The Code Book

Also by Simon Singh

Fermat's Enigma

The Code Book

The Evolution of Secrecy from Mary Queen of Scots to Quantum Cryptography

Simon Singh

Doubleday

New York London Toronto Sydney Auckland

PUBLISHED BY DOUBLEDAY
a division of Random House, Inc.
1540 Broadway, New York, New York 10036

DOUBLEDAY and the portrayal of an anchor with a dolphin are trademarks of
Doubleday, a division of Random House, Inc.

Book design by Jeffery Design

Library of Congress Cataloging-in-Publication Data
Singh, Simon.
The code book : the evolution of secrecy from Mary Queen of Scots to quantum cryptography / Simon Singh. –1st ed.
p. cm.
Includes bibliographical references and index.
1. Cryptography–History. 2. Data encryption (Computer science)–History.
I. Title.
Z103.S56 1999

652'.8'09–dc21 99-35261
CIP

ISBN 0-385-49531-5

Printed in the United States of America
October 1999
First Edition

10 9 8 7 6 5 4 3 2 1

For my mother and father,

Sawaran Kaur and Mehnga Singh

The urge to discover secrets is deeply ingrained in human nature; even the least curious mind is roused by the promise of sharing knowledge withheld from others. Some are fortunate enough to find a job which consists in the solution of mysteries, but most of us are driven to sublimate this urge by the solving of artificial puzzles devised for our entertainment. Detective stories or crossword puzzles cater for the majority; the solution of secret codes may be the pursuit of a few.

John Chadwick

The Decipherment of Linear B

Contents

Introduction

For thousands of years, kings, queens and generals have relied on efficient communication in order to govern their countries and command their armies. At the same time, they have all been aware of the consequences of their messages falling into the wrong hands, revealing precious secrets to rival nations and betraying vital information to opposing forces. It was the threat of enemy interception that motivated the development of codes and ciphers: techniques for disguising a message so that only the intended recipient can read it.

The desire for secrecy has meant that nations have operated codemaking departments, responsible for ensuring the security of communications by inventing and implementing the best possible codes. At the same time, enemy codebreakers have attempted to break these codes, and steal secrets. Codebreakers are linguistic alchemists, a mystical tribe attempting to conjure sensible words out of meaningless symbols. The history of codes and ciphers is the story of the centuries-old battle between codemakers and codebreakers, an intellectual arms race that has had a dramatic impact on the course of history.

In writing *The Code Book*, I have had two main objectives. The first is to chart the evolution of codes. Evolution is a wholly appropriate term, because the development of codes can be viewed as an evolutionary struggle. A code is constantly under attack from codebreakers. When the codebreakers have developed a new weapon that reveals a code's weakness, then the code is no longer useful. It either becomes extinct or it evolves into a new, stronger code. In turn, this new code thrives only until the codebreakers identify its weakness, and so on. This is analogous to the situation facing, for example, a strain of infectious bacteria. The bacteria live, thrive and survive until doctors discover an antibiotic that exposes a weakness in the bacteria and kills them. The bacteria are forced to evolve and outwit the antibiotic, and, if successful, they will thrive once again and re-establish themselves. The bacteria are continually forced to evolve in order to survive the onslaught of new antibiotics.

The ongoing battle between codemakers and codebreakers has inspired a whole series of remarkable scientific breakthroughs. The codemakers have continually striven to construct ever-stronger codes for defending communications, while codebreakers have continually invented more powerful methods for attacking them. In their efforts to destroy and preserve secrecy, both sides have drawn upon a diverse range of disciplines and technologies, from mathematics to linguistics, from information theory to quantum theory. In return, codemakers and codebreakers have enriched these subjects, and their work has accelerated technological development, most notably in the case of the modern computer.

History is punctuated with codes. They have decided the outcomes of battles and led to the deaths of kings and queens. I have therefore been able to call upon stories of political intrigue and tales of life and death to illustrate the key turning points in the evolutionary development of codes. The history of codes is so inordinately rich that I have been forced to leave out many fascinating stories, which in turn means that my account is not definitive. If you would like to find out more about your favourite tale or your favourite codebreaker then I would refer you to the list of further reading, which should help those readers who would like to study the subject in more detail.

Having discussed the evolution of codes and their impact on history, the book's second objective is to demonstrate how the subject is more relevant today than ever before. As information becomes an increasingly valuable commodity, and as the communications revolution changes society, so the process of encoding messages, known as encryption, will play an increasing role in everyday life. Nowadays our phone calls bounce off satellites and our e-mails pass through various computers, and both forms of communication can be intercepted with ease, so jeopardising our privacy. Similarly, as more and more business is conducted over the Internet, safeguards must be put in place to protect companies and their clients. Encryption is the only way to protect our privacy and guarantee the success of the digital marketplace. The art of secret communication, otherwise known as cryptography, will provide the locks and keys of the Information Age.

However, the public's growing demand for cryptography conflicts with the needs of law enforcement and national security. For decades, the police and the intelligence services have used wire-taps to gather evidence against terrorists and organised crime syndicates, but the recent development of ultra-strong codes threatens to undermine the value of wire-taps. As we enter the twenty-first century, civil libertarians are pressing for the widespread use of cryptography in order to protect the privacy of the individual. Arguing alongside them are businesses, who require strong cryptography in order to guarantee the security of transactions within the fast-growing world of Internet commerce. At the same time, the forces of law and order are lobbying governments to restrict the use of cryptography. The question is, which do we value more – our privacy or an effective police force? Or is there a compromise?

Although cryptography is now having a major impact on civilian activities, it should be noted that military cryptography remains an important subject. It has been said that the First World War was the chemists' war, because mustard gas and chlorine were employed for the first time, and that the Second World War was the physicists' war, because the atom bomb was detonated. Similarly, it has been argued that the Third World War would be the mathematicians' war, because mathematicians will have control over the next great weapon of war – information. Mathematicians have been responsible for developing the codes that are currently used to protect military information. Not surprisingly, mathematicians are also at the forefront of the battle to break these codes.

While describing the evolution of codes and their impact on history, I have allowed myself a minor detour. Chapter 5 describes the decipherment of various ancient scripts, including Linear B and Egyptian hieroglyphics. Technically, cryptography concerns communications that are deliberately designed to keep secrets from an enemy, whereas the writings of ancient civilisations were not intended to be indecipherable: it is merely that we have lost the ability to interpret them. However, the skills required to uncover the meaning of archaeological texts are closely related to the art of codebreaking. Ever since reading *The Decipherment of Linear B*, John Chadwick's description of how an ancient Mediterranean text was unravelled, I have been struck by the astounding intellectual achievements of those men and women who have been able to decipher the scripts of our ancestors, thereby allowing us to read about their civilisations, religions and everyday lives.

Turning to the purists, I should apologise for the title of this book. *The Code Book* is about more than just codes. The word 'code' refers to a very particular type of secret communication, one that has declined in use over the centuries. In a code, a word or phrase is replaced with a word, number or symbol. For example, secret agents have codenames, words that are used instead of their real names in order to mask their identities. Similarly, the phrase **Attack at dawn** could be replaced by the codeword **Jupiter**, and this word could be sent to a commander in the battlefield as a way of baffling the enemy. If headquarters and the commander have previously agreed on the code, then the meaning of **Jupiter** will be clear to the intended recipient, but it will mean nothing to an enemy who intercepts it. The alternative to a code is a cipher, a technique that acts at a more fundamental level, by replacing letters rather than whole words. For example, each letter in a phrase could be replaced by the next letter in the alphabet, so that **A** is replaced by **B**, **B** by **C**, and so on. **Attack at dawn** thus becomes **Buubdl bu ebxo**. Ciphers play an integral role in cryptography, and so this book should really have been called *The Code and Cipher Book*. I have, however, forsaken accuracy for snappiness.

As the need arises, I have defined the various technical terms used within cryptography. Although I have generally adhered to these definitions, there will be occasions when I use a term that is perhaps not technically accurate, but which I feel is more familiar to the non-specialist. For example, when describing a person attempting to break a cipher, I have often used *codebreaker* rather than the more accurate *cipherbreaker*. I have done this only when the meaning of the word is obvious from the context. There is a glossary of terms at the end of the book. More often than not, though, crypto-jargon is quite transparent: for example, *plaintext* is the message before encryption, and *ciphertext* is the message after encryption.

Before concluding this introduction, I must mention a problem that faces any author who tackles the subject of cryptography: the science of secrecy is largely a secret science. Many of the heroes in this book never gained recognition for their work during their lifetimes because their contribution could not be publicly acknowledged while their invention was still of diplomatic or military value. While researching this book, I was able to talk to experts at Britain's Government Communications Headquarters (GCHQ), who revealed details of extraordinary research done in the 1970s which has only just been declassified. As a result of this declassification, three of the world's greatest cryptographers can now receive the credit they deserve. However, this recent revelation has merely served to remind me that there is a great deal more going on, of which neither I nor any other science writer is aware. Organisations such as GCHQ and America's National Security Agency continue to conduct classified research into cryptography, which means that their breakthroughs remain secret and the individuals who make them remain anonymous.

Despite the problems of government secrecy and classified research, I have spent the final chapter of this book speculating about the future of codes and ciphers. Ultimately, this chapter is an attempt to see if we can predict who will win the evolutionary struggle between codemaker and codebreaker. Will codemakers ever design a truly unbreakable code and succeed in their quest for absolute secrecy? Or will codebreakers build a machine that can decipher any message? Bearing in mind that some of the greatest minds work in classified laboratories, and that they receive the bulk of research funds, it is clear that some of the statements in my final chapter may be inaccurate. For example, I state that quantum computers – machines potentially capable of breaking all today's ciphers – are at a very primitive stage, but it is possible that somebody has already built one. The only people who are in a position to point out my errors are also those who are not at liberty to reveal them.

1 The Cipher of Mary Queen of Scots

On the morning of Wednesday, 15 October 1586, Queen Mary entered the crowded courtroom at Fotheringhay Castle. Years of imprisonment and the onset of rheumatism had taken their toll, yet she remained dignified, composed and indisputably regal. Assisted by her physician, she made her way past the judges, officials and spectators, and approached the throne that stood halfway along the long, narrow chamber. Mary had assumed that the throne was a gesture of respect towards her, but she was mistaken. The throne symbolised the absent Queen Elizabeth, Mary's enemy and prosecutor. Mary was gently guided away from the throne and towards the opposite side of the room, to the defendant's seat, a crimson velvet chair.

Mary Queen of Scots was on trial for treason. She had been accused of plotting to assassinate Queen Elizabeth in order to take the English crown for herself. Sir Francis Walsingham, Elizabeth's Principal Secretary, had already arrested the other conspirators, extracted confessions, and executed them. Now he planned to prove that Mary was at the heart of the plot, and was therefore equally culpable and equally deserving of death.

Walsingham knew that before he could have Mary executed, he would have to convince Queen Elizabeth of her guilt. Although Elizabeth despised Mary, she had several reasons for being reluctant to see her put to death. First, Mary was a Scottish queen, and many questioned whether an English court had the authority to execute a foreign head of state. Second, executing Mary might establish an awkward precedent – if the state is allowed to kill one queen, then perhaps rebels might have fewer reservations about killing another, namely Elizabeth. Third, Elizabeth and Mary were cousins, and their blood tie made Elizabeth all the more squeamish about ordering her execution. In short, Elizabeth would

Figure 1 Mary Queen of Scots.

sanction Mary's execution only if Walsingham could prove beyond any hint of doubt that she had been part of the assassination plot.

The conspirators were a group of young English Catholic noblemen intent on removing Elizabeth, a Protestant, and replacing her with Mary, a fellow Catholic. It was apparent to the court that Mary was a figurehead for the conspirators, but it was not clear that she had actually given her blessing to the conspiracy. In fact, Mary had authorised the plot. The challenge for Walsingham was to demonstrate a palpable link between Mary and the plotters.

On the morning of her trial, Mary sat alone in the dock, dressed in sorrowful black velvet. In cases of treason, the accused was forbidden counsel and was not permitted to call witnesses. Mary was not even allowed secretaries to help her prepare her case. However, her plight was not hopeless because she had been careful to ensure that all her correspondence with the conspirators had been written in cipher. The cipher turned her words into a meaningless series of symbols, and Mary believed that even if Walsingham had captured the letters, then he could have no idea of the meaning of the words within them. If their contents were a mystery, then the letters could not be used as evidence against her. However, this all depended on the assumption that her cipher had not been broken.

Unfortunately for Mary, Walsingham was not merely Principal Secretary, he was also England's spymaster. He had intercepted Mary's letters to the plotters, and he knew exactly who might be capable of deciphering them. Thomas Phelippes was the nation's foremost expert on breaking codes, and for years he had been deciphering the messages of those who plotted against Queen Elizabeth, thereby providing the evidence needed to condemn them. If he could decipher the incriminating letters between Mary and the conspirators, then her death would be inevitable. On the other hand, if Mary's cipher was strong enough to conceal her secrets, then there was a chance that she might survive. Not for the first time, a life hung on the strength of a cipher.

The Evolution of Secret Writing

Some of the earliest accounts of secret writing date back to Herodotus, 'the father of history' according to the Roman philosopher and statesman

Cicero. In *The Histories*, Herodotus chronicled the conflicts between Greece and Persia in the fifth century BC, which he viewed as a confrontation between freedom and slavery, between the independent Greek states and the oppressive Persians. According to Herodotus, it was the art of secret writing that saved Greece from being conquered by Xerxes, King of Kings, the despotic leader of the Persians.

The long-running feud between Greece and Persia reached a crisis soon after Xerxes began constructing a city at Persepolis, the new capital for his kingdom. Tributes and gifts arrived from all over the empire and neighbouring states, with the notable exceptions of Athens and Sparta. Determined to avenge this insolence, Xerxes began mobilising a force, declaring that 'we shall extend the empire of Persia such that its boundaries will be God's own sky, so the sun will not look down upon any land beyond the boundaries of what is our own'. He spent the next five years secretly assembling the greatest fighting force in history, and then, in 480 BC, he was ready to launch a surprise attack.

However, the Persian military build-up had been witnessed by Demaratus, a Greek who had been expelled from his homeland and who lived in the Persian city of Susa. Despite being exiled he still felt some loyalty to Greece, so he decided to send a message to warn the Spartans of Xerxes' invasion plan. The challenge was how to dispatch the message without it being intercepted by the Persian guards. Herodotus wrote:

> As the danger of discovery was great, there was only one way in which he could contrive to get the message through: this was by scraping the wax off a pair of wooden folding tablets, writing on the wood underneath what Xerxes intended to do, and then covering the message over with wax again. In this way the tablets, being apparently blank, would cause no trouble with the guards along the road. When the message reached its destination, no one was able to guess the secret, until, as I understand, Cleomenes' daughter Gorgo, who was the wife of Leonides, divined and told the others that if they scraped the wax off, they would find something written on the wood underneath. This was done; the message was revealed and read, and afterwards passed on to the other Greeks.

As a result of this warning, the hitherto defenceless Greeks began to arm themselves. Profits from the state-owned silver mines, which were usually shared among the citizens, were instead diverted to the navy for the construction of two hundred warships.

Xerxes had lost the vital element of surprise and, on 23 September 480 BC, when the Persian fleet approached the Bay of Salamis near Athens, the Greeks were prepared. Although Xerxes believed he had trapped the Greek navy, the Greeks were deliberately enticing the Persian ships to enter the bay. The Greeks knew that their ships, smaller and fewer in number, would have been destroyed in the open sea, but they realised that within the confines of the bay they might outmanoeuvre the Persians. As the wind changed direction the Persians found themselves being blown into the bay, forced into an engagement on Greek terms. The Persian princess Artemisia became surrounded on three sides and attempted to head back out to sea, only to ram one of her own ships. Panic ensued, more Persian ships collided and the Greeks launched a full-blooded onslaught. Within a day, the formidable forces of Persia had been humbled.

Demaratus' strategy for secret communication relied on simply hiding the message. Herodotus also recounted another incident in which concealment was sufficient to secure the safe passage of a message. He chronicled the story of Histaiaeus, who wanted to encourage Aristagoras of Miletus to revolt against the Persian king. To convey his instructions securely, Histaiaeus shaved the head of his messenger, wrote the message on his scalp, and then waited for the hair to regrow. This was clearly a period of history that tolerated a certain lack of urgency. The messenger, apparently carrying nothing contentious, could travel without being harassed. Upon arriving at his destination he then shaved his head and pointed it at the intended recipient.

Secret communication achieved by hiding the existence of a message is known as *steganography*, derived from the Greek words *steganos*, meaning 'covered', and *graphein*, meaning 'to write'. In the two thousand years since Herodotus, various forms of steganography have been used throughout the world. For example, the ancient Chinese wrote messages on fine silk, which was then scrunched into a tiny ball and covered in wax. The messenger would then swallow the ball of wax. In the fifteenth century, the Italian scientist Giovanni Porta described how to conceal a message within a hard-boiled egg by making an ink from a mixture of one ounce of alum and a pint of vinegar, and then using it to write on the shell. The solution penetrates the porous shell, and leaves a message on the surface of the hardened egg albumen, which can be read only when the shell is removed. Steganography also includes the practice of writing in invisible ink. As far back as the first century AD, Pliny the Elder explained how the 'milk' of the thithymallus plant could be used as an invisible ink. Although transparent after drying, gentle heating chars the ink and turns it brown. Many organic fluids behave in a similar way, because they are rich in carbon and therefore char easily. Indeed, it is not unknown for modern spies who have run out of standard-issue invisible ink to improvise by using their own urine.

The longevity of steganography illustrates that it certainly offers a modicum of security, but it suffers from a fundamental weakness. If the messenger is searched and the message is discovered, then the contents of the secret communication are revealed at once. Interception of the message immediately compromises all security. A thorough guard might routinely search any person crossing a border, scraping any wax tablets, heating blank sheets of paper, shelling boiled eggs, shaving people's heads, and so on, and inevitably there will be occasions when the message is uncovered.

Hence, in parallel with the development of steganography, there was the evolution of *cryptography*, derived from the Greek word *kryptos*, meaning 'hidden'. The aim of cryptography is not to hide the existence of a message, but rather to hide its meaning, a process known as *encryption*. To render a message unintelligible, it is scrambled according to a particular protocol which is agreed beforehand between the sender and the intended recipient. Thus the recipient can reverse the scrambling protocol and make the message comprehensible. The advantage of cryptography is that if the enemy intercepts an encrypted message, then the message is unreadable. Without knowing the scrambling protocol, the enemy should find it difficult, if not impossible, to recreate the original message from the encrypted text.

Although cryptography and steganography are independent, it is possible to both scramble and hide a message to maximise security. For example, the microdot is a form of steganography that became popular during the Second World War. German agents in Latin America would photographically shrink a page of text down to a dot less than 1 millimetre in diameter, and then hide this microdot on top of a full stop in an apparently innocuous letter. The first microdot to be spotted by the FBI was in 1941, following a tip-off that the Americans should look for a tiny gleam from the surface of a letter, indicative of smooth film. Thereafter, the Americans could read the contents of most intercepted microdots, except when the German agents had taken the extra precaution of scrambling their message before reducing it. In such cases of cryptography combined with steganography, the Americans were sometimes able to intercept and block communications, but they were prevented from gaining any new information about German spying activity. Of the two branches of secret communication, cryptography is the more powerful because of this ability to prevent information from falling into enemy hands.

In turn, cryptography itself can be divided into two branches, known as *transposition* and *substitution*. In transposition, the letters of the message are simply rearranged, effectively generating an anagram. For very short messages, such as a single word, this method is relatively insecure because there are only a limited number of ways of rearranging a handful of letters. For example, three letters can be arranged in only six different ways, e.g. **cow, cwo, ocw, owc, wco, woc**. However, as the number of letters gradually increases, the number of possible arrangements rapidly explodes, making it impossible to get back to the original message unless the exact scrambling process is known. **For example, consider this short sentence.** It contains just 35 letters, and yet there are more than 50,000,000,000,000,000,000,000,000,000,000 distinct arrangements of them. If one person could check one arrangement per second, and if all the people in the world worked night and day, it would still take more than a thousand times the lifetime of the universe to check all the arrangements.

A random transposition of letters seems to offer a very high level of security, because it would be impractical for an enemy interceptor to unscramble even a short sentence. But there is a drawback. Transposition effectively generates an incredibly difficult anagram, and if the letters are randomly jumbled, with neither rhyme nor reason, then unscrambling the anagram is impossible for the intended recipient, as well as an enemy interceptor. In order for transposition to be effective, the rearrangement of letters needs to follow a straightforward system, one that has been previously agreed by sender and receiver, but kept secret from the enemy. For example, schoolchildren sometimes send messages using the 'rail fence' transposition, in which the message is written with alternate letters on separate upper and lower lines. The sequence of letters on the lower line is then tagged on at the end of the sequence on the upper line to create the final encrypted message. For example:

THY SECRET IS THY PRISONER; IF THOU LET IT GO, THOU ART A PRISONER TO IT

↓

T Y E R T S H P I O E I T O L T T O H U R A R S N R O T
H S C E I T Y R S N R F H U E I G T O A T P I O E T I

↓

TYERTSHPIOEITOLTTOHURARSNROTHSCEITYRSNRFHUEIGTOATPIOETI

The receiver can recover the message by simply reversing the process. There are various other forms of systematic transposition, including the three-line rail fence cipher, in which the message is first written on three separate lines instead of two. Alternatively, one could swap each pair of letters, so that the first and second letters switch places, the third and fourth letters switch places, and so on.

Another form of transposition is embodied in the first ever military cryptographic device, the Spartan *scytale*, dating back to the fifth century BC. The scytale is a wooden staff around which a strip of leather or parchment is wound, as shown in Figure 2. The sender writes the message along the length of the scytale, and then unwinds the strip, which now appears to

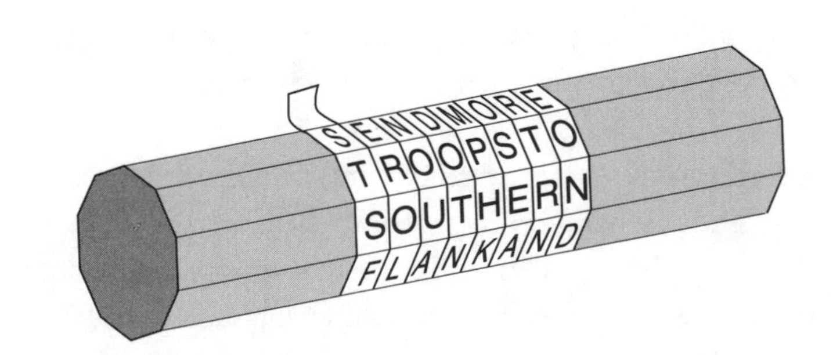

Figure 2 When it is unwound from the sender's scytale (wooden staff), the leather strip appears to carry a list of random letters; S, T, S, F, ... Only by rewinding the strip around another scytale of the correct diameter will the message reappear.

carry a list of meaningless letters. The message has been scrambled. The messenger would take the leather strip, and, as a steganographic twist, he would sometimes disguise it as a belt with the letters hidden on the inside. To recover the message, the receiver simply wraps the leather strip around a scytale of the same diameter as the one used by the sender. In 404 BC Lysander of Sparta was confronted by a messenger, bloody and battered, one of only five to have survived the arduous journey from Persia. The messenger handed his belt to Lysander, who wound it around his scytale to learn that Pharnabazus of Persia was planning to attack him. Thanks to the scytale, Lysander was prepared for the attack and repulsed it.

The alternative to transposition is substitution. One of the earliest descriptions of encryption by substitution appears in the *Kāma-sūtra*, a text written in the fourth century AD by the Brahmin scholar Vātsyāyana, but based on manuscripts dating back to the fourth century BC. The *Kāma-sūtra* recommends that women should study 64 arts, such as cooking, dressing, massage and the preparation of perfumes. The list also includes some less obvious arts, namely conjuring, chess, bookbinding and carpentry. Number 45 on the list is *mlecchita-vikalpā*, the art of secret writing, advocated in order to help women conceal the details of their liaisons. One of the recommended techniques is to pair letters of the alphabet at random, and then substitute each letter in the original message with its partner. If we apply the principle to the Roman alphabet, we could pair letters as follows:

A	D	H	I	K	M	O	R	S	U	W	Y	Z
↕	↕	↕	↕	↕	↕	↕	↕	↕	↕	↕	↕	↕
V	X	B	G	J	C	Q	L	N	E	F	P	T

Then, instead of **meet at midnight**, the sender would write **CUUZ VZ CGXSGIBZ**. This form of secret writing is called a substitution cipher because each letter in the plaintext is substituted for a different letter, thus acting in a complementary way to the transposition cipher. In transposition each letter retains its identity but changes its position, whereas in substitution each letter changes its identity but retains its position.

The first documented use of a substitution cipher for military purposes appears in Julius Caesar's *Gallic Wars*. Caesar describes how he sent a message to Cicero, who was besieged and on the verge of surrendering.

The substitution replaced Roman letters with Greek letters, rendering the message unintelligible to the enemy. Caesar described the dramatic delivery of the message:

> The messenger was instructed, if he could not approach, to hurl a spear, with the letter fastened to the thong, inside the entrenchment of the camp. Fearing danger, the Gaul discharged the spear, as he had been instructed. By chance it stuck fast in the tower, and for two days was not sighted by our troops; on the third day it was sighted by a soldier, taken down, and delivered to Cicero. He read it through and then recited it at a parade of the troops, bringing the greatest rejoicing to all.

Caesar used secret writing so frequently that Valerius Probus wrote an entire treatise on his ciphers, which unfortunately has not survived. However, thanks to Suetonius' *Lives of the Caesars LVI*, written in the second century AD, we do have a detailed description of one of the types of substitution cipher used by Caesar. The emperor simply replaced each letter in the message with the letter that is three places further down the alphabet. Cryptographers often think in terms of the *plain alphabet*, the alphabet used to write the original message, and the *cipher alphabet*, the letters that are substituted in place of the plain letters. When the plain alphabet is placed above the cipher alphabet, as shown in Figure 3, it is clear that the cipher alphabet has been shifted by three places, and hence this form of substitution is often called the *Caesar shift cipher*, or simply the Caesar cipher. A cipher is the name given to any form of

Plain alphabet	a	b	c	d	e	f	g	h	i	j	k	l	m	n	o	p	q	r	s	t	u	v	w	x	y	z
Cipher alphabet	D	E	F	G	H	I	J	K	L	M	N	O	P	Q	R	S	T	U	V	W	X	Y	Z	A	B	C

Plaintext	v e n i, v i d i, v i c i
Ciphertext	Y H Q L, Y L G L, Y L F L

Figure 3 The Caesar cipher applied to a short message. The Caesar cipher is based on a cipher alphabet that is shifted a certain number of places (in this case three), relative to the plain alphabet. The convention in cryptography is to write the plain alphabet in lower-case letters, and the cipher alphabet in capitals. Similarly, the original message, the plaintext, is written in lower case, and the encrypted message, the ciphertext, is written in capitals.

cryptographic substitution in which each letter is replaced by another letter or symbol.

Although Suetonius mentions only a Caesar shift of three places, it is clear that by using any shift between 1 and 25 places it is possible to generate 25 distinct ciphers. In fact, if we do not restrict ourselves to shifting the alphabet and permit the cipher alphabet to be any rearrangement of the plain alphabet, then we can generate an even greater number of distinct ciphers. There are over 400,000,000,000,000,000,000,000,000 such rearrangements, and therefore the same number of distinct ciphers.

Each distinct cipher can be considered in terms of a general encrypting method, known as the *algorithm*, and a *key*, which specifies the exact details of a particular encryption. In this case, the algorithm involves substituting each letter in the plain alphabet with a letter from a cipher alphabet, and the cipher alphabet is allowed to consist of any rearrangement of the plain alphabet. The key defines the exact cipher alphabet to be used for a particular encryption. The relationship between the algorithm and the key is illustrated in Figure 4.

An enemy studying an intercepted scrambled message may have a strong suspicion of the algorithm, but would not know the exact key. For example, they may well suspect that each letter in the plaintext has been

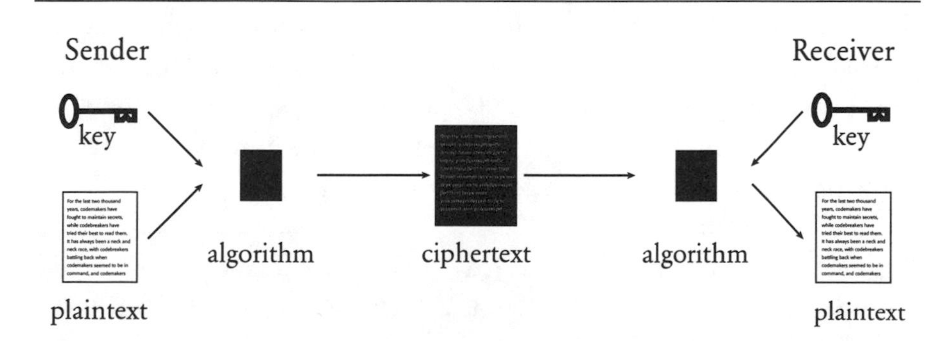

Figure 4 To encrypt a plaintext message, the sender passes it through an encryption algorithm. The algorithm is a general system for encryption, and needs to be specified exactly by selecting a key. Applying the key and algorithm together to a plaintext generates the encrypted message, or ciphertext. The ciphertext may be intercepted by an enemy while it is being transmitted to the receiver, but the enemy should not be able to decipher the message. However, the receiver, who knows both the key and the algorithm used by the sender, is able to turn the ciphertext back into the plaintext message.

replaced by a different letter according to a particular cipher alphabet, but they are unlikely to know which cipher alphabet has been used. If the cipher alphabet, the key, is kept a closely guarded secret between the sender and the receiver, then the enemy cannot decipher the intercepted message. The significance of the key, as opposed to the algorithm, is an enduring principle of cryptography. It was definitively stated in 1883 by the Dutch linguist Auguste Kerckhoffs von Nieuwenhof in his book *La Cryptographie militaire*: 'Kerckhoffs' Principle: The security of a crypto-system must not depend on keeping secret the crypto-algorithm. The security depends only on keeping secret the key.'

In addition to keeping the key secret, a secure cipher system must also have a wide range of potential keys. For example, if the sender uses the Caesar shift cipher to encrypt a message, then encryption is relatively weak because there are only 25 potential keys. From the enemy's point of view, if they intercept the message and suspect that the algorithm being used is the Caesar shift, then they merely have to check the 25 possibilities. However, if the sender uses the more general substitution algorithm, which permits the cipher alphabet to be any rearrangement of the plain alphabet, then there are 400,000,000,000,000,000,000,000,000 possible keys from which to choose. One such is shown in Figure 5. From the enemy's point of view, if the message is intercepted and the algorithm is known, there is still the horrendous task of checking all possible keys. If an enemy agent were able to check one of the 400,000,000,000,000,000,000,000,000 possible keys every second, it would take roughly a billion times the lifetime of the universe to check all of them and decipher the message.

Plain alphabet a b c d e f g h i j k l m n o p q r s t u v w x y z
Cipher alphabet J L P A W I Q B C T R Z Y D S K E G F X H U O N V M

Plaintext e t t u, b r u t e ?
Ciphertext W X X H, L G H X W ?

Figure 5 An example of the general substitution algorithm, in which each letter in the plaintext is substituted with another letter according to a key. The key is defined by the cipher alphabet, which can be any rearrangement of the plain alphabet.

The beauty of this type of cipher is that it is easy to implement, but provides a high level of security. It is easy for the sender to define the key, which consists merely of stating the order of the 26 letters in the rearranged cipher alphabet, and yet it is effectively impossible for the enemy to check all possible keys by the so-called brute-force attack. The simplicity of the key is important, because the sender and receiver have to share knowledge of the key, and the simpler the key, the less the chance of a misunderstanding.

In fact, an even simpler key is possible if the sender is prepared to accept a slight reduction in the number of potential keys. Instead of randomly rearranging the plain alphabet to achieve the cipher alphabet, the sender chooses a *keyword* or *keyphrase*. For example, to use **JULIUS CAESAR** as a keyphrase, begin by removing any spaces and repeated letters (**JULISCAER**), and then use this as the beginning of the jumbled cipher alphabet. The remainder of the cipher alphabet is merely the remaining letters of the alphabet, in their correct order, starting where the keyphrase ends. Hence, the cipher alphabet would read as follows.

Plain alphabet	a	b	c	d	e	f	g	h	i	j	k	l	m	n	o	p	q	r	s	t	u	v	w	x	y	z
Cipher alphabet	J	U	L	I	S	C	A	E	R	T	V	W	X	Y	Z	B	D	F	G	H	K	M	N	O	P	Q

The advantage of building a cipher alphabet in this way is that it is easy to memorise the keyword or keyphrase, and hence the cipher alphabet. This is important, because if the sender has to keep the cipher alphabet on a piece of paper, the enemy can capture the paper, discover the key, and read any communications that have been encrypted with it. However, if the key can be committed to memory it is less likely to fall into enemy hands. Clearly the number of cipher alphabets generated by keyphrases is smaller than the number of cipher alphabets generated without restriction, but the number is still immense, and it would be effectively impossible for the enemy to unscramble a captured message by testing all possible keyphrases.

This simplicity and strength meant that the substitution cipher dominated the art of secret writing throughout the first millennium AD. Codemakers had evolved a system for guaranteeing secure communication, so there was no need for further development – without necessity, there was no need for further invention. The onus had fallen upon the codebreakers, those who were attempting to crack the substitution cipher. Was there any way for an enemy interceptor to unravel an encrypted message? Many ancient scholars considered that the substitution cipher was unbreakable, thanks to the gigantic number of possible keys, and for centuries this seemed to be true. However, codebreakers would eventually find a shortcut to the process of exhaustively searching all keys. Instead of taking billions of years to crack a cipher, the shortcut could reveal the message in a matter of minutes. The breakthrough occurred in the East, and required a brilliant combination of linguistics, statistics and religious devotion.

The Arab Cryptanalysts

At the age of about forty, Muhammad began regularly visiting an isolated cave on Mount Hira just outside Mecca. This was a retreat, a place for prayer, meditation and contemplation. It was during a period of deep reflection, around AD 610, that he was visited by the archangel Gabriel, who proclaimed that Muhammad was to be the messenger of God. This was the first of a series of revelations which continued until Muhammad died some twenty years later. The revelations were recorded by various scribes during the Prophet's life, but only as fragments, and it was left to Abū Bakr, the first caliph of Islam, to gather them together into a single text. The work was continued by Umar, the second caliph, and his daughter Hafsa, and was eventually completed by Uthmān, the third caliph. Each revelation became one of the 114 chapters of the Koran.

The ruling caliph was responsible for carrying on the work of the Prophet, upholding his teachings and spreading his word. Between the appointment of Abū Bakr in 632 to the death of the fourth caliph, Alī, in 661, Islam spread until half of the known world was under Muslim rule. Then in 750, after a century of consolidation, the start of the Abbasid caliphate (or dynasty) heralded the golden age of Islamic civilisation. The arts and sciences flourished in equal measure. Islamic craftsmen bequeathed us magnificent paintings, ornate carvings, and the most elaborate textiles in history, while the legacy of Islamic scientists is evident from the number of Arabic words that pepper the lexicon of modern science such as *algebra*, *alkaline* and *zenith*.

The richness of Islamic culture was to a large part the result of a wealthy and peaceful society. The Abbasid caliphs were less interested than their predecessors in conquest, and instead concentrated on establishing an organised and affluent society. Lower taxes encouraged businesses to grow and gave rise to greater commerce and industry, while strict laws reduced corruption and protected the citizens. All of this relied on an effective system of administration, and in turn the administrators relied on secure communication achieved through the use of encryption. As well as encrypting sensitive affairs of state, it is documented that officials protected tax records, demonstrating a widespread and routine use of cryptography. Further evidence comes from many administrative manuals, such as the tenth-century *Adab al-Kuttāb* ('The Secretaries' Manual'), which include sections devoted to cryptography.

The administrators usually employed a cipher alphabet which was simply a rearrangement of the plain alphabet, as described earlier, but they also used cipher alphabets that contained other types of symbols. For example, a in the plain alphabet might be replaced by # in the cipher alphabet, b might be replaced by +, and so on. The *monoalphabetic substitution cipher* is the general name given to any substitution cipher in which the cipher alphabet consists of either letters or symbols, or a mix of both. All the substitution ciphers that we have met so far come within this general category.

Had the Arabs merely been familiar with the use of the monoalphabetic substitution cipher, they would not warrant a significant mention in any history of cryptography. However, in addition to employing ciphers, the Arab scholars were also capable of destroying ciphers. They in fact invented *cryptanalysis*, the science of unscrambling a message without knowledge of the key. While the cryptographer develops new methods of secret writing, it is the cryptanalyst who struggles to find weaknesses in these methods in order to break into secret messages. Arabian cryptanalysts succeeded in finding a method for breaking the monoalphabetic substitution cipher, a cipher that had remained invulnerable for several centuries.

Cryptanalysis could not be invented until a civilisation had reached a sufficiently sophisticated level of scholarship in several disciplines, including mathematics, statistics and linguistics. The Muslim civilisation provided an ideal cradle for cryptanalysis, because Islam demands justice in all spheres of human activity, and achieving this requires knowledge, or *ilm*. Every Muslim is obliged to pursue knowledge in all its forms, and the economic success of the Abbasid caliphate meant that scholars had the time, money and materials required to fulfil their duty. They endeavoured to acquire the knowledge of previous civilisations by obtaining Egyptian, Babylonian, Indian, Chinese, Farsi, Syriac, Armenian, Hebrew and Roman texts and translating them into Arabic. In 815, the Caliph al-Ma'mūn established in Baghdad the Bait al-Hikmah ('House of Wisdom'), a library and centre for translation.

At the same time as acquiring knowledge, the Islamic civilisation was able to disperse it, because it had procured the art of paper-making from the Chinese. The manufacture of paper gave rise to the profession of *warraqīn*, or 'those who handle paper', human photocopying machines who copied manuscripts and supplied the burgeoning publishing industry. At its peak, tens of thousands of books were published every year, and in just one suburb of Baghdad there were over a hundred bookshops. As well as such classics as *Tales from the Thousand and One Nights*, these bookshops also sold textbooks on every imaginable subject, and helped to support the most literate and learned society in the world.

In addition to a greater understanding of secular subjects, the invention of cryptanalysis also depended on the growth of religious scholarship. Major theological schools were established in Basra, Kufa and Baghdad, where theologians scrutinised the revelations of Muhammad as contained in the Koran. The theologians were interested in establishing the chronology of the revelations, which they did by counting the frequencies of words contained in each revelation. The theory was that certain words had evolved relatively recently, and hence if a revelation contained a high number of these newer words, this would indicate that it came later in the chronology. Theologians also studied the *Hadīth*, which consists of the Prophet's daily utterances. They tried to demonstrate that each statement was indeed attributable to Muhammad. This was done by studying the etymology of words and the structure of sentences, to test whether particular texts were consistent with the linguistic patterns of the Prophet.

Significantly, the religious scholars did not stop their scrutiny at the level of words. They also analysed individual letters, and in particular they discovered that some letters are more common than others. The letters a and l are the most common in Arabic, partly because of the definite article al-, whereas the letter j appears only a tenth as frequently. This apparently innocuous observation would lead to the first great breakthrough in cryptanalysis.

Although it is not known who first realised that the variation in the frequencies of letters could be exploited in order to break ciphers, the earliest known description of the technique is by the ninth-century scientist Abū Yūsūf Ya'qūb ibn Is-hāq ibn as-Sabbāh ibn 'omrān ibn Ismaīl al-Kindī. Known as 'the philosopher of the Arabs', al-Kindī was the author of 290 books on medicine, astronomy, mathematics, linguistics and music. His greatest treatise, which was rediscovered only in 1987 in the Sulaimaniyyah Ottoman Archive in Istanbul, is entitled *A Manuscript on Deciphering Cryptographic Messages*; the first page is shown in Figure 6. Although it contains detailed discussions on statistics, Arabic phonetics and Arabic syntax, al-Kindī's revolutionary system of cryptanalysis is encapsulated in two short paragraphs:

> One way to solve an encrypted message, if we know its language, is to find a different plaintext of the same language long enough to fill one sheet or so, and then we count the occurrences of each letter. We call the most frequently occurring letter the 'first', the next most occurring letter the 'second', the following most occurring letter the 'third', and so on, until we account for all the different letters in the plaintext sample.
>
> Then we look at the ciphertext we want to solve and we also classify its symbols. We find the most occurring symbol and change it to the form of the 'first' letter of the plaintext sample, the next most common symbol is changed to the form of the 'second' letter, and the following most common symbol is changed to the form of the 'third' letter, and so on, until we account for all symbols of the cryptogram we want to solve.

Al-Kindī's explanation is easier to explain in terms of the English alphabet. First of all, it is necessary to study a lengthy piece of normal English text, perhaps several, in order to establish the frequency of each letter of the alphabet. In English, e is the most common letter, followed by t, then a, and so on, as given in Table 1. Next, examine the ciphertext in question, and work out the frequency of each letter. If the most common letter in

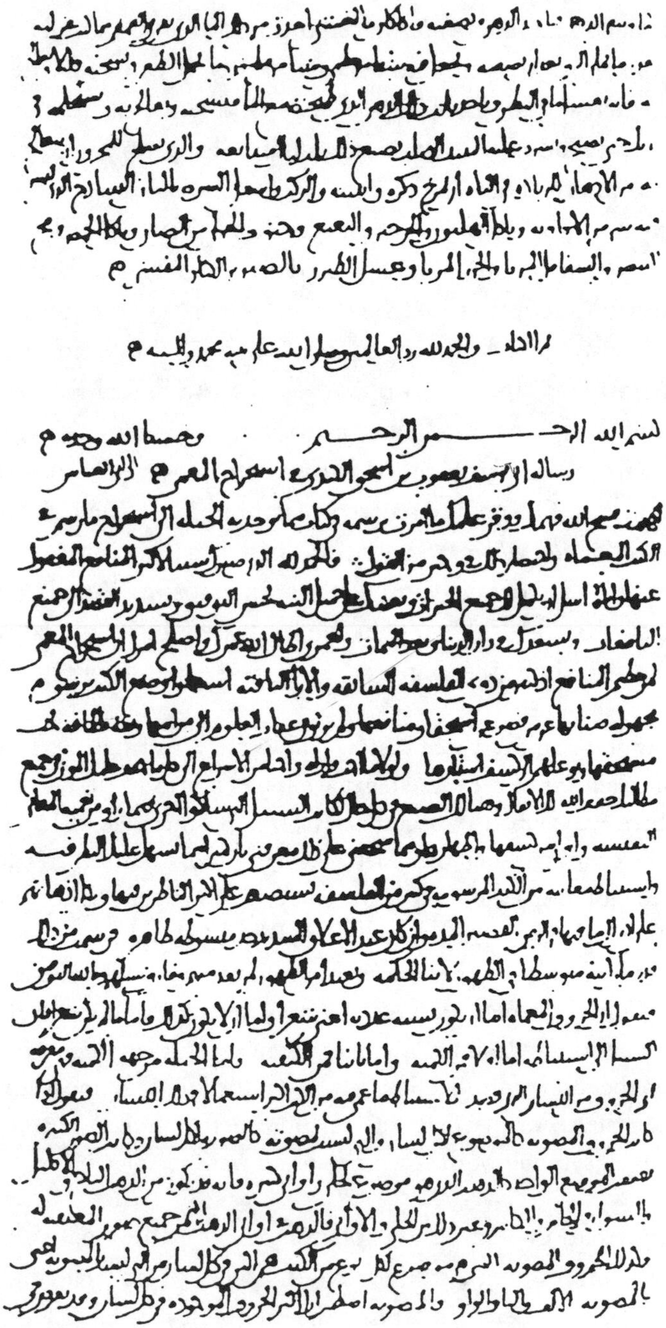

[illegible]

تم الكتاب والحمد لله رب العالمين وصلى الله على سيدنا محمد وآله [illegible]

بسم الله الرحمن الرحيم وحسبنا الله وحده

رسالة ابي يوسف يعقوب بن اسحق الكندي في استخراج المعمى [illegible]

[illegible]

Figure 6 The first page of al-Kindī's manuscript *On Deciphering Cryptographic Messages*, containing the oldest known description of cryptanalysis by frequency analysis.

the ciphertext is, for example, J then it would seem likely that this is a substitute for e. And if the second most common letter in the ciphertext is P, then this is probably a substitute for t, and so on. Al-Kindī's technique, known as *frequency analysis*, shows that it is unnecessary to check each of the billions of potential keys. Instead, it is possible to reveal the contents of a scrambled message simply by analysing the frequency of the characters in the ciphertext.

However, it is not possible to apply al-Kindī's recipe for cryptanalysis unconditionally, because the standard list of frequencies in Table 1 is only an average, and it will not correspond exactly to the frequencies of every text. For example, a brief message discussing the effect of the atmosphere on the movement of striped quadrupeds in Africa would not yield to straightforward frequency analysis: 'From Zanzibar to Zambia and Zaire, ozone zones make zebras run zany zigzags.' In general, short texts are likely to deviate significantly from the standard frequencies, and if there are less than a hundred letters, then decipherment will be very difficult. On the other hand, longer texts are more likely to follow the standard frequencies, although this is not always the case. In 1969, the French author

Table 1 This table of relative frequencies is based on passages taken from newspapers and novels, and the total sample was 100,362 alphabetic characters. The table was compiled by H. Beker and F. Piper, and originally published in *Cipher Systems: The Protection Of Communication.*

Letter	Percentage	Letter	Percentage
a	8.2	n	6.7
b	1.5	o	7.5
c	2.8	p	1.9
d	4.3	q	0.1
e	12.7	r	6.0
f	2.2	s	6.3
g	2.0	t	9.1
h	6.1	u	2.8
i	7.0	v	1.0
j	0.2	w	2.4
k	0.8	x	0.2
l	4.0	y	2.0
m	2.4	z	0.1

Georges Perec wrote *La Disparition*, a 200-page novel that did not use words that contain the letter e. Doubly remarkable is the fact that the English novelist and critic Gilbert Adair succeeded in translating *La Disparition* into English, while still following Perec's shunning of the letter e. Entitled *A Void*, Adair's translation is surprisingly readable (see Appendix A). If the entire book were encrypted via a monoalphabetic substitution cipher, then a naive attempt to decipher it might be stymied by the complete lack of the most frequently occurring letter in the English alphabet.

Having described the first tool of cryptanalysis, I shall continue by giving an example of how frequency analysis is used to decipher a ciphertext. I have avoided peppering the whole book with examples of cryptanalysis, but with frequency analysis I make an exception. This is partly because frequency analysis is not as difficult as it sounds, and partly because it is the primary cryptanalytic tool. Furthermore, the example that follows provides insight into the modus operandi of the cryptanalyst. Although frequency analysis requires logical thinking, you will see that it also demands guile, intuition, flexibility and guesswork.

Cryptanalysing a Ciphertext

PCQ VMJYPD LBYK LYSO KBXBJXWXV BXV ZCJPO EYPD KBXBJYUXJ LBJOO KCPK. CP LBO LBCMKXPV XPV IYJKL PYDBL, QBOP KBO BXV OPVOV LBO LXRO CI SX'XJMI, KBO JCKO XPV EYKKOV LBO DJCMPV ZOICJO BYS, KXUYPD: 'DJOXL EYPD, ICJ X LBCMKXPV XPV CPO PYDBLK Y BXNO ZOOP JOACMPLYPD LC UCM LBO IXZROK CI FXKL XDOK XPV LBO RODOPVK CI XPAYOPL EYPDK. SXU Y SXEO KC ZCRV XK LC AJXNO X IXNCMJ CI UCMJ SXGOKLU?'

OFYRCDMO, LXROK IJCS LBO LBCMKXPV XPV CPO PYDBLK

Imagine that we have intercepted this scrambled message. The challenge is to decipher it. We know that the text is in English, and that it has been scrambled according to a monoalphabetic substitution cipher, but we have no idea of the key. Searching all possible keys is impractical, so we must apply frequency analysis. What follows is a step-by-step guide to cryptanalysing the ciphertext, but if you feel confident then you might prefer to ignore this and attempt your own independent cryptanalysis.

The immediate reaction of any cryptanalyst upon seeing such a ciphertext is to analyse the frequency of all the letters, which results in Table 2. Not surprisingly, the letters vary in their frequency. The question is, can we identify what any of them represent, based on their frequencies? The ciphertext is relatively short, so we cannot slavishly apply frequency analysis. It would be naive to assume that the commonest letter in the ciphertext, O, represents the commonest letter in English, e, or that the eighth most frequent letter in the ciphertext, Y, represents the eighth most frequent letter in English, h. An unquestioning application of frequency analysis would lead to gibberish. For example, the first word PCQ would be deciphered as aov.

However, we can begin by focusing attention on the only three letters that appear more than thirty times in the ciphertext, namely O, X and P. It is fairly safe to assume that the commonest letters in the ciphertext probably represent the commonest letters in the English alphabet, but not necessarily in the right order. In other words, we cannot be sure that O = e, X = t, and P = a, but we can make the tentative assumption that:

O = e, t or a, X = e, t or a, P = e, t or a.

Table 2 Frequency analysis of enciphered message.

Letter	Frequency		Letter	Frequency	
	Occurrences	Percentage		Occurrences	Percentage
A	3	0.9	N	3	0.9
B	25	7.4	O	38	11.2
C	27	8.0	P	31	9.2
D	14	4.1	Q	2	0.6
E	5	1.5	R	6	1.8
F	2	0.6	S	7	2.1
G	1	0.3	T	0	0.0
H	0	0.0	U	6	1.8
I	11	3.3	V	18	5.3
J	18	5.3	W	1	0.3
K	26	7.7	X	34	10.1
L	25	7.4	Y	19	5.6
M	11	3.3	Z	5	1.5

In order to proceed with confidence, and pin down the identity of the three most common letters, O, X and P, we need a more subtle form of frequency analysis. Instead of simply counting the frequency of the three letters, we can focus on how often they appear next to all the other letters. For example, does the letter O appear before or after several other letters, or does is tend to neighbour just a few special letters? Answering this question will be a good indication of whether O represents a vowel or a consonant. If O represents a vowel it should appear before and after most of the other letters, whereas if it represents a consonant, it will tend to avoid many of the other letters. For example, the letter e can appear before and after virtually every other letter, but the letter t is rarely seen before or after b, d, g, j, k, m, q or v.

The table below takes the three most common letters in the ciphertext, O, X and P, and lists how frequently each appears before or after every letter. For example, O appears before A on 1 occasion, but never appears immediately after it, giving a total of 1 in the first box. The letter O neighbours the majority of letters, and there are only 7 that it avoids completely, represented by the 7 zeros in the O row. The letter X is equally sociable, because it too neighbours most of the letters, and avoids only 8 of them. However, the letter P is much less friendly. It tends to lurk around just a few letters, and avoids 15 of them. This evidence suggests that O and X represent vowels, while P represents a consonant.

	A	B	C	D	E	F	G	H	I	J	K	L	M	N	O	P	Q	R	S	T	U	V	W	X	Y	Z
O	1	9	0	3	1	1	1	0	1	4	6	0	1	2	2	8	0	4	1	0	0	3	0	1	1	2
X	0	7	0	1	1	1	1	0	2	4	6	3	0	3	1	9	0	2	4	0	3	3	2	0	0	1
P	1	0	5	6	0	0	0	0	0	1	1	2	2	0	8	0	0	0	0	0	0	11	0	9	9	0

Now we must ask ourselves which vowels are represented by O and X. They are probably e and a, the two most popular vowels in the English language, but does O = e and X = a, or does O = a and X = e? An interesting feature in the ciphertext is that the combination OO appears twice, whereas XX does not appear at all. Since the letters ee appear far more often than aa in plaintext English, it is likely that O = e and X = a.

At this point, we have confidently identified two of the letters in the ciphertext. Our conclusion that X = a is supported by the fact that X appears on its own in the ciphertext, and a is one of only two English words that consist of a single letter. The only other letter that appears on its own in the ciphertext is Y, and it seems highly likely that this represents the only other one-letter English word, which is i. Focusing on words with only one letter is a standard cryptanalytic trick, and I have included it among a list of cryptanalytic tips in Appendix B. This particular trick works only because this ciphertext still has spaces between the words. Often, a cryptographer will remove all the spaces to make it harder for an enemy interceptor to unscramble the message.

Although we have spaces between words, the following trick would also work where the ciphertext has been merged into a single string of characters. The trick allows us to spot the letter h, once we have already identified the letter e. In the English language, the letter h frequently goes before the letter e (as in the, then, they, etc.), but rarely after e. The table below shows how frequently the O, which we think represents e, goes before and after all the other letters in the ciphertext. The table suggests that B represents h, because it appears before O on 9 occasions, but it never goes after it. No other letter in the table has such an asymmetric relationship with O.

	A	B	C	D	E	F	G	H	I	J	K	L	M	N	O	P	Q	R	S	T	U	V	W	X	Y	Z
after O	1	0	0	1	0	1	0	0	1	0	4	0	0	0	2	5	0	0	0	0	0	2	0	1	0	0
before O	0	9	0	2	1	0	1	0	0	4	2	0	1	2	2	3	0	4	1	0	0	1	0	0	1	2

Each letter in the English language has its own unique personality, which includes its frequency and its relation to other letters. It is this personality that allows us to establish the true identity of a letter, even when it has been disguised by monoalphabetic substitution.

We have now confidently established four letters, O = e, X = a, Y = i and B = h, and we can begin to replace some of the letters in the ciphertext with their plaintext equivalents. I shall stick to the convention of keeping ciphertext letters in upper case, while putting plaintext letters in lower case. This will help to distinguish between those letters we still have to identify, and those that have already been established.

PCQ VMJiPD LhiK LiSe KhahJaWaV haV ZCJPe EiPD KhahJiUaJ LhJee KCPK. CP Lhe LhCMKaPV aPV IiJKL PiDhL, QheP Khe haV ePVeV Lhe LaRe CI Sa'aJMI, Khe JCKe aPV EiKKev Lhe DJCMPV ZeICJe hiS, KaUiPD: 'DJeaL EiPD, ICJ a LhCMKaPV aPV CPe PiDhLK i haNe ZeeP JeACMPLiPD LC UCM Lhe IaZReK CI FaKL aDeK aPV Lhe ReDePVK CI aPAiePL EiPDK. SaU i SaEe KC ZCRV aK LC AJaNe a IaNCMJ CI UCMJ SaGeKLU?'

eFiRCDMe, LaReK IJCS Lhe LhCMKaPV aPV CPe PiDhLK

This simple step helps us to identify several other letters, because we can guess some of the words in the ciphertext. For example, the most common three-letter words in English are **the** and **and**, and these are relatively easy to spot – **Lhe**, which appears six times, and **aPV**, which appears five times. Hence, **L** probably represents **t**, **P** probably represents **n**, and **V** probably represents **d**. We can now replace these letters in the ciphertext with their true values:

nCQ dMJinD thiK tiSe KhahJaWad had ZCJne EinD KhahJiUaJ thJee KCnK. Cn the thCMKand and IiJKt niDht, Qhen Khe had ended the taRe CI Sa'aJMI, Khe JCKe and EiKKed the DJCMnd ZeICJe hiS, KaUinD: 'DJeat EinD, ICJ a thCMKand and Cne niDhtK i haNe Zeen JeACMntinD tC UCM the IaZReK CI FaKt aDeK and the ReDendK CI anAient EinDK. SaU i SaEe KC ZCRd aK tC AJaNe a IaNCMJ CI UCMJ SaGeKtU?'

eFiRCDMe, taReK IJCS the thCMKand and Cne niDhtK

Once a few letters have been established, cryptanalysis progresses very rapidly. For example, the word at the beginning of the second sentence is **Cn**. Every word has a vowel in it, so **C** must be a vowel. There are only two vowels that remain to be identified, **u** and **o**; **u** does not fit, so **C** must represent **o**. We also have the word **Khe**, which implies that **K** represents either **t** or **s**. But we already know that **L** = **t**, so it becomes clear that **K** = **s**. Having identified these two letters, we insert them into the ciphertext, and there appears the phrase **thoMsand and one niDhts**. A sensible guess for this would be **thousand and one nights**, and it seems likely that the final line is telling us that this is a passage from *Tales from the Thousand and One Nights*. This implies that **M** = **u**, **I** = **f**, **J** = **r**, **D** = **g**, **R** = **l**, and **S** = **m**.

We could continue trying to establish other letters by guessing other words, but instead let us have a look at what we know about the plain alphabet and cipher alphabet. These two alphabets form the key, and they were used by the cryptographer in order to perform the substitution that scrambled the message. Already, by identifying the true values of letters in the ciphertext, we have effectively been working out the details of the cipher alphabet. A summary of our achievements, so far, is given in the plain and cipher alphabets below.

Plain alphabet	a	b	c	d	e	f	g	h	i	j	k	l	m	n	o	p	q	r	s	t	u	v	w	x	y	z
Cipher alphabet	X	–	–	V	O	I	D	B	Y	–	–	R	S	P	C	–	–	J	K	L	M	–	–	–	–	–

By examining the partial cipher alphabet, we can complete the cryptanalysis. The sequence **VOIDBY** in the cipher alphabet suggests that the cryptographer has chosen a keyphrase as the basis for the key. Some guesswork is enough to suggest the keyphrase might be **A VOID BY GEORGES PEREC**, which is reduced to **AVOIDBYGERSPC** after removing spaces and repetitions. Thereafter, the letters continue in alphabetical order, omitting any that have already appeared in the keyphrase. In this particular case, the cryptographer took the unusual step of not starting the keyphrase at the beginning of the cipher alphabet, but rather starting it three letters in. This is possibly because the keyphrase begins with the letter **A**, and the cryptographer wanted to avoid encrypting **a** as **A**. At last, having established the complete cipher alphabet, we can unscramble the entire ciphertext, and the cryptanalysis is complete.

Plain alphabet	a	b	c	d	e	f	g	h	i	j	k	l	m	n	o	p	q	r	s	t	u	v	w	x	y	z
Cipher alphabet	X	Z	A	V	O	I	D	B	Y	G	E	R	S	P	C	F	H	J	K	L	M	N	Q	T	U	W

Now during this time Shahrazad had borne King Shahriyar three sons. On the thousand and first night, when she had ended the tale of Ma'aruf, she rose and kissed the ground before him, saying: 'Great King, for a thousand and one nights I have been recounting to you the fables of past ages and the legends of ancient kings. May I make so bold as to crave a favour of your majesty?'

Epilogue, *Tales from the Thousand and One Nights*

Renaissance in the West

Between AD 800 and 1200 Arab scholars enjoyed a vigorous period of intellectual achievement. At the same time, Europe was firmly stuck in the Dark Ages. While al-Kindī was describing the invention of cryptanalysis, Europeans were still struggling with the basics of cryptography. The only European institutions to encourage the study of secret writing were the monasteries, where monks would study the Bible in search of hidden meanings, a fascination that has persisted through to modern times (see Appendix C).

Medieval monks were intrigued by the fact that the Old Testament contained deliberate and obvious examples of cryptography. For example, the Old Testament includes pieces of text encrypted with *atbash*, a traditional form of Hebrew substitution cipher. Atbash involves taking each letter, noting the number of places it is from the beginning of the alphabet, and replacing it with a letter that is an equal number of places from the end of the alphabet. In English this would mean that a, at the beginning of the alphabet, is replaced by Z, at the end of the alphabet, b is replaced by Y, and so on. The term atbash itself hints at the substitution it describes, because it consists of the first letter of the Hebrew alphabet, *aleph*, followed by the last letter *taw*, and then there is the second letter, *beth*, followed by the second to last letter *shin*. An example of atbash appears in Jeremiah 25: 26 and 51: 41, where 'Babel' is replaced by the word 'Sheshach'; the first letter of Babel is *beth*, the second letter of the Hebrew alphabet, and this is replaced by *shin*, the second-to-last letter; the second letter of Babel is also *beth*, and so it too is replaced by *shin*; and the last letter of Babel is *lamed*, the twelfth letter of the Hebrew alphabet, and this is replaced by *kaph*, the twelfth-to-last letter.

Atbash and other similar Biblical ciphers were probably intended only to add mystery, rather than to conceal meaning, but they were enough to spark an interest in serious cryptography. European monks began to rediscover old substitution ciphers, they invented new ones, and, in due course, they helped to reintroduce cryptography into Western civilisation. The first known European book to describe the use of cryptography was written in the thirteenth century by the English Franciscan monk and polymath Roger Bacon. *Epistle on the Secret Works of Art and the Nullity of*

Magic included seven methods for keeping messages secret, and cautioned: 'A man is crazy who writes a secret in any other way than one which will conceal it from the vulgar.'

By the fourteenth century the use of cryptography had become increasingly widespread, with alchemists and scientists using it to keep their discoveries secret. Although better known for his literary achievements, Geoffrey Chaucer was also an astronomer and a cryptographer, and he is responsible for one of the most famous examples of early European encryption. In his *Treatise on the Astrolabe* he provided some additional notes entitled 'The Equatorie of the Planetis', which included several encrypted paragraphs. Chaucer's encryption replaced plaintext letters with symbols, for example b with δ. A ciphertext consisting of strange symbols rather than letters may at first sight seem more complicated, but it is essentially equivalent to the traditional letter-for-letter substitution. The process of encryption and the level of security are exactly the same.

By the fifteenth century, European cryptography was a burgeoning industry. The revival in the arts, sciences and scholarship during the Renaissance nurtured the capacity for cryptography, while an explosion in political machinations offered ample motivation for secret communication. Italy, in particular, provided the ideal environment for cryptography. As well as being at the heart of the Renaissance, it consisted of independent city states, each trying to outmanoeuvre the others. Diplomacy flourished, and each state would send ambassadors to the courts of the others. Each ambassador received messages from his respective head of state, describing details of the foreign policy he was to implement. In response, each ambassador would send back any information that he had gleaned. Clearly there was a great incentive to encrypt communications in both directions, so each state established a cipher office, and each ambassador had a cipher secretary.

At the same time that cryptography was becoming a routine diplomatic tool, the science of cryptanalysis was beginning to emerge in the West. Diplomats had only just familiarised themselves with the skills required to establish secure communications, and already there were individuals attempting to destroy this security. It is quite probable that cryptanalysis was independently discovered in Europe, but there is also the possibility that it was introduced from the Arab world. Islamic discoveries in science and mathematics strongly influenced the rebirth of science in Europe, and cryptanalysis might have been among the imported knowledge.

Arguably the first great European cryptanalyst was Giovanni Soro, appointed as Venetian cipher secretary in 1506. Soro's reputation was known throughout Italy, and friendly states would send intercepted messages to Venice for cryptanalysis. Even the Vatican, probably the second most active centre of cryptanalysis, would send Soro seemingly impenetrable messages that had fallen into its hands. In 1526, Pope Clement VII sent him two encrypted messages, and both were returned having been successfully cryptanalysed. And when one of the Pope's own encrypted messages was captured by the Florentines, the Pope sent a copy to Soro in the hope that he would be reassured that it was unbreakable. Soro claimed that he could not break the Pope's cipher, implying that the Florentines would also be unable to decipher it. However, this may have been a ploy to lull the Vatican cryptographers into a false sense of security – Soro might have been reluctant to point out the weaknesses of the Papal cipher, because this would only have encouraged the Vatican to switch to a more secure cipher, one that Soro might not have been able to break.

Elsewhere in Europe, other courts were also beginning to employ skilled cryptanalysts, such as Philibert Babou, cryptanalyst to King Francis I of France. Babou gained a reputation for being incredibly persistent, working day and night and persevering for weeks on end in order to crack an intercepted message. Unfortunately for Babou, this gave the king ample opportunity to carry on a long-term affair with his wife. Towards the end of the sixteenth century the French consolidated their codebreaking prowess with the arrival of François Viète, who took particular pleasure in cracking Spanish ciphers. Spain's cryptographers, who appear to have been naive compared with their rivals elsewhere in Europe, could not believe it when they discovered that their messages were transparent to the French. King Philip II of Spain went as far as petitioning the Vatican, claiming that the only explanation for Viète's cryptanalysis was that he was an 'archfiend in league with the devil'. Philip argued that Viète should be tried before a Cardinal's Court for his demonic deeds; but the Pope, who was aware that his own cryptanalysts had been reading Spanish ciphers for years, rejected the Spanish petition.

News of the petition soon reached cipher experts in various countries, and Spanish cryptographers became the laughing stock of Europe.

The Spanish embarrassment was symptomatic of the state of the battle between cryptographers and cryptanalysts. This was a period of transition, with cryptographers still relying on the monoalphabetic substitution cipher, while cryptanalysts were beginning to use frequency analysis to break it. Those yet to discover the power of frequency analysis continued to trust monoalphabetic substitution, ignorant of the extent to which cryptanalysts such as Soro, Babou and Viète were able to read their messages.

Meanwhile, countries that were alert to the weakness of the straightforward monoalphabetic substitution cipher were anxious to develop a better cipher, something that would protect their own nation's messages from being unscrambled by enemy cryptanalysts. One of the simplest improvements to the security of the monoalphabetic substitution cipher was the introduction of *nulls*, symbols or letters that were not substitutes for actual letters, merely blanks that represented nothing. For example, one could substitute each plain letter with a number between 1 and 99, which would leave 73 numbers that represent nothing, and these could be randomly sprinkled throughout the ciphertext with varying frequencies. The nulls would pose no problem to the intended recipient, who would know that they were to be ignored. However, the nulls would baffle an enemy interceptor because they would confuse an attack by frequency analysis. An equally simple development was that cryptographers would sometimes deliberately misspell words before encrypting the message. **Thys haz thi ifekkt off diztaughting thi ballans off frikwenseas** – making it harder for the cryptanalyst to apply frequency analysis. However, the intended recipient, who knows the key, can unscramble the message and then deal with the bad, but not unintelligible, spelling.

Another attempt to shore up the monoalphabetic substitution cipher involved the introduction of codewords. The term *code* has a very broad meaning in everyday language, and it is often used to describe any method for communicating in secret. However, as mentioned in the Introduction, it actually has a very specific meaning, and applies only to a certain form of substitution. So far we have concentrated on the idea of a substitution cipher, whereby each letter is replaced by a different letter, number or symbol. However, it is also possible to have substitution at a much higher level, whereby each word is represented by another word or symbol – this would be a code. For example,

assassinate	= D	general	= Σ	immediately	= 08
blackmail	= P	king	= Ω	today	= 73
capture	= J	minister	= ψ	tonight	= 28
protect	= Z	prince	= θ	tomorrow	= 43

Plain message = assassinate the king tonight
Encoded message = D–Ω–28

Technically, a *code* is defined as substitution at the level of words or phrases, whereas a *cipher* is defined as substitution at the level of letters. Hence the term *encipher* means to scramble a message using a cipher, while *encode* means to scramble a message using a code. Similarly, the term *decipher* applies to unscrambling an enciphered message, and *decode* to unscrambling an encoded message. The terms *encrypt* and *decrypt* are more general, and cover scrambling and unscrambling with respect to both codes and ciphers. Figure 7 presents a brief summary of these definitions. In general, I shall keep to these definitions, but when the sense is clear, I might use a term such as 'codebreaking' to describe a process that is really 'cipher breaking' – the latter phrase might be technically accurate, but the former phrase is widely accepted.

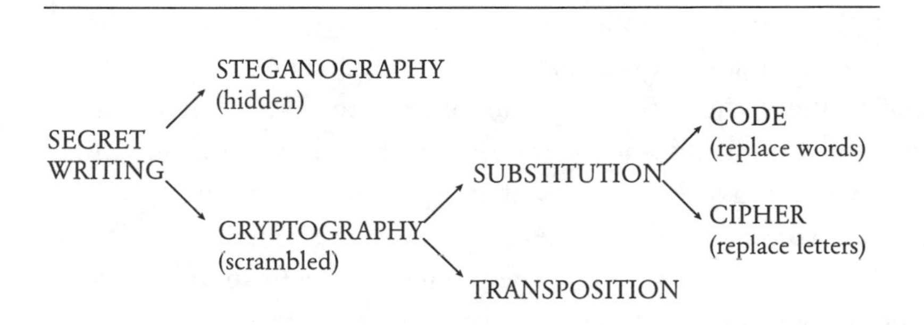

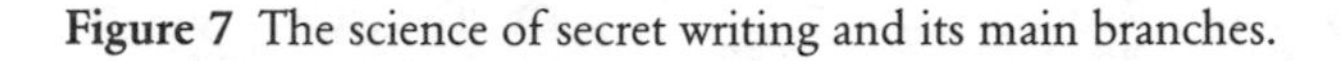
Figure 7 The science of secret writing and its main branches.

At first sight, codes seem to offer more security than ciphers, because words are much less vulnerable to frequency analysis than letters. To decipher a monoalphabetic cipher you need only identify the true value of each of the 26 characters, whereas to decipher a code you need to identify the true value of hundreds or even thousands of codewords. However, if we examine codes in more detail, we see that they suffer from two major practical failings when compared with ciphers. First, once the sender and receiver have agreed upon the 26 letters in the cipher alphabet (the key), they can encipher any message, but to achieve the same level of flexibility using a code they would need to go through the painstaking task of defining a codeword for every one of the thousands of possible plaintext words. The codebook would consist of hundreds of pages, and would look something like a dictionary. In other words, compiling a codebook is a major task, and carrying it around is a major inconvenience.

Second, the consequences of having a codebook captured by the enemy are devastating. Immediately, all the encoded communications would become transparent to the enemy. The senders and receivers would have to go through the painstaking process of having to compile an entirely new codebook, and then this hefty new tome would have to be distributed to everyone in the communications network, which might mean securely transporting it to every ambassador in every state. In comparison, if the enemy succeeds in capturing a cipher key, then it is relatively easy to compile a new cipher alphabet of 26 letters, which can be memorised and easily distributed.

Even in the sixteenth century, cryptographers appreciated the inherent weaknesses of codes, and instead relied largely on ciphers, or sometimes *nomenclators*. A nomenclator is a system of encryption that relies on a cipher alphabet, which is used to encrypt the majority of a message, and a limited list of codewords. For example, a nomenclator book might consist of a front page containing the cipher alphabet, and then a second page containing a list of codewords. Despite the addition of codewords, a nomenclator is not much more secure than a straightforward cipher, because the bulk of a message can be deciphered using frequency analysis, and the remaining encoded words can be guessed from the context.

As well as coping with the introduction of the nomenclator, the best cryptanalysts were also capable of dealing with badly spelt messages and the presence of nulls. In short, they were able to break the majority of encrypted messages. Their skills provided a steady flow of uncovered secrets, which influenced the decisions of their masters and mistresses, thereby affecting Europe's history at critical moments.

Nowhere is the impact of cryptanalysis more dramatically illustrated than in the case of Mary Queen of Scots. The outcome of her trial depended wholly on the battle between her codemakers and Queen Elizabeth's codebreakers. Mary was one of the most significant figures of the sixteenth century – Queen of Scotland, Queen of France, pretender to the English throne – yet her fate would be decided by a slip of paper, the message it bore, and whether or not that message could be deciphered.

The Babington Plot

On 24 November 1542, the English forces of Henry VIII demolished the Scottish army at the Battle of Solway Moss. It appeared that Henry was on the verge of conquering Scotland and stealing the crown of King James V. After the battle, the distraught Scottish king suffered a complete mental and physical breakdown, and withdrew to the palace at Falkland. Even the birth of a daughter, Mary, just two weeks later could not revive the ailing king. It was as if he had been waiting for news of an heir so that he could die in peace, safe in the knowledge that he had done his duty. Just a week after Mary's birth, King James V, still only thirty years old, died. The baby princess had become Mary Queen of Scots.

Mary was born prematurely, and initially there was considerable concern that she would not survive. Rumours in England suggested that the baby had died, but this was merely wishful thinking at the English court, which was keen to hear any news that might destabilise Scotland. In fact, Mary soon grew strong and healthy, and at the age of nine months, on 9 September 1543, she was crowned in the chapel of Stirling Castle, surrounded by three earls, bearing on her behalf the royal crown, sceptre and sword.

The fact that Queen Mary was so young offered Scotland a respite from English incursions. It would have been deemed unchivalrous had Henry VIII attempted to invade the country of a recently dead king, now under the rule of an infant queen. Instead, the English king decided on a policy of wooing Mary in the hope of arranging a marriage between her and his son Edward, thereby uniting the two nations under a Tudor sovereign. He began his manoeuvring by releasing the Scottish nobles captured at Solway Moss, on the condition that they campaign in favour of a union with England.

However, after considering Henry's offer, the Scottish court rejected it in favour of a marriage to Francis, the dauphin of France. Scotland was choosing to ally itself with a fellow Roman Catholic nation, a decision which pleased Mary's mother, Mary of Guise, whose own marriage with James V had been intended to cement the relationship between Scotland and France. Mary and Francis were still children, but the plan for the future was that they would eventually marry, and Francis would ascend the throne of France with Mary as his queen, thereby uniting Scotland and France. In the meantime, France would defend Scotland against any English onslaught.

The promise of protection was reassuring, particularly as Henry VIII had switched from diplomacy to intimidation in order to persuade the Scots that his own son was a more worthy groom for Mary Queen of Scots. His forces committed acts of piracy, destroyed crops, burnt villages and attacked towns and cities along the border. The 'rough wooing', as it is known, continued even after Henry's death in 1547. Under the auspices of his son, King Edward VI (the would-be suitor), the attacks culminated in the Battle of Pinkie Cleugh, in which the Scottish army was routed. As a result of this slaughter it was decided that, for her own safety, Mary should leave for France, beyond the reach of the English threat, where she could prepare for her marriage to Francis. On 7 August 1548, at the age of six, she set sail for the port of Roscoff.

Mary's first few years in the French court would be the most idyllic time of her life. She was surrounded by luxury, protected from harm, and she grew to love her future husband, the dauphin. At the age of sixteen they married, and the following year Francis and Mary became King and Queen of France. Everything seemed set for her triumphant return to Scotland, until her husband, who had always suffered from poor health, fell gravely ill. An ear infection that he had nursed since a child had worsened, the inflammation spread towards his brain, and an abscess began to develop. In 1560, within a year of being crowned, Francis was dead and Mary was widowed.

From this point onwards, Mary's life would be repeatedly struck by tragedy. She returned to Scotland in 1561, where she discovered a transformed nation. During her long absence Mary had confirmed her Catholic faith, while her Scottish subjects had increasingly moved towards the Protestant church. Mary tolerated the wishes of the majority and at first reigned with relative success, but in 1565 she married her cousin, Henry Stewart, the Earl of Darnley, an act that led to a spiral of decline. Darnley was a vicious and brutal man whose ruthless greed for power lost Mary the loyalty of the Scottish nobles. The following year Mary witnessed for herself the full horror of her husband's barbaric nature when he murdered David Riccio, her secretary, in front of her. It became clear to everyone that for the sake of Scotland it was necessary to get rid of Darnley. Historians debate whether it was Mary or the Scottish nobles who instigated the plot, but on the night of 9 February 1567, Darnley's house was blown up and, as he attempted to escape, he was strangled. The only good to come from the marriage was a son and heir, James.

Mary's next marriage, to James Hepburn, the Fourth Earl of Bothwell, was hardly more successful. By the summer of 1567 the Protestant Scottish nobles had become completely disillusioned with their Catholic Queen, and they exiled Bothwell and imprisoned Mary, forcing her to abdicate in favour of her fourteen-month-old son, James VI, while her half-brother, the Earl of Moray, acted as regent. The next year, Mary escaped from her prison, gathered an army of six thousand royalists, and made a final attempt to regain her crown. Her soldiers confronted the regent's army at the small village of Langside, near Glasgow, and Mary witnessed the battle from a nearby hilltop. Although her troops were greater in number, they lacked discipline, and Mary watched as they were torn apart. When defeat was inevitable, she fled. Ideally she would have headed east to the coast, and then on to France, but this would have meant crossing territory loyal to her half-brother, and so instead she headed south to England, where she hoped that her cousin Queen Elizabeth I would provide refuge.

Mary had made a terrible misjudgement. Elizabeth offered Mary nothing more than another prison. The official reason for her arrest was in connection with the murder of Darnley, but the true reason was that Mary posed a threat to Elizabeth, because English Catholics considered

Mary to be the true queen of England. Through her grandmother, Margaret Tudor, the elder sister of Henry VIII, Mary did indeed have a claim to the throne, but Henry's last surviving offspring, Elizabeth I, would seem to have a prior claim. However, according to Catholics, Elizabeth was illegitimate because she was the daughter of Anne Boleyn, Henry's second wife after he had divorced Catherine of Aragon in defiance of the Pope. English Catholics did not recognise Henry VIII's divorce, they did not acknowledge his ensuing marriage to Anne Boleyn, and they certainly did not accept their daughter Elizabeth as Queen. Catholics saw Elizabeth as a bastard usurper.

Mary was imprisoned in a series of castles and manors. Although Elizabeth thought of her as one of the most dangerous figures in England, many Englishmen admitted that they admired her gracious manner, her obvious intelligence and her great beauty. William Cecil, Elizabeth's Great Minister, commented on 'her cunning and sugared entertainment of all men', and Nicholas White, Cecil's emissary, made a similar observation: 'She hath withal an alluring grace, a pretty Scotch accent, and a searching wit, clouded with mildness.' But, as each year passed, her appearance waned, her health deteriorated and she began to lose hope. Her jailer, Sir Amyas Paulet, a Puritan, was immune to her charms, and treated her with increasing harshness.

By 1586, after 18 years of imprisonment, she had lost all her privileges. She was confined to Chartley Hall in Staffordshire, and was no longer allowed to take the waters at Buxton, which had previously helped to alleviate her frequent illnesses. On her last visit to Buxton she used a diamond to inscribe a message on a window-pane: 'Buxton, whose warm waters have made thy name famous, perchance I shall visit thee no more – Farewell.' It appears that she suspected that she was about to lose what little freedom she had. Mary's growing sorrow was compounded by the actions of her nineteen-year-old son, King James VI of Scotland. She had always hoped that one day she would escape and return to Scotland to share power with her son, whom she had not seen since he was one year old. However, James felt no such affection for his mother. He had been brought up by Mary's enemies, who had taught James that his mother had murdered his father in order to marry her lover. James despised her, and feared that if she returned then she might seize his crown. His hatred towards Mary was demonstrated by the fact that he had no qualms in seeking a marriage with Elizabeth I, the woman responsible for his mother's imprisonment (and who was also thirty years his senior). Elizabeth declined the offer.

Mary wrote to her son in an attempt to win him over, but her letters never reached the Scottish border. By this stage, Mary was more isolated then ever before: all her outgoing letters were confiscated, and any incoming correspondence was kept by her jailer. Mary's morale was at its lowest, and it seemed that all hope was lost. It was under these severe and desperate circumstances that, on 6 January 1586, she received an astonishing package of letters.

The letters were from Mary's supporters on the Continent, and they had been smuggled into her prison by Gilbert Gifford, a Catholic who had left England in 1577 and trained as a priest at the English College in Rome. Upon returning to England in 1585, apparently keen to serve Mary, he immediately approached the French Embassy in London, where a pile of correspondence had accumulated. The Embassy had known that if they forwarded the letters by the formal route, Mary would never see them. However Gifford claimed that he could smuggle the letters into Chartley Hall, and sure enough he lived up to his word. This delivery was the first of many, and Gifford began a career as a courier, not only passing messages to Mary but also collecting her replies. He had a rather cunning way of sneaking letters into Chartley Hall. He took the messages to a local brewer, who wrapped them in a leather packet, which was then hidden inside a hollow bung used to seal a barrel of beer. The brewer would deliver the barrel to Chartley Hall, whereupon one of Mary's servants would open the bung and take the contents to the Queen of Scots. The process worked equally well for getting messages out of Chartley Hall.

Meanwhile, unknown to Mary, a plan to rescue her was being hatched in the taverns of London. At the centre of the plot was Anthony Babington, aged just twenty-four but already well known in the city as a handsome, charming and witty bon viveur. What his many admiring contemporaries failed to appreciate was that Babington deeply resented the establishment, which had persecuted him, his family and his faith. The state's anti-Catholic policies had reached new heights of horror, with priests being accused of treason, and anybody caught harbouring them punished by the rack, mutilation and disembowelling while still alive. The Catholic mass was officially banned, and families who remained loyal to the Pope were forced to pay crippling taxes. Babington's animosity was fuelled by the death of Lord Darcy, his great-grandfather, who was beheaded for his involvement in the Pilgrimage of Grace, a Catholic uprising against Henry VIII.

The conspiracy began one evening in March 1586, when Babington and six confidants gathered in The Plough, an inn outside Temple Bar. As the historian Philip Caraman observed, 'He drew to himself by the force of his exceptional charm and personality many young Catholic gentlemen of his own standing, gallant, adventurous and daring in defence of the Catholic faith in its day of stress; and ready for any arduous enterprise whatsoever that might advance the common Catholic cause.' Over the next few months an ambitious plan emerged to free Mary Queen of Scots, assassinate Queen Elizabeth and incite a rebellion supported by an invasion from abroad.

The conspirators were agreed that the Babington Plot, as it became known, could not proceed without the blessing of Mary, but there was no apparent way to communicate with her. Then, on 6 July 1586, Gifford arrived on Babington's doorstep. He delivered a letter from Mary, explaining that she had heard about Babington via her supporters in Paris, and looked forward to hearing from him. In reply, Babington compiled a detailed letter in which he outlined his scheme, including a reference to the excommunication of Elizabeth by Pope Pius V in 1570, which he believed legitimised her assassination.

> Myself with ten gentlemen and a hundred of our followers will undertake the delivery of your royal person from the hands of your enemies. For the dispatch of the usurper, from the obedience of whom we are by the excommunication of her made free, there be six noble gentlemen, all my private friends, who for the zeal they bear to the Catholic cause and your Majesty's service will undertake that tragical execution.

As before, Gifford used his trick of putting the message in the bung of a beer barrel in order to sneak it past Mary's guards. This can be considered a form of steganography, because the letter was being hidden. As an extra precaution, Babington enciphered his letter so that even if it was intercepted by Mary's jailer, it would be indecipherable and the plot would not be uncovered. He used a cipher which was not a simple monoalphabetic substitution, but rather a nomenclator, as shown in Figure 8. It consisted of 23 symbols that were to be substituted for the letters of the alphabet (excluding j, v and w), along with 36 symbols representing words or phrases. In addition, there were four nulls (ff.⌐.⊣.d.) and a symbol σ which signified that the next symbol represents a double letter ('dowbleth').

Gifford was still a youth, even younger than Babington, and yet he conducted his deliveries with confidence and guile. His aliases, such as Mr Colerdin, Pietro and Cornelys, enabled him to travel the country without suspicion, and his contacts within the Catholic community provided him with a series of safe houses between London and Chartley Hall. However, each time Gifford travelled to or from Chartley Hall, he would make a detour. Although Gifford was apparently acting as an agent for Mary, he was actually a double agent. Back in 1585, before his return to England, Gifford had written to Sir Francis Walsingham, Principal Secretary to Queen Elizabeth, offering his services. Gifford realised that his Catholic background would act as a perfect mask for infiltrating plots

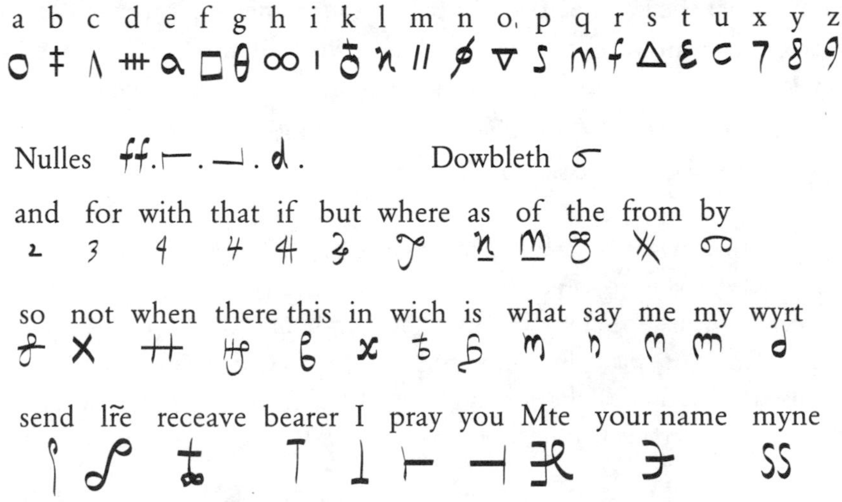

a b c d e f g h i k l m n o p q r s t u x y z

Nulles ff.⌐.⊣.d.   Dowbleth σ

and for with that if but where as of the from by

so not when there this in wich is what say me my wyrt

send lr̃e receave bearer I pray you Mte your name myne

Figure 8 The nomenclator of Mary Queen of Scots, consisting of a cipher alphabet and codewords.

against Queen Elizabeth. In the letter to Walsingham, he wrote, 'I have heard of the work you do and I want to serve you. I have no scruples and no fear of danger. Whatever you order me to do I will accomplish.'

Walsingham was Elizabeth's most ruthless minister. He was a machiavellian figure, a spymaster who was responsible for the security of the monarch. He had inherited a small network of spies, which he rapidly expanded into the Continent, where many of the plots against Elizabeth were being hatched. After his death it was discovered that he had been receiving regular reports from twelve locations in France, nine in Germany, four in Italy, four in Spain and three in the Low Countries, as well as having informants in Constantinople, Algiers and Tripoli.

Walsingham recruited Gifford as a spy, and in fact it was Walsingham who ordered Gifford to approach the French Embassy and offer himself as a courier. Each time Gifford collected a message to or from Mary, he would first take it to Walsingham. The vigilant spymaster would then pass it to his counterfeiters, who would break the seal on each letter, make a copy, and then reseal the original letter with an identical stamp before handing it back to Gifford. The apparently untouched letter could then be delivered to Mary or her correspondents, who remained oblivious to what was going on.

When Gifford handed Walsingham a letter from Babington to Mary, the first objective was to decipher it. Walsingham had originally encountered codes and ciphers while reading a book written by the Italian mathematician and cryptographer Girolamo Cardano (who, incidentally, proposed a form of writing for the blind based on touch, a precursor of Braille). Cardano's book aroused Walsingham's interest, but it was a decipherment by the Flemish cryptanalyst Philip van Marnix that really convinced him of the power of having a codebreaker at his disposal. In 1577, Philip of Spain was using ciphers to correspond with his half-brother and fellow Catholic, Don John of Austria, who was in control of much of the Netherlands. Philip's letter described a plan to invade England, but it was intercepted by William of Orange, who passed it to Marnix, his cipher secretary. Marnix deciphered the plan, and William passed the information to Daniel Rogers, an English agent working on the Continent, who in turn warned Walsingham of the invasion. The English reinforced their defences, which was enough to deter the invasion attempt.

Now fully aware of the value of cryptanalysis, Walsingham established a cipher school in London and employed Thomas Phelippes as his cipher secretary, a man 'of low stature, slender every way, dark yellow haired on the head, and clear yellow bearded, eaten in the face with smallpox, of short sight, thirty years of age by appearance'. Phelippes was a linguist who could speak French, Italian, Spanish, Latin and German, and, more importantly, he was one of Europe's finest cryptanalysts.

Upon receiving any message to or from Mary, Phelippes devoured it. He was a master of frequency analysis, and it would be merely a matter of time before he found a solution. He established the frequency of each character, and tentatively proposed values for those that appeared most often. When a particular approach hinted at absurdity, he would backtrack and try alternative substitutions. Gradually he would identify the nulls, the cryptographic red herrings, and put them to one side. Eventually all that remained were the handful of codewords, whose meaning could be guessed from the context.

When Phelippes deciphered Babington's message to Mary, which clearly proposed the assassination of Elizabeth, he immediately forwarded the damning text to his master. At this point Walsingham could have pounced on Babington, but he wanted more than the execution of a handful of rebels. He bided his time in the hope that Mary would reply and authorise the plot, thereby incriminating herself. Walsingham had long wished for the death of Mary Queen of Scots, but he was aware of Elizabeth's reluctance to execute her cousin. However, if he could prove that Mary was endorsing an attempt on the life of Elizabeth, then surely his queen would permit the execution of her Catholic rival. Walsingham's hopes were soon fulfilled.

On 17 July, Mary replied to Babington, effectively signing her own death warrant. She explicitly wrote about the 'design', showing particular concern that she should be released simultaneously with, or before, Elizabeth's assassination, otherwise news might reach her jailer, who might then murder her. Before reaching Babington, the letter made the usual detour to Phelippes. Having cryptanalysed the earlier message, he deciphered this one with ease, read its contents, and marked it with a 'Π' – the sign of the gallows.

Walsingham had all the evidence he needed to arrest Mary and Babington, but still he was not satisfied. In order to destroy the conspiracy completely, he needed the names of all those involved. He asked Phelippes to forge a postscript to Mary's letter, which would entice Babington to name names. One of Phelippes's additional talents was as a forger, and it was said that he had the ability 'to write any man's hand, if he had once seen it, as if the man himself had writ it'. Figure 9 shows the postscript that was added at the end of Mary's letter to Babington. It can be deciphered using Mary's nomenclator, as shown in Figure 8, to reveal the following plaintext:

> I would be glad to know the names and qualities of the six gentlemen which are to accomplish the designment; for it may be that I shall be able, upon knowledge of the parties, to give you some further advice necessary to be followed therein, as also from time to time particularly how you proceed: and as soon as you may, for the same purpose, who be already, and how far everyone is privy hereunto.

The cipher of Mary Queen of Scots clearly demonstrates that a weak encryption can be worse than no encryption at all. Both Mary and Babington wrote explicitly about their intentions because they believed that their communications were secure, whereas if they had been communicating openly they would have referred to their plan in a more discreet manner. Furthermore, their faith in their cipher made them particularly vulnerable to accepting Phelippes's forgery. Sender and receiver often have such confidence in the strength of their cipher that

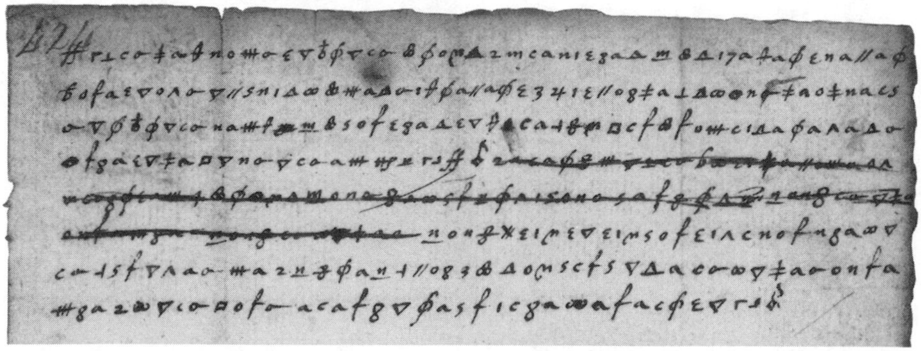

Figure 9 The forged postscript added by Thomas Phelippes to Mary's message. It can be deciphered by referring to Mary's nomenclator (Figure 8).

they consider it impossible for the enemy to mimic the cipher and insert forged text. The correct use of a strong cipher is a clear boon to sender and receiver, but the misuse of a weak cipher can generate a very false sense of security.

Soon after receiving the message and its postscript, Babington needed to go abroad to organise the invasion, and had to register at Walsingham's department in order to acquire a passport. This would have been an ideal time to capture the traitor, but the bureaucrat who was manning the office, John Scudamore, was not expecting the most wanted traitor in England to turn up at his door. Scudamore, with no support to hand, took the unsuspecting Babington to a nearby tavern, stalling for time while his assistant organised a group of soldiers. A short while later a note arrived at the tavern, informing Scudamore that it was time for the arrest. Babington, however, caught sight of it. He casually said that he would pay for the beer and meal and rose to his feet, leaving his sword and coat at the table, implying that he would return in an instant. Instead, he slipped out of the back door and escaped, first to St John's Wood and then on to Harrow. He attempted to disguise himself, cutting his hair short and staining his skin with walnut juice to mask his aristocratic background. He managed to elude capture for ten days, but by 15 August Babington and his six colleagues were captured and brought to London. Church bells across the city rang out in triumph. Their executions were horrid in the extreme. In the words of the Elizabethan historian William Camden, 'they were all cut down, their privities were cut off, bowelled alive and seeing, and quartered'.

Meanwhile, on 11 August, Mary Queen of Scots and her entourage had been allowed the exceptional privilege of riding in the grounds of Chartley Hall. As Mary crossed the moors she spied some horsemen approaching, and immediately thought that these must be Babington's men coming to rescue her. It soon became clear that these men had come to arrest her, not release her. Mary had been implicated in the Babington Plot, and was charged under the Act of Association, an Act of Parliament passed in 1584 specifically designed to convict anybody involved in a conspiracy against Elizabeth.

The trial was held in Fotheringhay Castle, a bleak, miserable place in the middle of the featureless fens of East Anglia. It began on Wednesday 15 October, in front of two chief justices, four other judges, the Lord

Chancellor, the Lord Treasurer, Walsingham, and various earls, knights and barons. At the back of the courtroom there was space for spectators, such as local villagers and the servants of the commissioners, all eager to see the humiliated Scottish queen beg forgiveness and plead for her life. However, Mary remained dignified and composed throughout the trial. Mary's main defence was to deny any connection with Babington. 'Can I be responsible for the criminal projects of a few desperate men', she proclaimed, 'which they planned without my knowledge or participation?' Her statement had little impact in the face of the evidence against her.

Mary and Babington had relied on a cipher to keep their plans secret, but they lived during a period when cryptography was being weakened by advances in cryptanalysis. Although their cipher would have been sufficient protection against the prying eyes of an amateur, it stood no chance against an expert in frequency analysis. In the spectators' gallery sat Phelippes, quietly watching the presentation of the evidence that he had conjured from the enciphered letters.

The trial went into a second day, and Mary continued to deny any knowledge of the Babington Plot. When the trial finished, she left the judges to decide her fate, pardoning them in advance for the inevitable decision. Ten days later, the Star Chamber met in Westminster and concluded that Mary had been guilty of 'compassing and imagining since June 1st matters tending to the death and destruction of the Queen of England'. They recommended the death penalty, and Elizabeth signed the death warrant.

On 8 February 1587, in the Great Hall of Fotheringhay Castle, an audience of three hundred gathered to watch the beheading. Walsingham was determined to minimise Mary's influence as a martyr, and he ordered that the block, Mary's clothing, and everything else relating to the execution be burned in order to avoid the creation of any holy relics. He also planned a lavish funeral procession for his son-in-law, Sir Philip Sidney, to take place the following week. Sidney, a popular and heroic figure, had died fighting Catholics in the Netherlands, and Walsingham believed that a magnificent parade in his honour would dampen sympathy for Mary. However, Mary was equally determined that her final appearance should be a defiant gesture, an opportunity to reaffirm her Catholic faith and inspire her followers.

While the Dean of Peterborough led the prayers, Mary spoke aloud her own prayers for the salvation of the English Catholic Church, for her son and for Elizabeth. With her family motto, 'In my end is my beginning', in her mind, she composed herself and approached the block. The executioners requested her forgiveness, and she replied, 'I forgive you with all my heart, for now I hope you shall make an end of all my troubles'. Richard Wingfield, in his *Narration of the Last Days of the Queen of Scots*, describes her final moments:

> Then she laide herself upon the blocke most quietlie, & stretching out her armes & legges cryed out In manus tuas domine three or foure times, & at the laste while one of the executioners held her slightlie with one of his handes, the other gave two strokes with an axe before he cutt of her head, & yet lefte a little gristle behinde at which time she made verie small noyse & stirred not any parte of herself from the place where she laye . . . Her lipps stirred up & downe almost a quarter of an hower after her head was cutt of. Then one of her executioners plucking of her garters espied her little dogge which was crept under her clothes which could not be gotten forth but with force & afterwardes could not depart from her dead corpse, but came and laye betweene her head & shoulders a thing dilligently noted.

Figure 10 The execution of Mary Queen of Scots.

2 Le Chiffre Indéchiffrable

For centuries, the simple monoalphabetic substitution cipher had been sufficient to ensure secrecy. The subsequent development of frequency analysis, first in the Arab world and then in Europe, destroyed its security. The tragic execution of Mary Queen of Scots was a dramatic illustration of the weaknesses of monoalphabetic substitution, and in the battle between cryptographers and cryptanalysts it was clear that the cryptanalysts had gained the upper hand. Anybody sending an encrypted message had to accept that an expert enemy codebreaker might intercept and decipher their most precious secrets.

The onus was clearly on the cryptographers to concoct a new, stronger cipher, something that could outwit the cryptanalysts. Although this cipher would not emerge until the end of the sixteenth century, its origins can be traced back to the fifteenth-century Florentine polymath Leon Battista Alberti. Born in 1404, Alberti was one the leading figures of the Renaissance – a painter, composer, poet and philosopher, as well as the author of the first scientific analysis of perspective, a treatise on the housefly and a funeral oration for his dog. He is probably best known as an architect, having designed Rome's first Trevi Fountain and having written *De re aedificatoria*, the first printed book on architecture, which acted as a catalyst for the transition from Gothic to Renaissance design.

Sometime in the 1460s, Alberti was wandering through the gardens of the Vatican when he bumped into his friend Leonardo Dato, the pontifical secretary, who began chatting to him about some of the finer points of cryptography. This casual conversation prompted Alberti to write an essay on the subject, outlining what he believed to be a new form of cipher. At the time, all substitution ciphers required a single cipher alphabet for encrypting each message. However, Alberti proposed using two or more cipher alphabets, switching between them during encipherment, thereby confusing potential cryptanalysts.

Plain alphabet	a	b	c	d	e	f	g	h	i	j	k	l	m	n	o	p	q	r	s	t	u	v	w	x	y	z
Cipher alphabet 1	F	Z	B	V	K	I	X	A	Y	M	E	P	L	S	D	H	J	O	R	G	N	Q	C	U	T	W
Cipher alphabet 2	G	O	X	B	F	W	T	H	Q	I	L	A	P	Z	J	D	E	S	V	Y	C	R	K	U	H	N

For example, here we have two possible cipher alphabets, and we could encrypt a message by alternating between them. To encrypt the message hello, we would encrypt the first letter according to the first cipher alphabet, so that h becomes A, but we would encrypt the second letter according to the second cipher alphabet, so that e becomes F. To encrypt the third letter we return to the first cipher alphabet, and to encrypt the fourth letter we return to the second alphabet. This means that the first l is enciphered as P, but the second l is enciphered as A. The final letter, o, is enciphered according to the first cipher alphabet and becomes D. The complete ciphertext reads AFPAD. The crucial advantage of Alberti's system is that the same letter in the plaintext does not necessarily appear as the same letter in the ciphertext, so the repeated l in hello is enciphered differently in each case. Similarly, the repeated A in the ciphertext represents a different plaintext letter in each case, first h and then l.

Although he had hit upon the most significant breakthrough in encryption for over a thousand years, Alberti failed to develop his concept into a fully formed system of encryption. That task fell to a diverse group of intellectuals, who built on his initial idea. First came Johannes Trithemius, a German abbot born in 1462, then Giovanni Porta, an Italian scientist born in 1535, and finally Blaise de Vigenère, a French diplomat born in 1523. Vigenère became acquainted with the writings of Alberti, Trithemius and Porta when, at the age of twenty-six, he was sent to Rome on a two-year diplomatic mission. To start with, his interest in cryptography was purely practical and was linked to his diplomatic work. Then, at the age of thirty-nine, Vigenère decided that he had accumulated enough money for him to be able to abandon his career and concentrate on a life of study. It was only then that he examined in detail the ideas of Alberti, Trithemius and Porta, weaving them into a coherent and powerful new cipher.

Figure 11 Blaise de Vigenère.

Although Alberti, Trithemius and Porta all made vital contributions, the cipher is known as the Vigenère cipher in honour of the man who developed it into its final form. The strength of the Vigenère cipher lies in its using not one, but 26 distinct cipher alphabets to encrypt a message. The first step in encipherment is to draw up a so-called Vigenère square, as shown in Table 3, a plaintext alphabet followed by 26 cipher alphabets, each shifted by one letter with respect to the previous alphabet. Hence, row 1 represents a cipher alphabet with a Caesar shift of 1, which means that it could be used to implement a Caesar shift cipher in which every letter of the plaintext is replaced by the letter one place further on in the alphabet. Similarly, row 2 represents a cipher alphabet with a Caesar shift

Table 3 A Vigenère square.

Plain	a	b	c	d	e	f	g	h	i	j	k	l	m	n	o	p	q	r	s	t	u	v	w	x	y	z
1	B	C	D	E	F	G	H	I	J	K	L	M	N	O	P	Q	R	S	T	U	V	W	X	Y	Z	A
2	C	D	E	F	G	H	I	J	K	L	M	N	O	P	Q	R	S	T	U	V	W	X	Y	Z	A	B
3	D	E	F	G	H	I	J	K	L	M	N	O	P	Q	R	S	T	U	V	W	X	Y	Z	A	B	C
4	E	F	G	H	I	J	K	L	M	N	O	P	Q	R	S	T	U	V	W	X	Y	Z	A	B	C	D
5	F	G	H	I	J	K	L	M	N	O	P	Q	R	S	T	U	V	W	X	Y	Z	A	B	C	D	E
6	G	H	I	J	K	L	M	N	O	P	Q	R	S	T	U	V	W	X	Y	Z	A	B	C	D	E	F
7	H	I	J	K	L	M	N	O	P	Q	R	S	T	U	V	W	X	Y	Z	A	B	C	D	E	F	G
8	I	J	K	L	M	N	O	P	Q	R	S	T	U	V	W	X	Y	Z	A	B	C	D	E	F	G	H
9	J	K	L	M	N	O	P	Q	R	S	T	U	V	W	X	Y	Z	A	B	C	D	E	F	G	H	I
10	K	L	M	N	O	P	Q	R	S	T	U	V	W	X	Y	Z	A	B	C	D	E	F	G	H	I	J
11	L	M	N	O	P	Q	R	S	T	U	V	W	X	Y	Z	A	B	C	D	E	F	G	H	I	J	K
12	M	N	O	P	Q	R	S	T	U	V	W	X	Y	Z	A	B	C	D	E	F	G	H	I	J	K	L
13	N	O	P	Q	R	S	T	U	V	W	X	Y	Z	A	B	C	D	E	F	G	H	I	J	K	L	M
14	O	P	Q	R	S	T	U	V	W	X	Y	Z	A	B	C	D	E	F	G	H	I	J	K	L	M	N
15	P	Q	R	S	T	U	V	W	X	Y	Z	A	B	C	D	E	F	G	H	I	J	K	L	M	N	O
16	Q	R	S	T	U	V	W	X	Y	Z	A	B	C	D	E	F	G	H	I	J	K	L	M	N	O	P
17	R	S	T	U	V	W	X	Y	Z	A	B	C	D	E	F	G	H	I	J	K	L	M	N	O	P	Q
18	S	T	U	V	W	X	Y	Z	A	B	C	D	E	F	G	H	I	J	K	L	M	N	O	P	Q	R
19	T	U	V	W	X	Y	Z	A	B	C	D	E	F	G	H	I	J	K	L	M	N	O	P	Q	R	S
20	U	V	W	X	Y	Z	A	B	C	D	E	F	G	H	I	J	K	L	M	N	O	P	Q	R	S	T
21	V	W	X	Y	Z	A	B	C	D	E	F	G	H	I	J	K	L	M	N	O	P	Q	R	S	T	U
22	W	X	Y	Z	A	B	C	D	E	F	G	H	I	J	K	L	M	N	O	P	Q	R	S	T	U	V
23	X	Y	Z	A	B	C	D	E	F	G	H	I	J	K	L	M	N	O	P	Q	R	S	T	U	V	W
24	Y	Z	A	B	C	D	E	F	G	H	I	J	K	L	M	N	O	P	Q	R	S	T	U	V	W	X
25	Z	A	B	C	D	E	F	G	H	I	J	K	L	M	N	O	P	Q	R	S	T	U	V	W	X	Y
26	A	B	C	D	E	F	G	H	I	J	K	L	M	N	O	P	Q	R	S	T	U	V	W	X	Y	Z

of 2, and so on. The top row of the square, in lower case, represents the plaintext letters. You could encipher each plaintext letter according to any one of the 26 cipher alphabets. For example, if cipher alphabet number 2 is used, then the letter a is enciphered as C, but if cipher alphabet number 12 is used, then a is enciphered as M.

If the sender were to use just one of the cipher alphabets to encipher an entire message, this would effectively be a simple Caesar cipher, which would be a very weak form of encryption, easily deciphered by an enemy interceptor. However, in the Vigenère cipher a different row of the Vigenère square (a different cipher alphabet) is used to encrypt different letters of the message. In other words, the sender might encrypt the first letter according to row 5, the second according to row 14, the third according to row 21, and so on.

To unscramble the message, the intended receiver needs to know which row of the Vigenère square has been used to encipher each letter, so there must be an agreed system of switching between rows. This is achieved by using a keyword. To illustrate how a keyword is used with the Vigenère square to encrypt a short message, let us encipher **divert troops to east ridge**, using the keyword WHITE. First of all, the keyword is spelt out above the message, and repeated over and over again so that each letter in the message is associated with a letter from the keyword. The ciphertext is then generated as follows. To encrypt the first letter, d, begin by identifying the key letter above it, W, which in turn defines a particular row in the Vigenère square. The row beginning with W, row 22, is the cipher alphabet that will be used to find the substitute letter for the plaintext d. We look to see where the column headed by d intersects the row beginning with W, which turns out to be at the letter Z. Consequently, the letter d in the plaintext is represented by Z in the ciphertext.

Keyword W H I T E W H I T E W H I T E W H I T E W H I
Plaintext d i v e r t t r o o p s t o e a s t r i d g e
Ciphertext Z P D X V P A Z H S L Z B H I W Z B K M Z N M

To encipher the second letter of the message, i, the process is repeated. The key letter above i is H, so it is encrypted via a different row in the Vigenère square: the H row (row 7) which is a new cipher alphabet. To encrypt i, we look to see where the column headed by i intersects the row beginning with H, which turns out to be at the letter P. Consequently, the letter i in the plaintext is represented by P in the ciphertext. Each letter of the keyword indicates a particular cipher alphabet within the Vigenère square, and because the keyword contains five letters, the sender encrypts the message by cycling through five rows of the Vigenère square. The fifth letter of the message is enciphered according to the fifth letter of the keyword, E, but to encipher the sixth letter of the message we have to return to the first letter of the keyword. A longer keyword, or perhaps a

Table 4 A Vigenère square with the rows defined by the keyword WHITE highlighted. Encryption is achieved by switching between the five highlighted cipher alphabets, defined by W, H, I, T and E.

Plain	a	b	c	d	e	f	g	h	i	j	k	l	m	n	o	p	q	r	s	t	u	v	w	x	y	z
1	B	C	D	E	F	G	H	I	J	K	L	M	N	O	P	Q	R	S	T	U	V	W	X	Y	Z	A
2	C	D	E	F	G	H	I	J	K	L	M	N	O	P	Q	R	S	T	U	V	W	X	Y	Z	A	B
3	D	E	F	G	H	I	J	K	L	M	N	O	P	Q	R	S	T	U	V	W	X	Y	Z	A	B	C
4	E	F	G	H	I	J	K	L	M	N	O	P	Q	R	S	T	U	V	W	X	Y	Z	A	B	C	D
5	F	G	H	I	J	K	L	M	N	O	P	Q	R	S	T	U	V	W	X	Y	Z	A	B	C	D	E
6	G	H	I	J	K	L	M	N	O	P	Q	R	S	T	U	V	W	X	Y	Z	A	B	C	D	E	F
7	H	I	J	K	L	M	N	O	P	Q	R	S	T	U	V	W	X	Y	Z	A	B	C	D	E	F	G
8	I	J	K	L	M	N	O	P	Q	R	S	T	U	V	W	X	Y	Z	A	B	C	D	E	F	G	H
9	J	K	L	M	N	O	P	Q	R	S	T	U	V	W	X	Y	Z	A	B	C	D	E	F	G	H	I
10	K	L	M	N	O	P	Q	R	S	T	U	V	W	X	Y	Z	A	B	C	D	E	F	G	H	I	J
11	L	M	N	O	P	Q	R	S	T	U	V	W	X	Y	Z	A	B	C	D	E	F	G	H	I	J	K
12	M	N	O	P	Q	R	S	T	U	V	W	X	Y	Z	A	B	C	D	E	F	G	H	I	J	K	L
13	N	O	P	Q	R	S	T	U	V	W	X	Y	Z	A	B	C	D	E	F	G	H	I	J	K	L	M
14	O	P	Q	R	S	T	U	V	W	X	Y	Z	A	B	C	D	E	F	G	H	I	J	K	L	M	N
15	P	Q	R	S	T	U	V	W	X	Y	Z	A	B	C	D	E	F	G	H	I	J	K	L	M	N	O
16	Q	R	S	T	U	V	W	X	Y	Z	A	B	C	D	E	F	G	H	I	J	K	L	M	N	O	P
17	R	S	T	U	V	W	X	Y	Z	A	B	C	D	E	F	G	H	I	J	K	L	M	N	O	P	Q
18	S	T	U	V	W	X	Y	Z	A	B	C	D	E	F	G	H	I	J	K	L	M	N	O	P	Q	R
19	T	U	V	W	X	Y	Z	A	B	C	D	E	F	G	H	I	J	K	L	M	N	O	P	Q	R	S
20	U	V	W	X	Y	Z	A	B	C	D	E	F	G	H	I	J	K	L	M	N	O	P	Q	R	S	T
21	V	W	X	Y	Z	A	B	C	D	E	F	G	H	I	J	K	L	M	N	O	P	Q	R	S	T	U
22	W	X	Y	Z	A	B	C	D	E	F	G	H	I	J	K	L	M	N	O	P	Q	R	S	T	U	V
23	X	Y	Z	A	B	C	D	E	F	G	H	I	J	K	L	M	N	O	P	Q	R	S	T	U	V	W
24	Y	Z	A	B	C	D	E	F	G	H	I	J	K	L	M	N	O	P	Q	R	S	T	U	V	W	X
25	Z	A	B	C	D	E	F	G	H	I	J	K	L	M	N	O	P	Q	R	S	T	U	V	W	X	Y
26	A	B	C	D	E	F	G	H	I	J	K	L	M	N	O	P	Q	R	S	T	U	V	W	X	Y	Z

keyphrase, would bring more rows into the encryption process and increase the complexity of the cipher. Table 4 shows a Vigenère square, highlighting the five rows (i.e. the five cipher alphabets) defined by the keyword WHITE.

The great advantage of the Vigenère cipher is that it is impregnable to the frequency analysis described in Chapter 1. For example, a cryptanalyst applying frequency analysis to a piece of ciphertext would usually begin by identifying the most common letter in the ciphertext, which in this case is Z, and then assume that this represents the most common letter in English, e. In fact, the letter Z represents three different letters, d, r and s, but not e. This is clearly a problem for the cryptanalyst. The fact that a letter which appears several times in the ciphertext can represent a different plaintext letter on each occasion generates tremendous ambiguity for the cryptanalyst. Equally confusing is the fact that a letter which appears several times in the plaintext can be represented by different letters in the ciphertext. For example, the letter o is repeated in **troops**, but it is substituted by two different letters – the oo is enciphered as HS.

As well as being invulnerable to frequency analysis, the Vigenère cipher has an enormous number of keys. The sender and receiver can agree on any word in the dictionary, any combination of words, or even fabricate words. A cryptanalyst would be unable to crack the message by searching all possible keys because the number of options is simply too great.

Vigenère's work culminated in his *Traicté des Chiffres* ('A Treatise on Secret Writing'), published in 1586. Ironically, this was the same year that Thomas Phelippes was breaking the cipher of Mary Queen of Scots. If only Mary's secretary had read this treatise, he would have known about the Vigenère cipher, Mary's messages to Babington would have baffled Phelippes, and her life might have been spared.

Because of its strength and its guarantee of security, it would seem natural that the Vigenère cipher would be rapidly adopted by cipher secretaries around Europe. Surely they would be relieved to have access, once again, to a secure form of encryption? On the contrary, cipher secretaries seem to have spurned the Vigenère cipher. This apparently flawless system would remain largely neglected for the next two centuries.

From Shunning Vigenère to the Man in the Iron Mask

The traditional forms of substitution cipher, those that existed before the Vigenère cipher, were called monoalphabetic substitution ciphers because they used only one cipher alphabet per message. In contrast, the Vigenère cipher belongs to a class known as *polyalphabetic*, because it employs several cipher alphabets per message. The polyalphabetic nature of the Vigenère cipher is what gives it its strength, but it also makes it much more complicated to use. The additional effort required in order to implement the Vigenère cipher discouraged many people from employing it.

For many seventeenth-century purposes, the monoalphabetic substitution cipher was perfectly adequate. If you wanted to ensure that your servant was unable to read your private correspondence, or if you wanted to protect your diary from the prying eyes of your spouse, then the old-fashioned type of cipher was ideal. Monoalphabetic substitution was quick, easy to use, and secure against people unschooled in cryptanalysis. In fact, the simple monoalphabetic substitution cipher endured in various forms for many centuries (see Appendix D). For more serious applications, such as military and government communications, where security was paramount, the straightforward monoalphabetic cipher was clearly inadequate. Professional cryptographers in combat with professional cryptanalysts needed something better, yet they were still reluctant to adopt the polyalphabetic cipher because of its complexity. Military communications, in particular, required speed and simplicity, and a diplomatic office might be sending and receiving hundreds of messages each day, so time was of the essence. Consequently, cryptographers searched for an intermediate cipher, one that was harder to crack than a straightforward monoalphabetic cipher, but one that was simpler to implement than a polyalphabetic cipher.

The various candidates included the remarkably effective *homophonic substitution cipher*. Here, each letter is replaced with a variety of substitutes, the number of potential substitutes being proportional to the frequency of the letter. For example, the letter a accounts for roughly 8 per cent of all letters in written English, and so we would assign eight symbols to represent it. Each time a appears in the plaintext it would be replaced in the ciphertext by one of the eight symbols chosen at random, so that by the end of the encipherment each symbol would constitute roughly 1 per cent of the enciphered text. By comparison, the letter b accounts for only 2 per cent of all letters, and so we would assign only two symbols to represent it. Each time b appears in the plaintext either of the two symbols could be chosen, and by the end of the encipherment each symbol would also constitute roughly 1 per cent of the enciphered text. This process of allotting varying numbers of symbols to act as substitutes for each letter continues throughout the alphabet, until we get to z, which is so rare that it has only one symbol to act as a substitute. In the example given in Table 5, the substitutes in the cipher alphabet happen to be two-digit numbers, and there are between one and twelve substitutes for each letter in the plain alphabet, depending on each letter's relative abundance.

We can think of all the two-digit numbers that correspond to the plaintext letter a as effectively representing the same sound in the ciphertext, namely the sound of the letter a. Hence the origin of the term homophonic substitution, *homos* meaning 'same' and *phone* meaning 'sound' in Greek. The point of offering several substitution options for popular letters is to balance out the frequencies of symbols in the

Table 5 An example of a homophonic substitution cipher. The top row represents the plain alphabet, while the numbers below represent the cipher alphabet, with several options for frequently occurring letters.

a	b	c	d	e	f	g	h	i	j	k	l	m	n	o	p	q	r	s	t	u	v	w	x	y	z
09	48	13	01	14	10	06	23	32	15	04	26	22	18	00	38	94	29	11	17	08	34	60	28	21	02
12	81	41	03	16	31	25	39	70			37	27	58	05	95		35	19	20	61		89		52	
33		62	45	24			50	73			51		59	07			40	36	30	63					
47			79	44			56	83			84		66	54			42	76	43						
53				46			65	88					71	72			77	86	49						
67				55			68	93					91	90			80	96	69						
78				57										99					75						
92				64															85						
				74															97						
				82																					
				87																					
				98																					

ciphertext. If we enciphered a message using the cipher alphabet in Table 5, then every number would constitute roughly 1 per cent of the entire text. If no symbol appears more frequently than any other, then this would appear to defy any potential attack via frequency analysis. Perfect security? Not quite.

The ciphertext still contains many subtle clues for the clever cryptanalyst. As we saw in Chapter 1, each letter in the English language has its own personality, defined according to its relationship with all the other letters, and these traits can still be discerned even if the encryption is by homophonic substitution. In English, the most extreme example of a letter with a distinct personality is the letter q, which is only followed by one letter, namely u. If we were attempting to decipher a ciphertext, we might begin by noting that q is a rare letter, and is therefore likely to be represented by just one symbol, and we know that u, which accounts for roughly 3 per cent of all letters, is probably represented by three symbols. So, if we find a symbol in the ciphertext that is only ever followed by three particular symbols, then it would be sensible to assume that the first symbol represents q and the other three symbols represent u. Other letters are harder to spot, but are also betrayed by their relationships to one another. Although the homophonic cipher is breakable, it is much more secure than a straightforward monoalphabetic cipher.

A homophonic cipher might seem similar to a polyalphabetic cipher inasmuch as each plaintext letter can be enciphered in many ways, but there is a crucial difference, and the homophonic cipher is in fact a type of monoalphabetic cipher. In the table of homophones shown above, the letter a can be represented by eight numbers. Significantly, these eight numbers represent only the letter a. In other words, a plaintext letter can be represented by several symbols, but each symbol can only represent one letter. In a polyalphabetic cipher, a plaintext letter will also be represented by different symbols, but, even more confusingly, these symbols will represent different letters during the course of an encipherment.

Perhaps the fundamental reason why the homophonic cipher is considered monoalphabetic is that once the cipher alphabet has been established, it remains constant throughout the process of encryption. The fact that the cipher alphabet contains several options for encrypting each letter is irrelevant. However, a cryptographer who is using a polyalphabetic cipher must continually switch between distinctly different cipher alphabets during the process of encryption.

By tweaking the basic monoalphabetic cipher in various ways, such as adding homophones, it became possible to encrypt messages securely, without having to resort to the complexities of the polyalphabetic cipher. One of the strongest examples of an enhanced monoalphabetic cipher was the Great Cipher of Louis XIV. The Great Cipher was used to encrypt the king's most secret messages, protecting details of his plans, plots and political schemings. One of these messages mentioned one of the most enigmatic characters in French history, the Man in the Iron Mask, but the strength of the Great Cipher meant that the message and its remarkable contents would remain undeciphered and unread for two centuries.

The Great Cipher was invented by the father-and-son team of Antoine and Bonaventure Rossignol. Antoine had first come to prominence in 1626 when he was given a coded letter captured from a messenger leaving the besieged city of Réalmont. Before the end of the day he had deciphered the letter, revealing that the Huguenot army which held the city was on the verge of collapse. The French, who had previously been unaware of the Huguenots' desperate plight, returned the letter accompanied by a decipherment. The Huguenots, who now knew that their enemy would not back down, promptly surrendered. The decipherment had resulted in a painless French victory.

The power of codebreaking became obvious, and the Rossignols were appointed to senior positions in the court. After serving Louis XIII, they then acted as cryptanalysts for Louis XIV, who was so impressed that he moved their offices next to his own apartments so that Rossignol *père et fils* could play a central role in shaping French diplomatic policy. One of the greatest tributes to their abilities is that the word *rossignol* became French slang for a device that picks locks, a reflection of their ability to unlock ciphers.

The Rossignols' prowess at cracking ciphers gave them an insight into how to create a stronger form of encryption, and they invented the so-called Great Cipher. The Great Cipher was so secure that it defied the efforts of all enemy cryptanalysts attempting to steal French secrets. Unfortunately, after the death of both father and son, the Great Cipher fell into disuse and its exact details were rapidly lost, which meant that enciphered papers in the French archives could no longer be read. The Great Cipher was so strong that it even defied the efforts of subsequent generations of codebreakers.

Historians knew that the papers encrypted by the Great Cipher would offer a unique insight into the intrigues of seventeenth-century France, but even by the end of the nineteenth century they were still unable to decipher them. Then, in 1890, Victor Gendron, a military historian researching the campaigns of Louis XIV, unearthed a new series of letters enciphered with the Great Cipher. Unable to make sense of them, he passed them on to Commandant Étienne Bazeries, a distinguished expert in the French Army's Cryptographic Department. Bazeries viewed the letters as the ultimate challenge, and he spent the next three years of his life attempting to decipher them.

The encrypted pages contained thousands of numbers, but only 587 different ones. It was clear that the Great Cipher was more complicated than a straightforward substitution cipher, because this would require just 26 different numbers, one for each letter. Initially, Bazeries thought that the surplus of numbers represented homophones, and that several numbers represented the same letter. Exploring this avenue took months of painstaking effort, all to no avail. The Great Cipher was not a homophonic cipher.

Next, he hit upon the idea that each number might represent a pair of letters, or a *digraph*. There are only 26 individual letters, but there are 676 possible pairs of letters, and this is roughly equal to the variety of numbers in the ciphertexts. Bazeries attempted a decipherment by looking for the most frequent numbers in the ciphertexts (22, 42, 124, 125 and 341), assuming that these probably stood for the commonest French digraphs (es, en, ou, de, nt). In effect, he was applying frequency analysis at the level of pairs of letters. Unfortunately, again after months of work, this theory also failed to yield any meaningful decipherments.

Bazeries must have been on the point of abandoning his obsession, when a new line of attack occurred to him. Perhaps the digraph idea was not so far from the truth. He began to consider the possibility that each number represented not a pair of letters, but rather a whole syllable. He attempted to match each number to a syllable, the most frequently occurring numbers presumably representing the commonest French syllables. He tried various tentative permutations, but they all resulted in gibberish – until he succeeded in identifying one particular word. A cluster of numbers (124-22-125-46-345) appeared several times on each page, and Bazeries postulated that they represented les-en-ne-mi-s, that is, 'les ennemis'. This proved to be a crucial breakthrough.

Bazeries was then able to continue by examining other parts of the ciphertexts where these numbers appeared within different words. He then inserted the syllabic values derived from 'les enemis', which revealed parts of other words. As crossword addicts know, when a word is partly completed it is often possible to guess the remainder of the word. As Bazeries completed new words, he also identified further syllables, which in turn led to other words, and so on. Frequently he would be stumped, partly because the syllabic values were never obvious, partly because some of the numbers represented single letters rather than syllables, and partly because the Rossignols had laid traps within the cipher. For example, one number represented neither a syllable nor a letter, but instead deviously deleted the previous number.

When the decipherment was eventually completed, Bazeries became the first person for two hundred years to witness the secrets of Louis XIV. The newly deciphered material fascinated historians, who focused on one tantalising letter in particular. It seemed to solve one of the great mysteries of the seventeenth century: the true identity of the Man in the Iron Mask.

The Man in the Iron Mask has been the subject of much speculation ever since he was first imprisoned at the French fortress of Pignerole in Savoy. When he was transferred to the Bastille in 1698, peasants tried to catch a glimpse of him, and variously reported him as being short or tall, fair or dark, young or old. Some even claimed that he was a she. With so few facts, everyone from Voltaire to Benjamin Franklin concocted their own theory to explain the case of the Man in the Iron Mask. The most popular conspiracy theory relating to the Mask (as he is sometimes called) suggests that he was the twin of Louis XIV, condemned to imprisonment in order to avoid any controversy over who was the rightful heir to the throne. One version of this theory argues that there existed descendants of the Mask and an associated hidden royal bloodline. A pamphlet published in 1801 said that Napoleon himself was a descendant of the

Mask, a rumour which, since it enhanced his position, the emperor did not deny.

The myth of the Mask even inspired poetry, prose and drama. In 1848 Victor Hugo had begun writing a play entitled *Twins*, but when he found that Alexandre Dumas had already plumped for the same plot, he abandoned the two acts he had written. Ever since, it has been Dumas's name that we associate with the story of the Man in the Iron Mask. The success of his novel reinforced the idea that the Mask was related to the king, and this theory has persisted despite the evidence revealed in one of Bazeries's decipherments.

Bazeries had deciphered a letter written by François de Louvois, Louis XIV's Minister of War, which began by recounting the crimes of Vivien de Bulonde, the commander responsible for leading an attack on the town of Cuneo, on the French–Italian border. Although he was ordered to stand his ground, Bulonde became concerned about the arrival of enemy troops from Austria and fled, leaving behind his munitions and abandoning many of his wounded soldiers. According to the Minister of War, these actions jeopardised the whole Piedmont campaign, and the letter made it clear that the king viewed Bulonde's actions as an act of extreme cowardice:

> His Majesty knows better than any other person the consequences of this act, and he is also aware of how deeply our failure to take the place will prejudice our cause, a failure which must be repaired during the winter. His Majesty desires that you immediately arrest General Bulonde and cause him to be conducted to the fortress of Pignerole, where he will be locked in a cell under guard at night, and permitted to walk the battlements during the day with a mask.

This was an explicit reference to a masked prisoner at Pignerole, and a sufficiently serious crime, with dates that seem to fit the myth of the Man in the Iron Mask. Does this solve the mystery? Not surprisingly, those favouring more conspiratorial solutions have found flaws in Bulonde as a candidate. For example, there is the argument that if Louis XIV was actually attempting to secretly imprison his unacknowledged twin, then he would have left a series of false trails. Perhaps the encrypted letter was meant to be deciphered. Perhaps the nineteenth-century codebreaker Bazeries had fallen into a seventeenth-century trap.

The Black Chambers

Reinforcing the monoalphabetic cipher by applying it to syllables or adding homophones might have been sufficient during the 1600s, but by the 1700s cryptanalysis was becoming industrialised, with teams of government cryptanalysts working together to crack many of the most complex monoalphabetic ciphers. Each European power had its own so-called Black Chamber, a nerve centre for deciphering messages and gathering intelligence. The most celebrated, disciplined and efficient Black Chamber was the Geheime Kabinets-Kanzlei in Vienna.

It operated according to a rigorous timetable, because it was vital that its nefarious activities should not interrupt the smooth running of the postal service. Letters which were supposed to be delivered to embassies in Vienna were first routed via the Black Chamber, arriving at 7 a.m. Secretaries melted seals, and a team of stenographers worked in parallel to make copies of the letters. If necessary, a language specialist would take responsibility for duplicating unusual scripts. Within three hours the letters had been resealed in their envelopes and returned to the central post office, so that they could be delivered to their intended destination. Mail merely in transit through Austria would arrive at the Black Chamber at 10 a.m., and mail leaving Viennese embassies for destinations outside Austria would arrive at 4 p.m. All these letters would also be copied before being allowed to continue on their journey. Each day a hundred letters would filter through the Viennese Black Chamber.

The copies were passed to the cryptanalysts, who sat in little kiosks, ready to tease out the meanings of the messages. As well as supplying the emperors of Austria with invaluable intelligence, the Viennese Black Chamber sold the information it harvested to other powers in Europe. In 1774 an arrangement was made with Abbot Georgel, the secretary at the French Embassy, which gave him access to a twice-weekly package of information in exchange for 1,000 ducats. He then sent these letters, which contained the supposedly secret plans of various monarchs, straight to Louis XV in Paris.

The Black Chambers were effectively making all forms of monoalphabetic cipher insecure. Confronted with such professional cryptanalytic opposition, cryptographers were at last forced to adopt the more complex but more secure Vigenère cipher. Gradually, cipher secretaries began to switch to using polyalphabetic ciphers. In addition to more effective cryptanalysis, there was another pressure that was encouraging the move towards securer forms of encryption: the development of the telegraph, and the need to protect telegrams from interception and decipherment.

Although the telegraph, together with the ensuing telecommunications revolution, came in the nineteenth century, its origins can be traced all the way back to 1753. An anonymous letter in a Scottish magazine described how a message could be sent across large distances by connecting the sender and receiver with 26 cables, one for each letter of the alphabet. The sender could then spell out the message by sending pulses of electricity along each wire. For example, to spell out hello, the sender would begin by sending a signal down the h wire, then down the e wire, and so on. The receiver would somehow sense the electrical current emerging from each wire and read the message. However, this 'expeditious method of conveying intelligence', as the inventor called it, was never constructed, because there were several technical obstacles that had to be overcome.

For example, engineers needed a sufficiently sensitive system for detecting electrical signals. In England, Sir Charles Wheatstone and William Fothergill Cooke built detectors from magnetised needles, which would be deflected in the presence of an incoming electric current. By 1839, the Wheatstone–Cooke system was being used to send messages between railway stations in West Drayton and Paddington, a distance of 29 km. The reputation of the telegraph and its remarkable speed of communication soon spread, and nothing did more to popularise its power than the birth of Queen Victoria's second son, Prince Alfred, at Windsor on 6 August 1844. News of the birth was telegraphed to London, and within the hour *The Times* was on the streets announcing the news. It credited the technology that had enabled this feat, mentioning that it was 'indebted to the extraordinary power of the Electro-Magnetic Telegraph'. The following year, the telegraph gained further fame when it helped capture John Tawell, who had murdered his mistress in Slough, and who had attempted to escape by jumping on to a London-bound train. The local police telegraphed Tawell's description to London, and he was arrested as soon as he arrived at Paddington.

Meanwhile, in America, Samuel Morse had just built his first telegraph line, a system spanning the 60 km between Baltimore and Washington. Morse used an electromagnet to enhance the signal, so that upon arriving at the receiver's end it was strong enough to make a series of short and long marks, dots and dashes, on a piece of paper. He also developed the now familiar Morse code for translating each letter of the alphabet into a series of dots and dashes, as given in Table 6. To complete his system he designed a sounder, so that the receiver would hear each letter as a series of audible dots and dashes.

Back in Europe, Morse's approach gradually overtook the Wheatstone–Cooke system in popularity, and in 1851 a European form of Morse Code, which included accented letters, was adopted throughout the Continent. As each year passed, Morse code and the telegraph had an increasing influence on the world, enabling the police to capture more criminals, helping newspapers to bring the very latest news, providing valuable information for businesses, and allowing distant companies to make instantaneous deals.

However, guarding these often sensitive communications was a major concern. The Morse code itself is not a form of cryptography, because there is no concealment of the message. The dots and dashes are merely a convenient way to represent letters for the telegraphic medium; Morse code is effectively nothing more than an alternative alphabet. The problem of security arose primarily because anyone wanting to send a message would have to deliver it to a Morse code operator, who would then have to read it in order to transmit it. The telegraph operators had access to every message, and hence there was a risk that one company might bribe an operator in order to gain access to a rival's communications. This problem was outlined in an article on telegraphy published in 1853 in England's *Quarterly Review*:

> Means should also be taken to obviate one great objection, at present felt with respect to sending private communications by telegraph – the violation of all secrecy – for in any case half-a-dozen people must be cognisant of every word addressed by one person to another. The clerks of the English Telegraph Company are sworn to secrecy, but we often write things that it would be intolerable to see strangers read before our eyes. This is a grievous fault in the telegraph, and it must be remedied by some means or other.

The solution was to encipher a message before handing it to the telegraph operator. The operator would then turn the ciphertext into Morse code before transmitting it. As well as preventing the operators from seeing sensitive material, encryption also stymied the efforts of any spy who might be tapping the telegraph wire. The polyalphabetic Vigenère cipher was clearly the best way to ensure secrecy for important business

Table 6 International Morse Code symbols.

Symbol	Code	Symbol	Code
A	· –	W	· – –
B	– · ·	X	– · · –
C	– · – ·	Y	– · – –
D	– · ·	Z	– – · ·
E	·	1	· – – – –
F	· · – ·	2	· · – – –
G	– – ·	3	· · · – –
H	· · · ·	4	· · · –
I	· ·	5	· · · · ·
J	· – – –	6	– · · · ·
K	– · –	7	– – · · ·
L	· – · ·	8	– – – · ·
M	– –	9	– – – – ·
N	– ·	10	– – – – –
O	– – –	full stop	· – · – · –
P	· – – ·	comma	– – · · – –
Q	– – · –	question mark	· · – – · ·
R	· – ·	colon	– – – · · ·
S	· · ·	semicolon	– · – · – ·
T	–	hyphen	– · · · –
U	· · –	slash	– · · – ·
V	· · · –	quotation mark	· – · · – ·

communications. It was considered unbreakable, and became known as *le chiffre indéchiffrable*. Cryptographers had, for the time being at least, a clear lead over the cryptanalysts.

Mr Babbage Versus the Vigenère Cipher

The most intriguing figure in nineteenth-century cryptanalysis is Charles Babbage, the eccentric British genius best known for developing the blueprint for the modern computer. He was born in 1791, the son of Benjamin Babbage, a wealthy London banker. When Charles married without his father's permission, he no longer had access to the Babbage fortune, but he still had enough money to be financially secure, and he pursued the life of a roving scholar, applying his mind to whatever problem tickled his fancy. His inventions include the speedometer and the cowcatcher, a device that could be fixed to the front of steam locomotives to clear cattle from railway tracks. In terms of scientific breakthroughs, he was the first to realise that the width of a tree ring depended on that year's weather, and he deduced that it was possible to determine past climates by studying ancient trees. He was also intrigued by statistics, and as a diversion he drew up a set of mortality tables, a basic tool for today's insurance industry.

Babbage did not restrict himself to tackling scientific and engineering problems. The cost of sending a letter used to depend on the distance the letter had to travel, but Babbage pointed out that the cost of the labour required to calculate the price for each letter was more than the cost of the postage. Instead, he proposed the system we still use today – a single price for all letters, regardless of where in the country the addressee lives. He was also interested in politics and social issues, and towards the end of his life he began a campaign to get rid of the organ-grinders and street musicians who roamed London. He complained that the music 'not infrequently gives rise to a dance by little ragged urchins, and sometimes half-intoxicated men, who occasionally accompany the noise with their own discordant voices. Another class who are great supporters of street music consists of ladies of elastic virtue and cosmopolitan tendencies, to whom it affords a decent excuse for displaying their fascinations at their open windows.' Unfortunately for

Babbage, the musicians fought back by gathering in large groups around his house and playing as loud as possible.

The turning point in Babbage's scientific career came in 1821, when he and the astronomer John Herschel were examining a set of mathematical tables, the sort used as the basis for astronomical, engineering and navigational calculations. The two men were disgusted by the number of errors in the tables, which in turn would generate flaws in important calculations. One set of tables, the *Nautical Ephemeris for Finding Latitude and Longitude at Sea*, contained over a thousand errors. Indeed, many shipwrecks and engineering disasters were blamed on faulty tables.

These mathematical tables were calculated by hand, and the mistakes were simply the result of human error. This caused Babbage to exclaim, 'I wish to God these calculations had been executed by steam!' This marked the beginning of an extraordinary endeavour to build a machine capable of faultlessly calculating the tables to a high degree of accuracy. In 1823 Babbage designed 'Difference Engine No. 1', a magnificent calculator consisting of 25,000 precision parts, to be built with government funding. Although Babbage was a brilliant innovator, he was not a great implementer. After ten years of toil, he abandoned 'Difference Engine No. 1', cooked up an entirely new design, and set to work building 'Difference Engine No. 2'.

When Babbage abandoned his first machine, the government lost confidence in him and decided to cut its losses by withdrawing from the project – it had already spent £17,470, enough to build a pair of battleships. It was probably this withdrawal of support that later prompted Babbage to make the following complaint: 'Propose to an Englishman any principle, or any instrument, however admirable, and you will observe that the whole effort of the English mind is directed to find a difficulty, a defect, or an impossibility in it. If you speak to him of a machine for peeling a potato, he will pronounce it impossible: if you peel a potato with it before his eyes, he will declare it useless, because it will not slice a pineapple.'

Lack of government funding meant that Babbage never completed Difference Engine No. 2. The scientific tragedy was that Babbage's machine would have offered the unique feature of being programmable. Rather than merely calculating a specific set of tables, Difference Engine No. 2

Figure 12 Charles Babbage.

would have been able to solve a variety of mathematical problems depending on the instructions that it was given. In fact, Difference Engine No. 2 provided the template for modern computers. The design included a 'store' (memory) and a 'mill' (processor), which would allow it to make decisions and repeat instructions, which are equivalent to the 'IF . . . THEN . . .' and 'LOOP' commands in modern programming.

A century later, during the course of the Second World War, the first electronic incarnations of Babbage's machine would have a profound effect on cryptanalysis, but, in his own lifetime, Babbage made an equally important contribution to codebreaking: he succeeded in breaking the Vigenère cipher, and in so doing he made the greatest breakthrough in cryptanalysis since the Arab scholars of the ninth century broke the monoalphabetic cipher by inventing frequency analysis. Babbage's work required no mechanical calculations or complex computations. Instead, he employed nothing more than sheer cunning.

Babbage had become interested in ciphers at a very young age. In later life, he recalled how his childhood hobby occasionally got him into trouble: 'The bigger boys made ciphers, but if I got hold of a few words, I usually found out the key. The consequence of this ingenuity was occasionally painful: the owners of the detected ciphers sometimes thrashed me, though the fault lay in their own stupidity.' These beatings did not discourage him, and he continued to be enchanted by cryptanalysis. He wrote in his autobiography that 'deciphering is, in my opinion, one of the most fascinating of arts'.

He soon gained a reputation within London society as a cryptanalyst prepared to tackle any encrypted message, and strangers would approach him with all sorts of problems. For example, Babbage helped a desperate biographer attempting to decipher the shorthand notes of John Flamsteed, England's first Astronomer Royal. He also came to the rescue of a historian, solving a cipher of Henrietta Maria, wife of Charles I. In 1854, he collaborated with a barrister and used cryptanalysis to reveal crucial evidence in a legal case. Over the years, he accumulated a thick file of encrypted messages, which he planned to use as the basis for an authoritative book on cryptanalysis, entitled *The Philosophy of Decyphering*. The book would contain two examples of every kind of cipher, one that would be broken as a demonstration and one that would be left as an exercise for the reader. Unfortunately, as with many other of his grand plans, the book was never completed.

While most cryptanalysts had given up all hope of ever breaking the Vigenère cipher, Babbage was inspired to attempt a decipherment by an exchange of letters with John Hall Brock Thwaites, a dentist from Bristol with a rather innocent view of ciphers. In 1854, Thwaites claimed to have invented a new cipher, which, in fact, was equivalent to the Vigenère cipher. He wrote to the *Journal of the Society of Arts* with the intention of patenting his idea, apparently unaware that he was several centuries too late. Babbage wrote to the Society, pointing out that 'the cypher . . . is a very old one, and to be found in most books'. Thwaite was unapologetic and challenged Babbage to break his cipher. Whether or not it was breakable was irrelevant to whether or not it was new, but Babbage's curiosity was sufficiently aroused for him to embark on a search for a weakness in the Vigenère cipher.

Cracking a difficult cipher is akin to climbing a sheer cliff face. The cryptanalyst is seeking any nook or cranny which could provide the slightest purchase. In a monoalphabetic cipher the cryptanalyst will latch on to the frequency of the letters, because the commonest letters, such as e, t and a, will stand out no matter how they have been disguised. In the polyalphabetic Vigenère cipher the frequencies are much more balanced, because the keyword is used to switch between cipher alphabets. Hence, at first sight, the rock face seems perfectly smooth.

Remember, the great strength of the Vigenère cipher is that the same letter will be enciphered in different ways. For example, if the keyword is KING, then every letter in the plaintext can potentially be enciphered in four different ways, because the keyword contains four letters. Each letter of the keyword defines a different cipher alphabet in the Vigenère square, as shown in Table 7. The e column of the square has been highlighted to show how it is enciphered differently, depending on which letter of the keyword is defining the encipherment:

If the K of KING is used to encipher e, then the resulting ciphertext letter is O.
If the I of KING is used to encipher e, then the resulting ciphertext letter is M.
If the N of KING is used to encipher e, then the resulting ciphertext letter is R.
If the G of KING is used to encipher e, then the resulting ciphertext letter is K.

Table 7 A Vigenère square used in combination with the keyword **KING**. The keyword defines four separate cipher alphabets, so that the letter **e** may be encrypted as **O**, **M**, **R** or **K**.

Plain	a	b	c	d	e	f	g	h	i	j	k	l	m	n	o	p	q	r	s	t	u	v	w	x	y	z
1	B	C	D	E	F	G	H	I	J	K	L	M	N	O	P	Q	R	S	T	U	V	W	X	Y	Z	A
2	C	D	E	F	G	H	I	J	K	L	M	N	O	P	Q	R	S	T	U	V	W	X	Y	Z	A	B
3	D	E	F	G	H	I	J	K	L	M	N	O	P	Q	R	S	T	U	V	W	X	Y	Z	A	B	C
4	E	F	G	H	I	J	K	L	M	N	O	P	Q	R	S	T	U	V	W	X	Y	Z	A	B	C	D
5	F	G	H	I	J	K	L	M	N	O	P	Q	R	S	T	U	V	W	X	Y	Z	A	B	C	D	E
6	G	H	I	J	K	L	M	N	O	P	Q	R	S	T	U	V	W	X	Y	Z	A	B	C	D	E	F
7	H	I	J	K	L	M	N	O	P	Q	R	S	T	U	V	W	X	Y	Z	A	B	C	D	E	F	G
8	I	J	K	L	M	N	O	P	Q	R	S	T	U	V	W	X	Y	Z	A	B	C	D	E	F	G	H
9	J	K	L	M	N	O	P	Q	R	S	T	U	V	W	X	Y	Z	A	B	C	D	E	F	G	H	I
10	K	L	M	N	O	P	Q	R	S	T	U	V	W	X	Y	Z	A	B	C	D	E	F	G	H	I	J
11	L	M	N	O	P	Q	R	S	T	U	V	W	X	Y	Z	A	B	C	D	E	F	G	H	I	J	K
12	M	N	O	P	Q	R	S	T	U	V	W	X	Y	Z	A	B	C	D	E	F	G	H	I	J	K	L
13	N	O	P	Q	R	S	T	U	V	W	X	Y	Z	A	B	C	D	E	F	G	H	I	J	K	L	M
14	O	P	Q	R	S	T	U	V	W	X	Y	Z	A	B	C	D	E	F	G	H	I	J	K	L	M	N
15	P	Q	R	S	T	U	V	W	X	Y	Z	A	B	C	D	E	F	G	H	I	J	K	L	M	N	O
16	Q	R	S	T	U	V	W	X	Y	Z	A	B	C	D	E	F	G	H	I	J	K	L	M	N	O	P
17	R	S	T	U	V	W	X	Y	Z	A	B	C	D	E	F	G	H	I	J	K	L	M	N	O	P	Q
18	S	T	U	V	W	X	Y	Z	A	B	C	D	E	F	G	H	I	J	K	L	M	N	O	P	Q	R
19	T	U	V	W	X	Y	Z	A	B	C	D	E	F	G	H	I	J	K	L	M	N	O	P	Q	R	S
20	U	V	W	X	Y	Z	A	B	C	D	E	F	G	H	I	J	K	L	M	N	O	P	Q	R	S	T
21	V	W	X	Y	Z	A	B	C	D	E	F	G	H	I	J	K	L	M	N	O	P	Q	R	S	T	U
22	W	X	Y	Z	A	B	C	D	E	F	G	H	I	J	K	L	M	N	O	P	Q	R	S	T	U	V
23	X	Y	Z	A	B	C	D	E	F	G	H	I	J	K	L	M	N	O	P	Q	R	S	T	U	V	W
24	Y	Z	A	B	C	D	E	F	G	H	I	J	K	L	M	N	O	P	Q	R	S	T	U	V	W	X
25	Z	A	B	C	D	E	F	G	H	I	J	K	L	M	N	O	P	Q	R	S	T	U	V	W	X	Y
26	A	B	C	D	E	F	G	H	I	J	K	L	M	N	O	P	Q	R	S	T	U	V	W	X	Y	Z

Similarly, whole words will be deciphered in different ways: the word **the**, for example, could be enciphered as **DPR**, **BUK**, **GNO** or **ZRM**, depending on its position relative to the keyword. Although this makes cryptanalysis difficult, it is not impossible. The important point to note is that if there are only four ways to encipher the word **the**, and the original message contains several instances of the word **the**, then it is highly likely that some of the four possible encipherments will be repeated in the ciphertext. This is demonstrated in the following example, in which the line **The Sun and the Man in the Moon** has been enciphered using the Vigenère cipher and the keyword **KING**.

Keyword	K I N G K I N G K I N G K I N G K I N G K I N G
Plaintext	t h e s u n a n d t h e m a n i n t h e m o o n
Ciphertext	D P R Y E V N T N B U K W I A O X B U K W W B T

The word **the** is enciphered as **DPR** in the first instance, and then as **BUK** on the second and third occasions. The reason for the repetition of **BUK** is that the second **the** is displaced by eight letters with respect to the third **the**, and eight is a multiple of the length of the keyword, which is four letters long. In other words, the second **the** was enciphered according to its relationship to the key word (**the** is directly below **ING**), and by the time we reach the third **the**, the keyword has cycled round exactly twice, to repeat the relationship, and hence repeat the encipherment.

Babbage realised that this sort of repetition provided him with exactly the foothold he needed in order to conquer the Vigenère cipher. He was able to define a series of relatively simple steps which could be followed by any cryptanalyst to crack the hitherto *chiffre indéchiffrable*. To demonstrate his brilliant technique, let us imagine that we have intercepted the ciphertext shown in Figure 13. We know that it was enciphered using the Vigenère cipher, but we know nothing about the original message, and the keyword is a mystery.

The first stage in Babbage's cryptanalysis is to look for sequences of letters that appear more than once in the ciphertext. There are two ways that such repetitions could arise. The most likely is that the same sequence of letters in the plaintext has been enciphered using the same part of the key. Alternatively, there is a slight possibility that two different sequences of letters in the plaintext have been enciphered using different parts of the key, coincidentally leading to the identical sequence in the ciphertext. If we restrict ourselves to long sequences, then we largely discount the second possibility, and, in this case, we shall consider repeated sequences only if they are of four letters or more. Table 8 is a log of such repetitions, along with the spacing between the repetition. For example, the sequence E-F-I-Q appears in the first line of the ciphertext and then in the fifth line, shifted forward by 95 letters.

As well as being used to encipher the plaintext into ciphertext, the keyword is also used by the receiver to decipher the ciphertext back into plaintext. Hence, if we could identify the keyword, deciphering the text would be easy. At this stage we do not have enough information to work out the keyword, but Table 8 does provide some very good clues as to its length. Having listed which sequences repeat themselves and the spacing between these repetitions, the rest of the table is given over to identifying the *factors* of the spacing – the numbers that will divide into the spacing.

W U B E F I Q L Z U R M V O F E H M Y M W T
I X C G T M P I F K R Z U P M V O I R Q M M
W O Z M P U L M B N Y V Q Q Q M V M V J L E
Y M H F E F N Z P S D L P P S D L P E V Q M
W C X Y M D A V Q E E F I Q C A Y T Q O W C
X Y M W M S E M E F C F W Y E Y Q E T R L I
Q Y C G M T W C W F B S M Y F P L R X T Q Y
E E X M R U L U K S G W F P T L R Q A E R L
U V P M V Y Q Y C X T W F Q L M T E L S F J
P Q E H M O Z C I W C I W F P Z S L M A E Z
I Q V L Q M Z V P P X A W C S M Z M O R V G
V V Q S Z E T R L Q Z P B J A Z V Q I Y X E
W W O I C C G D W H Q M M V O W S G N T J P
F P P A Y B I Y B J U T W R L Q K L L L M D
P Y V A C D C F Q N Z P I F P P K S D V P T
I D G X M Q Q V E B M Q A L K E Z M G C V K
U Z K I Z B Z L I U A M M V Z

Figure 13 The ciphertext, enciphered using the Vigenère cipher.

For example, the sequence **W-C-X-Y-M** repeats itself after 20 letters, and the numbers 1, 2, 4, 5, 10 and 20 are factors, because they divide perfectly into 20 without leaving a remainder. These factors suggest six possibilities:

(1) The key is 1 letter long and is recycled 20 times between encryptions.
(2) The key is 2 letters long and is recycled 10 times between encryptions.
(3) The key is 4 letters long and is recycled 5 times between encryptions.
(4) The key is 5 letters long and is recycled 4 times between encryptions.
(5) The key is 10 letters long and is recycled 2 times between encryptions.
(6) The key is 20 letters long and is recycled 1 time between encryptions.

The first possibility can be excluded, because a key that is only 1 letter long gives rise to a monoalphabetic cipher – only one row of the Vigenère square would be used for the entire encryption, and the cipher alphabet would remain unchanged; it is unlikely that a cryptographer would do this. To indicate each of the other possiblities, a ✓ is placed in the appropriate column of Table 8. Each ✓ indicates a potential key length.

To identify whether the key is 2, 4, 5, 10 or 20 letters long, we need to look at the factors of all the other spacings. Because the keyword seems to be 20 letters or smaller, Table 8 lists those factors that are 20 or smaller for each of the other spacings. There is a clear propensity for a spacing divisible by 5. In fact, every spacing is divisible by 5. The first repeated sequence, **E-F-I-Q**, can be explained by a keyword of length 5 recycled nineteen times between the first and second encryptions. The second repeated sequence, **P-S-D-L-P**, can be explained by a keyword of length 5 recycled just once between the first and second encryptions. The third

Table 8 Repetitions and spacings in the ciphertext.

Repeated sequence	Repeat spacing	Possible length of key (or factors)																		
		2	3	4	5	6	7	8	9	10	11	12	13	14	15	16	17	18	19	20
E-F-I-Q	95				✓														✓	
P-S-D-L-P	5				✓															
W-C-X-Y-M	20	✓		✓	✓					✓										✓
E-T-R-L	120	✓	✓	✓	✓	✓		✓		✓		✓			✓					✓

repeated sequence, W-C-X-Y-M, can be explained by a keyword of length 5 recycled four times between the first and second encryptions. The fourth repeated sequence, E-T-R-L, can be explained by a keyword of length 5 recycled twenty-four times between the first and second encryptions. In short, everything is consistent with a five-letter keyword.

Assuming that the keyword is indeed 5 letters long, the next step is to work out the actual letters of the keyword. For the time being, let us call the keyword L_1-L_2-L_3-L_4-L_5, such that L_1 represents the first letter of the keyword, and so on. The process of encipherment would have begun with enciphering the first letter of the plaintext according to the first letter of the keyword, L_1. The letter L_1 defines one row of the Vigenère square, and effectively provides a monoalphabetic substitution cipher alphabet for the first letter of the plaintext. However, when it comes to encrypting the second letter of the plaintext, the cryptographer would have used L_2 to define a different row of the Vigenère square, effectively providing a different monoalphabetic substitution cipher alphabet. The third letter of plaintext would be encrypted according to L_3, the fourth according to L_4, and the fifth according to L_5. Each letter of the keyword is providing a different cipher alphabet for encryption. However, the sixth letter of the plaintext would once again be encrypted according to L_1, the seventh letter of the plaintext would once again be encrypted according to L_2, and the cycle repeats itself thereafter. In other words, the polyalphabetic cipher consists of five monoalphabetic ciphers, each monoalphabetic cipher is responsible for encrypting one-fifth of the entire message, and, most importantly, we already know how to cryptanalyse monoalphabetic ciphers.

We proceed as follows. We know that one of the rows of the Vigenère square, defined by L_1, provided the cipher alphabet to encrypt the 1st, 6th, 11th, 16th, . . . letters of the message. Hence, if we look at the 1st, 6th, 11th, 16th, . . . letters of the ciphertext, we should be able to use old-fashioned frequency analysis to work out the cipher alphabet in question. Figure 14 shows the frequency distribution of the letters that appear in the 1st, 6th, 11th, 16th, . . . positions of the ciphertext, which are W, I, R, E, ... At this point, remember that each cipher alphabet in the Vigenère square is simply a standard alphabet shifted by a value between 1 and 26. Hence, the frequency distribution in Figure 14 should have similar features to the frequency distribution of a standard alphabet, except that it will have been shifted by some distance. By comparing the L_1 distribution with the standard distribution, it should be possible to work out the shift. Figure 15 shows the standard frequency distribution for a piece of English plaintext.

The standard distribution has peaks, plateaus and valleys, and to match it with the L_1 cipher distribution we look for the most outstanding combination of features. For example, the three spikes at **R-S-T** in the

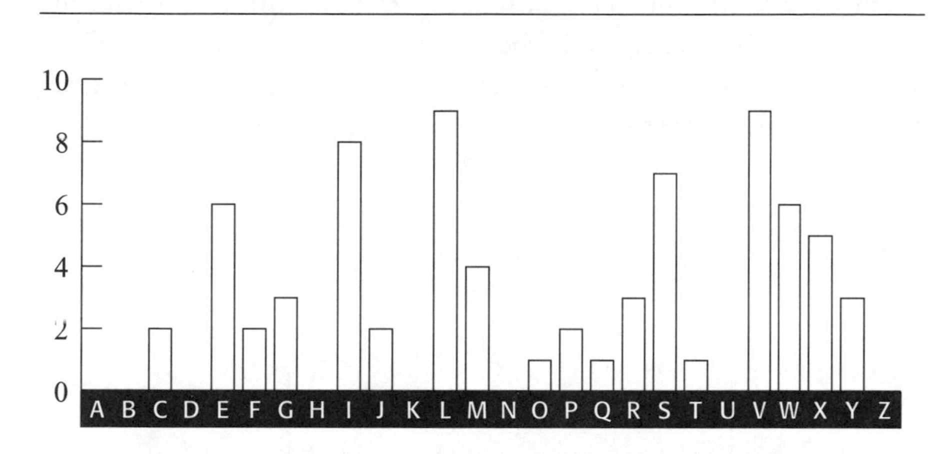

Figure 14 Frequency distribution for letters in the ciphertext encrypted using the L_1 cipher alphabet (number of occurrences).

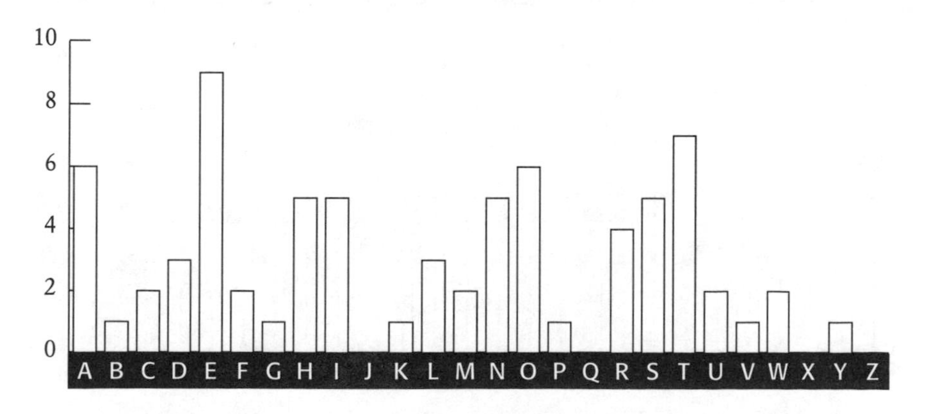

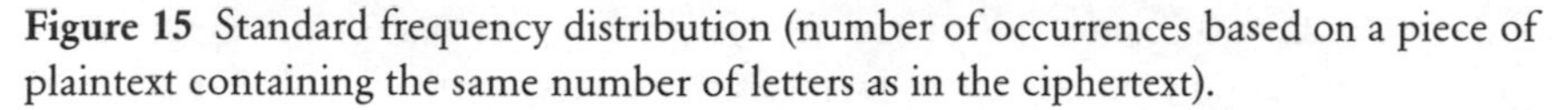

Figure 15 Standard frequency distribution (number of occurrences based on a piece of plaintext containing the same number of letters as in the ciphertext).

standard distribution (Figure 15) and the long depression to its right that stretches across six letters from U to Z together form a very distinctive pair of features. The only similar features in the L_1 distribution (Figure 14) are the three spikes at V-W-X, followed by the depression stretching six letters from Y to D. This would suggest that all the letters encrypted according to L_1 have been shifted four places, or that L_1 defines a cipher alphabet which begins E, F, G, H, ... In turn, this means that the first letter of the keyword, L_1, is probably E. This hypothesis can be tested by shifting the L_1 distribution back four letters and comparing it with the standard distribution. Figure 16 shows both distributions for comparison. The match between the major peaks is very strong, implying that it is safe to assume that the keyword does indeed begin with E.

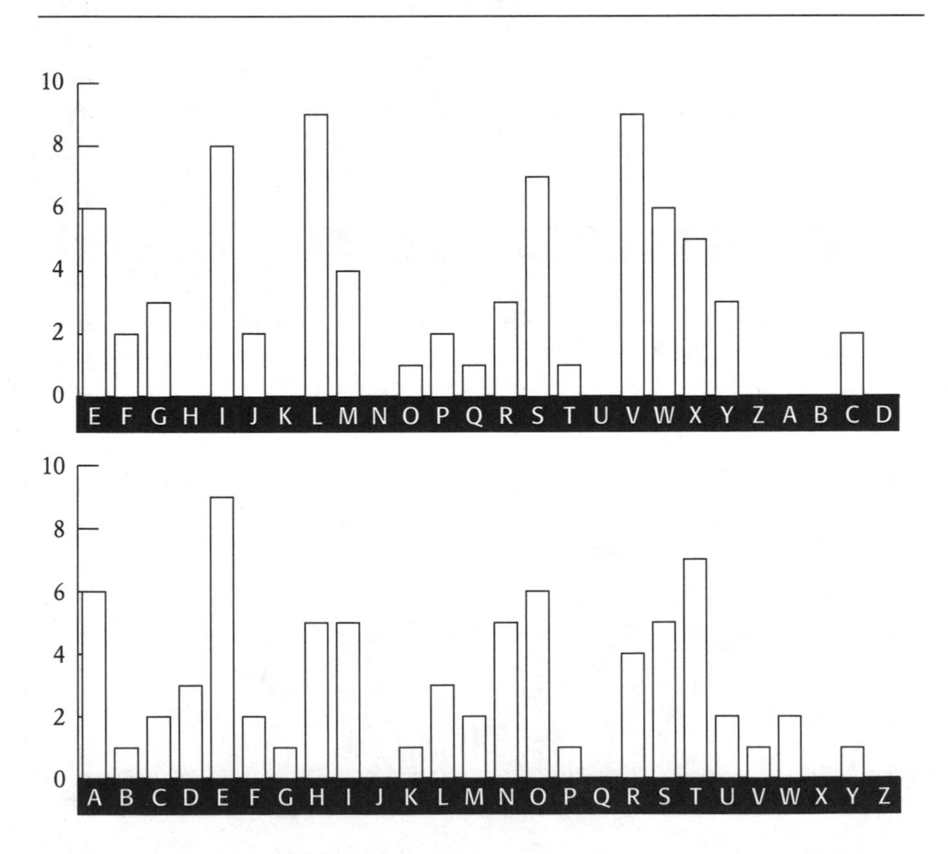

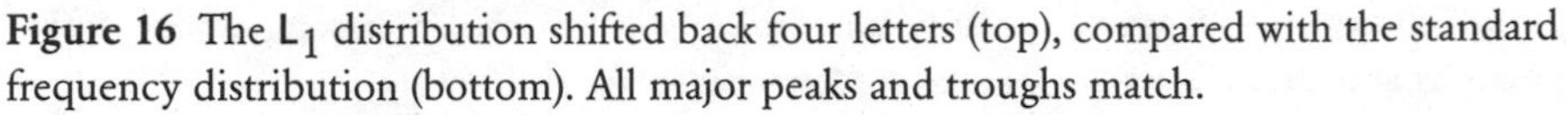
Figure 16 The L_1 distribution shifted back four letters (top), compared with the standard frequency distribution (bottom). All major peaks and troughs match.

To summarise, searching for repetitions in the ciphertext has allowed us to identify the length of the keyword, which turned out to be five letters long. This allowed us to split the ciphertext into five parts, each one enciphered according to a monoalphabetic substitution as defined by one letter of the keyword. By analysing the fraction of the ciphertext that was enciphered according to the first letter of the keyword, we have been able to show that this letter, L_1, is probably E. This process is repeated in order to identify the second letter of the keyword. A frequency distribution is established for the 2nd, 7th, 12th, 17th, . . . letters in the ciphertext. Again, the resulting distribution, shown in Figure 17, is compared with the standard distribution in order to deduce the shift.

This distribution is harder to analyse. There are no obvious candidates for the three neighbouring peaks that correspond to R-S-T. However, the depression that stretches from G to L is very distinct, and probably corresponds to the depression we expect to see stretching from U to Z in the standard distribution. If this were the case, we would expect the three R S T peaks to appear at D, E and F, but the peak at E is missing. For the time being, we shall dismiss the missing peak as a statistical glitch, and go

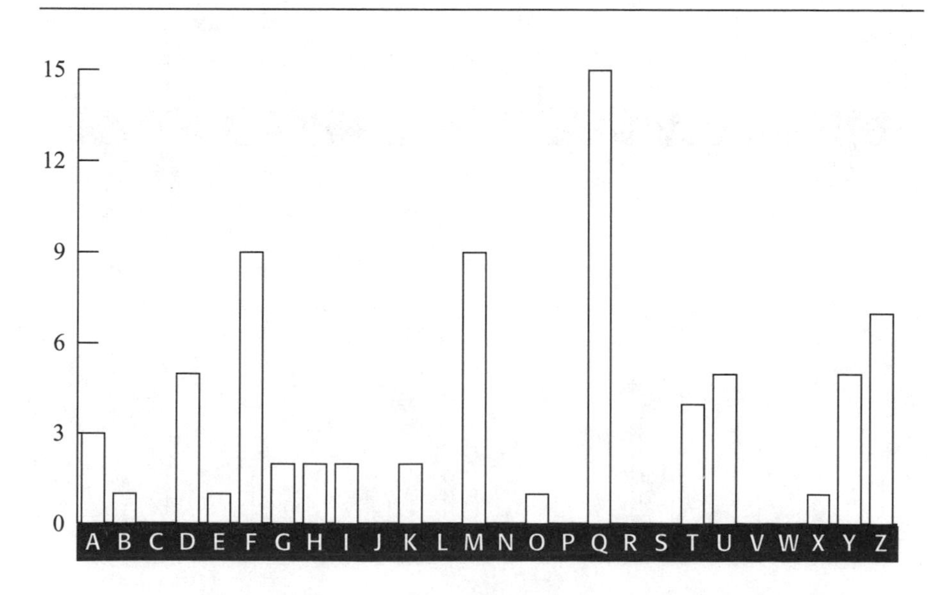

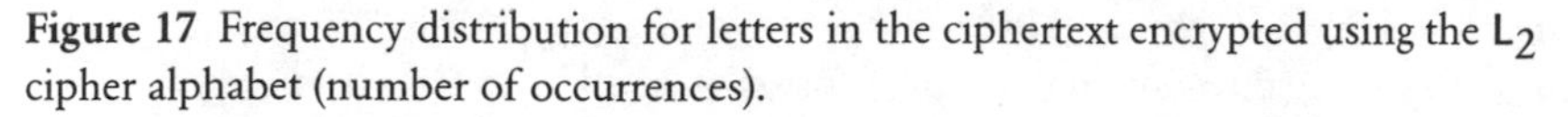

Figure 17 Frequency distribution for letters in the ciphertext encrypted using the L_2 cipher alphabet (number of occurrences).

with our initial reaction, which is that the depression from G to L is a recognisably shifted feature. This would suggest that all the letters encrypted according to L_2 have been shifted twelve places, or that L_2 defines a cipher alphabet which begins M, N, O, P, . . . and that the second letter of the keyword, L_2, is M. Once again, this hypothesis can be tested by shifting the L_2 distribution back twelve letters and comparing it with the standard distribution. Figure 18 shows both distributions, and the

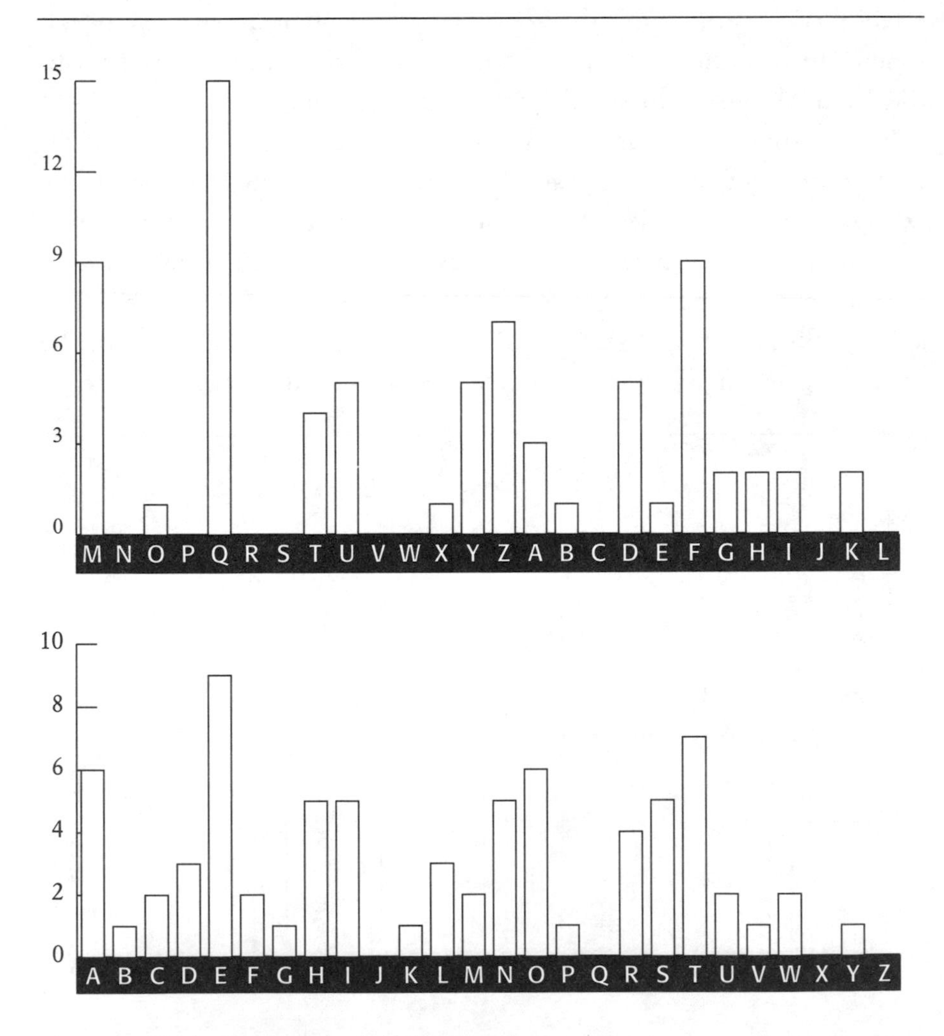

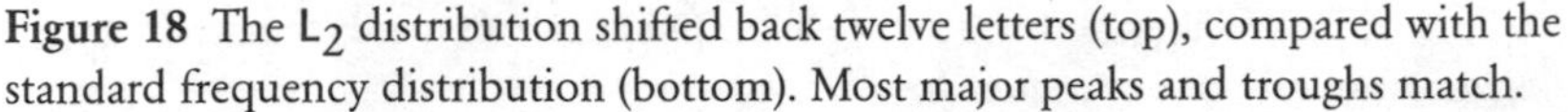

Figure 18 The L_2 distribution shifted back twelve letters (top), compared with the standard frequency distribution (bottom). Most major peaks and troughs match.

match between the major peaks is very strong, implying that it is safe to assume that the second letter of the keyword is indeed **M**.

I shall not continue the analysis; suffice to say that analysing the 3rd, 8th, 13th, . . . letters implies that the third letter of the keyword is **I**, analysing the 4th, 9th, 14th, . . . letters implies that the fourth letter is **L**, and analysing the 5th, 10th, 15th, . . . letters implies that the fifth letter is **Y**. The keyword is **EMILY**. It is now possible to reverse the Vigenère cipher and complete the cryptanalysis. The first letter of the ciphertext is **W**, and it was encrypted according to the first letter of the keyword, **E**. Working backwards, we look at the Vigenère square, and find **W** in the row beginning with **E**, and then we find which letter is at the top of that column. The letter is s, which must make it the first letter of the plaintext. By repeating this process, we see that the plaintext begins sittheedownandhavenoshamecheekbyjowl . . . . By inserting suitable word-breaks and punctuation, we eventually get:

Sit thee down, and have no shame,
Cheek by jowl, and knee by knee:
What care I for any name?
What for order or degree?

Let me screw thee up a peg:
Let me loose thy tongue with wine:
Callest thou that thing a leg?
Which is thinnest? thine or mine?

Thou shalt not be saved by works:
Thou hast been a sinner too:
Ruined trunks on withered forks,
Empty scarecrows, I and you!

Fill the cup, and fill the can:
Have a rouse before the morn:
Every moment dies a man,
Every moment one is born.

These are verses from a poem by Alfred Tennyson entitled 'The Vision of Sin'. The keyword happens to be the first name of Tennyson's wife, Emily Sellwood. I chose to use a section from this particular poem as an example for cryptanalysis because it inspired some curious correspondence between Babbage and the great poet. Being a keen statistician and compiler of mortality tables, Babbage was irritated by the lines 'Every moment dies a man, Every moment one is born', which are the last lines of the plaintext above. Consequently, he offered a correction to Tennyson's 'otherwise beautiful' poem:

> It must be manifest that if this were true, the population of the world would be at a standstill . . . I would suggest that in the next edition of your poem you have it read – 'Every moment dies a man, Every moment $1\frac{1}{16}$ is born.' . . . The actual figure is so long I cannot get it onto a line, but I believe the figure $1\frac{1}{16}$ will be sufficiently accurate for poetry.
>
> I am, Sir, yours, etc.,
>
> Charles Babbage.

Babbage's successful cryptanalysis of the Vigenère cipher was probably achieved in 1854, soon after his spat with Thwaites, but his discovery went completely unrecognised because he never published it. The discovery came to light only in the twentieth century, when scholars examined Babbage's extensive notes. In the meantime, his technique was independently discovered by Friedrich Wilhelm Kasiski, a retired officer in the Prussian army. Ever since 1863, when he published his cryptanalytic breakthrough in *Die Geheimschriften und die Dechiffrir-kunst* ('Secret Writing and the Art of Deciphering'), the technique has been known as the Kasiski Test, and Babbage's contribution has been largely ignored.

And why did Babbage fail to publicise his cracking of such a vital cipher? He certainly had a habit of not finishing projects and not publishing his discoveries, which might suggest that this is just one more example of his lackadaisical attitude. However, there is an alternative explanation. His discovery occurred soon after the outbreak of the Crimean War, and one theory is that it gave the British a clear advantage over their Russian enemy. It is quite possible that British Intelligence demanded that Babbage keep his work secret, thus providing them with a nine-year head start over the rest of the world. If this was the case, then it would fit in with the long-standing tradition of hushing up codebreaking achievements in the interests of national security, a practice that has continued into the twentieth century.

From Agony Columns to Buried Treasure

Thanks to the breakthroughs by Charles Babbage and Friedrich Kasiski, the Vigenère cipher was no longer secure. Cryptographers could no longer guarantee secrecy, now that cryptanalysts had fought back to regain control in the communications war. Although cryptographers attempted to design new ciphers, nothing of great significance emerged during the latter half of the nineteenth century, and professional cryptography was in disarray. However, this same period witnessed an enormous growth of interest in ciphers among the general public.

The development of the telegraph, which had driven a commercial interest in cryptography, was also responsible for generating public interest in cryptography. The public became aware of the need to protect personal messages of a highly sensitive nature, and if necessary they would use encryption, even though this took more time to send, thus adding to the cost of the telegram. Morse operators could send plain English at speeds of up to 35 words per minute because they could memorise entire phrases and transmit them in a single burst, whereas the jumble of letters that make up a ciphertext was considerably slower to transmit, because the operator had to continually refer back to the sender's written message to check the sequence of letters. The ciphers used by the general public would not have withstood attack by a professional cryptanalyst, but they were sufficient to guard against the casual snooper.

As people became comfortable with encipherment, they began to express their cryptographic skills in a variety of ways. For example, young lovers in Victorian England were often forbidden from publicly expressing their affection, and could not even communicate by letter in case their parents intercepted and read the contents. This resulted in lovers sending encrypted messages to each other via the personal columns of newspapers. These 'agony columns', as they became known, provoked the curiosity of cryptanalysts, who would scan the notes and try to decipher their titillating contents. Charles Babbage is known to have indulged in this activity, along with his friends Sir Charles Wheatstone and Baron Lyon Playfair, who together were responsible for developing the deft *Playfair cipher* (described in Appendix E). On one occasion,

Wheatstone deciphered a note in *The Times* from an Oxford student, suggesting to his true love that they elope. A few days later, Wheatstone inserted his own message, encrypted in the same cipher, advising the couple against this rebellious and rash action. Shortly afterwards there appeared a third message, this time unencrypted and from the lady in question: 'Dear Charlie, Write no more. Our cipher is discovered.'

In due course a wider variety of encrypted notes appeared in the newspapers. Cryptographers began to insert blocks of ciphertext merely to challenge their colleagues. On other occasions, encrypted notes were used to criticise public figures or organisations. *The Times* once unwittingly carried the following encrypted notice: '*The Times* is the Jeffreys of the press'. The newspaper was being likened to the notorious seventeenth-century Judge Jeffreys, implying that it was a ruthless, bullying publication which acted as a mouthpiece for the government.

Another example of the public's familiarity with cryptography was the widespread use of pinprick encryption. Ṭhe ancient Greek ḥistorian Aeneas thẹ Tactic̣ian suggested cọnveying a secret message by pricking tiny holes unḍer particular lettẹrs in an apparently innocuous page of text, just as there are dots under some letters in this paragraph. Those letters would spell out a secret message, easily read by the intended receiver. However, if an intermediary stared at the page, they would proḅably be obliviọus tọ the barely perceptible pinpricḳs, and would probably be unaware of the secret message. Two thousand years later, British letter-writers used exactly the same method, not to achieve secrecy but to avoid paying excessive postage costs. Before the overhaul of the postage system in the mid-1800s, sending a letter cost about a shilling for every hundred miles, beyond the means of most people. However, newspapers could be posted free of charge, and this provided a loophole for thrifty Victorians. Instead of writing and sending letters, people began to use pinpricks to spell out a message on the front page of a newspaper. They could then send the newspaper through the post without having to pay a penny.

The public's growing fascination with cryptographic techniques meant that codes and ciphers soon found their way into nineteenth-century literature. In Jules Verne's *Journey to the Centre of the Earth*, the decipherment of a parchment filled with runic characters prompts the first step on the epic journey. The characters are part of a substitution cipher which generates a Latin script, which in turn makes sense only when the letters are reversed: 'Descend the crater of the volcano of Sneffels when the shadow of Scartaris comes to caress it before the calends of July, audacious voyager, and you will reach the centre of the Earth.' In 1885, Verne also used a cipher as a pivotal element in his novel *Mathias Sandorff.* In Britain, one of the finest writers of cryptographic fiction was Sir Arthur Conan Doyle. Not surprisingly, Sherlock Holmes was an expert in cryptography and, as he explained to Dr Watson, was 'the author of a trifling monograph upon the subject in which I analyse one hundred and sixty separate ciphers'. The most famous of Holmes's decipherments is told in *The Adventure of the Dancing Men*, which involves a cipher consisting of stickmen, each pose representing a distinct letter.

On the other side of the Atlantic, Edgar Allan Poe was also developing an interest in cryptanalysis. Writing for Philadelphia's *Alexander Weekly Messenger*, he issued a challenge to readers, claiming that he could decipher any monoalphabetic substitution cipher. Hundreds of readers sent in their ciphertexts, and he successfully deciphered them all. Although this required nothing more than frequency analysis, Poe's readers were astonished by his achievements. One adoring fan proclaimed him 'the most profound and skilful cryptographer who ever lived'.

In 1843, keen to exploit the interest he had generated, Poe wrote a short story about ciphers, which is widely acknowledged by professional cryptographers to be the finest piece of fictional literature on the subject. *The Gold Bug* tells the story of William Legrand, who discovers an unusual beetle, the gold bug, and collects it using a scrap of paper lying nearby. That evening he sketches the gold bug upon the same piece of paper, and then holds his drawing up to the light of the fire to check its accuracy. However,

Figure 19 A section of the ciphertext from *The Adventure of the Dancing Men*, a Sherlock Holmes adventure by Sir Arthur Conan Doyle.

his sketch is obliterated by an invisible ink, which has been developed by the heat of the flames. Legrand examines the characters that have emerged and becomes convinced that he has in his hands the encrypted directions for finding Captain Kidd's treasure. The remainder of the story is a classic demonstration of frequency analysis, resulting in the decipherment of Captain Kidd's clues and the discovery of his buried treasure.

Although *The Gold Bug* is pure fiction, there is a true nineteenth-century story containing many of the same elements. The case of the Beale ciphers involves Wild West escapades, a cowboy who amassed a vast fortune, a buried treasure worth $20 million and a mysterious set of encrypted papers describing its whereabouts. Much of what we know about this story, including the encrypted papers, is contained in a pamphlet published in 1885. Although only 23 pages long, the pamphlet has baffled generations of cryptanalysts and captivated hundreds of treasure hunters.

The story begins at the Washington Hotel in Lynchburg, Virginia, sixty-five years before the publication of the pamphlet. According to the pamphlet, the hotel and its owner, Robert Morriss, were held in high regard: 'His kind disposition, strict probity, excellent management, and well ordered household, soon rendered him famous as a host, and his reputation extended even to other States. His was the house par excellence of the town, and no fashionable assemblages met at any other.' In January 1820 a stranger by the name of Thomas J. Beale rode into Lynchburg and checked into the Washington Hotel. 'In person, he was about six feet in height', recalled Morriss, 'with jet black eyes and hair of the same color, worn longer than was the style at the time. His form was symmetrical, and gave evidence of unusual strength and activity; but his distinguishing feature was a dark and swarthy complexion, as if much exposure to the sun and weather had thoroughly tanned and discolored him; this, however, did not detract from his appearance, and I thought him the handsomest man I had ever seen.' Although Beale spent the rest of the winter with Morriss and was 'extremely popular with every one, particularly the ladies,' he never spoke about his background, his family or the purpose of his visit. Then, at the end of March, he left as suddenly as he had arrived.

THE

BEALE PAPERS,

CONTAINING

AUTHENTIC STATEMENTS

REGARDING THE

TREASURE BURIED

IN

1819 AND 1821,

NEAR

BUFORDS, IN BEDFORD COUNTY, VIRGINIA,

AND

WHICH HAS NEVER BEEN RECOVERED.

PRICE FIFTY CENTS.

LYNCHBURG:
VIRGINIAN BOOK AND JOB PRINT,
1885.

Figure 20 The title page of *The Beale Papers*, the pamphlet that contains all that we know about the mystery of the Beale treasure.

Two years later, in January 1822, Beale returned to the Washington Hotel, 'darker and swarthier than ever'. Once again, he spent the rest of the winter in Lynchburg and disappeared in the spring, but not before he entrusted Morriss with a locked iron box, which he said contained 'papers of value and importance'. Morriss placed the box in a safe, and thought nothing more about it and its contents until he received a letter from Beale, dated 9 May 1822 and sent from St Louis. After a few pleasantries and a paragraph about an intended trip to the plains 'to hunt the buffalo and encounter the savage grizzlies', Beale's letter revealed the significance of the box:

> It contains papers vitally affecting the fortunes of myself and many others engaged in business with me, and in the event of my death, its loss might be irreparable. You will, therefore, see the necessity of guarding it with vigilance and care to prevent so great a catastrophe. Should none of us ever return you will please preserve carefully the box for the period of ten years from the date of this letter, and if I, or no one with authority from me, during that time demands its restoration, you will open it, which can be done by removing the lock. You will find, in addition to the papers addressed to you, other papers which will be unintelligible without the aid of a key to assist you. Such a key I have left in the hand of a friend in this place, sealed and addressed to yourself, and endorsed not to be delivered until June 1832. By means of this you will understand fully all you will be required to do.

Morriss dutifully continued to guard the box, waiting for Beale to collect it, but the swarthy man of mystery never returned to Lynchburg. He disappeared without explanation, never to be seen again. Ten years later, Morriss could have followed the letter's instructions and opened the box, but he seems to have been reluctant to break the lock. Beale's letter had mentioned that a note would be sent to Morriss in June 1832, and this was supposed to explain how to decipher the contents of the box. However, the note never arrived, and perhaps Morriss felt that there was no point opening the box if he could not decipher what was inside it. Eventually, in 1845, Morriss's curiosity got the better of him and he cracked open the lock. The box contained three sheets of enciphered characters, and a note written by Beale in plain English.

The intriguing note revealed the truth about Beale, the box, and the ciphers. It explained that in April 1817, almost three years before his first meeting with Morriss, Beale and 29 others had embarked on a journey across America. After travelling through the rich hunting grounds of the Western plains, they arrived in Santa Fé, and spent the winter in the 'little Mexican town'. In March they headed north and began tracking an 'immense herd of buffaloes', picking off as many as possible along the way. Then, according to Beale, they struck lucky:

> One day, while following them, the party encamped in a small ravine, some 250 or 300 miles north of Santa Fé, and, with their horses tethered, were preparing their evening meal, when one of the men discovered in a cleft of the rocks something that had the appearance of gold. Upon showing it to the others it was pronounced to be gold, and much excitement was the natural consequence.

The letter went on to explain that Beale and his men, with help from the local tribe, mined the site for the next eighteen months, by which time they had accumulated a large quantity of gold, as well as some silver which was found nearby. In due course they agreed that their new-found wealth should be moved to a secure place, and decided to take it back home to Virginia, where they would hide it in a secret location. In 1820, Beale travelled to Lynchburg with the gold and silver, found a suitable location, and buried it. It was on this occasion that he first lodged at the Washington Hotel and made the acquaintance of Morriss. When Beale left at the end of the winter, he rejoined his men who had continued to work the mine during his absence.

After another eighteen months Beale revisited Lynchburg with even more to add to his stash. This time there was an additional reason for his trip:

> Before leaving my companions on the plains it was suggested that, in case of an accident to ourselves, the treasure so concealed would be lost to their relatives, without some provision against such a contingency. I was, therefore, instructed to select some perfectly reliable person, if such could be found, who should, in the event of this proving acceptable to the party, be confided in to carry out their wishes in regard to their respective shares.

Beale believed that Morriss was a man of integrity, which is why he trusted him with the box containing the three enciphered sheets, the so-called Beale ciphers. Each enciphered sheet contained an array of numbers (reprinted here as Figures 21, 22 and 23), and deciphering the numbers would reveal all the relevant details; the first sheet described the treasure's location, the second outlined the contents of the treasure, and the third listed the relatives of the men who should receive a share of the treasure. When Morriss read all of this, it was some 23 years after he had last seen Thomas Beale. Working on the assumption that Beale and his men were dead, Morriss felt obliged to find the gold and share it among their relatives. However, without the promised key he was forced to decipher the ciphers from scratch, a task that troubled his mind for the next twenty years, and which ended in failure.

In 1862, at the age of eighty-four, Morriss knew that he was coming to the end of his life, and that he had to share the secret of the Beale ciphers, otherwise any hope of carrying out Beale's wishes would die with him. Morriss confided in a friend, but unfortunately the identity of this person remains a mystery. All we know about Morriss's friend is that it was he who wrote the pamphlet in 1885, so hereafter I will refer to him simply as *the author*. The author explained the reasons for his anonymity within the pamphlet:

> I anticipate for these papers a large circulation, and, to avoid the multitude of letters with which I should be assailed from all sections of the Union, propounding all sorts of questions, and requiring answers which, if attended to, would absorb my entire time, and only change the character of my work, I have decided upon withdrawing my name from the publication, after assuring all interested that I have given all that I know of the matter, and that I cannot add one word to the statements herein contained.

To protect his identity, the author asked James B. Ward, a respected member of the local community and the county's road surveyor, to act as his agent and publisher.

Everything we know about the strange tale of the Beale ciphers is published in the pamphlet, and so it is thanks to the author that we have the ciphers and Morriss's account of the story. In addition to this, the author is also responsible for successfully deciphering the second Beale cipher. Like the first and third ciphers, the second cipher consists of a page of numbers, and the author assumed that each number represented a letter. However, the range of numbers far exceeds the number of letters in the alphabet, so the author realised that he was dealing with a cipher that uses several numbers to represent the same letter. One cipher that fulfils

71, 194, 38, 1701, 89, 76, 11, 83, 1629, 48, 94, 63, 132, 16, 111, 95, 84, 341, 975, 14, 40, 64, 27, 81, 139, 213, 63, 90, 1120, 8, 15, 3, 126, 2018, 40, 74, 758, 485, 604, 230, 436, 664, 582, 150, 251, 284, 308, 231, 124, 211, 486, 225, 401, 370, 11, 101, 305, 139, 189, 17, 33, 88, 208, 193, 145, 1, 94, 73, 416, 918, 263, 28, 500, 538, 356, 117, 136, 219, 27, 176, 130, 10, 460, 25, 485, 18, 436, 65, 84, 200, 283, 118, 320, 138, 36, 416, 280, 15, 71, 224, 961, 44, 16, 401, 39, 88, 61, 304, 12, 21, 24, 283, 134, 92, 63, 246, 486, 682, 7, 219, 184, 360, 780, 18, 64, 463, 474, 131, 160, 79, 73, 440, 95, 18, 64, 581, 34, 69, 128, 367, 460, 17, 81, 12, 103, 820, 62, 116, 97, 103, 862, 70, 60, 1317, 471, 540, 208, 121, 890, 346, 36, 150, 59, 568, 614, 13, 120, 63, 219, 812, 2160, 1780, 99, 35, 18, 21, 136, 872, 15, 28, 170, 88, 4, 30, 44, 112, 18, 147, 436, 195, 320, 37, 122, 113, 6, 140, 8, 120, 305, 42, 58, 461, 44, 106, 301, 13, 408, 680, 93, 86, 116, 530, 82, 568, 9, 102, 38, 416, 89, 71, 216, 720, 965, 010, 2, 30, 121, 195, 14, 326, 148, 234, 18, 55, 131, 234, 361, 824, 5, 81, 623, 48, 961, 19, 26, 33, 10, 1101, 365, 92, 88, 181, 275, 346, 201, 206, 86, 36, 219, 324, 829, 840, 64, 326, 19, 48, 122, 85, 216, 284, 919, 861, 326, 985, 233, 64, 68, 232, 431, 960, 50, 29, 81, 216, 321, 603, 14, 612, 81, 360, 36, 51, 62, 194, 78, 60, 200, 314, 676, 112, 4, 28, 18, 61, 136, 247, 819, 921, 1060, 464, 895, 10, 6, 66, 119, 38, 41, 49, 602, 423, 962, 302, 294, 875, 78, 14, 23, 111, 109, 62, 31, 501, 823, 216, 280, 34, 24, 150, 1000, 162, 286, 19, 21, 17, 340, 19, 242, 31, 86, 234, 140, 607, 115, 33, 191, 67, 104, 86, 52, 88, 16, 80, 121, 67, 95, 122, 216, 548, 96, 11 , 201, 77, 364, 218, 65, 667, 890, 236, 154, 211, 10, 98, 34, 119, 56, 216, 119, 71, 218, 1164, 1496, 1817, 51, 39, 210, 36, 3, 19, 540, 232, 22, 141, 617, 84, 290, 80, 46, 207, 411, 150, 29, 38, 46, 172, 85, 194, 39, 261, 543, 897, 624, 18, 212, 416, 127, 931, 19, 4, 63, 96, 12, 101, 418, 16, 140, 230, 460, 538, 19, 27, 88, 612, 1431, 90, 716, 275, 74, 83, 11, 426, 89, 72, 84, 1300, 1706, 814, 221, 132, 40, 102, 34, 868, 975, 1101, 84, 16, 79, 23, 16, 81, 122, 324, 403, 912, 227, 936, 447, 55, 86, 34, 43, 212, 107, 96, 314, 264, 1065, 323, 428, 601, 203, 124, 95, 216, 814, 2906, 654, 820, 2, 301, 112, 176, 213, 71, 87, 96, 202, 35, 10, 2, 41, 17, 84, 221, 736, 820, 214, 11, 60, 760.

Figure 21 The first Beale cipher.

115, 73, 24, 807, 37, 52, 49, 17, 31, 62, 647, 22, 7, 15, 140, 47, 29, 107, 79, 84, 56, 239, 10, 26, 811, 5, 196, 308, 85, 52, 160, 136, 59, 211, 36, 9, 46, 316, 554, 122, 106, 95, 53, 58, 2, 42, 7, 35, 122, 53, 31, 82, 77, 250, 196, 56, 96, 118, 71, 140, 287, 28, 353, 37, 1005, 65, 147, 807, 24, 3, 8, 12, 47, 43, 59, 807, 45, 316, 101, 41, 78, 154, 1005, 122, 138, 191, 16, 77, 49, 102, 57, 72, 34, 73, 85, 35, 371, 59, 196, 81, 92, 191, 106, 273, 60, 394, 620, 270, 220, 106, 388, 287, 63, 3, 6, 191, 122, 43, 234, 400, 106, 290, 314, 47, 48, 81, 96, 26, 115, 92, 158, 191, 110, 77, 85, 197, 46, 10, 113, 140, 353, 48, 120, 106, 2, 607, 61, 420, 811, 29, 125, 14, 20, 37, 105, 28, 248, 16, 159, 7, 35, 19, 301, 125, 110, 486, 287, 98, 117, 511, 62, 51, 220, 37, 113, 140, 807, 138, 540, 8, 44, 287, 388, 117, 18, 79, 344, 34, 20, 59, 511, 548, 107, 603, 220, 7, 66, 154, 41, 20, 50, 6, 575, 122, 154, 248, 110, 61, 52, 33, 30, 5, 38, 8, 14, 84, 57, 540, 217, 115, 71, 29, 84, 63, 43, 131, 29, 138, 47, 73, 239, 540, 52, 53, 79, 118, 51, 44, 63, 196, 12, 239, 112, 3, 49, 79, 353, 105, 56, 371, 557, 211, 515, 125, 360, 133, 143, 101, 15, 284, 540, 252, 14, 205, 140, 344, 26, 811, 138, 115, 48, 73, 34, 205, 316, 607, 63, 220, 7, 52, 150, 44, 52, 16, 40, 37, 158, 807, 37, 121, 12, 95, 10, 15, 35, 12, 131, 62, 115, 102, 807, 49, 53, 135, 138, 30, 31, 62, 67, 41, 85, 63, 10, 106, 807, 138, 8, 113, 20, 32, 33, 37, 353, 287, 140, 47, 85, 50, 37, 49, 47, 64, 6, 7, 71, 33, 4, 43, 47, 63, 1, 27, 600, 208, 230, 15, 191, 246, 85, 94, 511, 2, 270, 20, 39, 7, 33, 44, 22, 40, 7, 10, 3, 811, 106, 44, 486, 230, 353, 211, 200, 31, 10, 38, 140, 297, 61, 603, 320, 302, 666, 287, 2, 44, 33, 32, 511, 548, 10, 6, 250, 557, 246, 53, 37, 52, 83, 47, 320, 38, 33, 807, 7, 44, 30, 31, 250, 10, 15, 35, 106, 160, 113, 31, 102, 406, 230, 540, 320, 29, 66, 33, 101, 807, 138, 301, 316, 353, 320, 220, 37, 52, 28, 540, 320, 33, 8, 48, 107, 50, 811, 7, 2, 113, 73, 16, 125, 11, 110, 67, 102, 807, 33, 59, 81, 158, 38, 43, 581, 138, 19, 85, 400, 38, 43, 77, 14, 27, 8, 47, 138, 63, 140, 44, 35, 22, 177, 106, 250, 314, 217, 2, 10, 7, 1005, 4, 20, 25, 44, 48, 7, 26, 46, 110, 230, 807, 191, 34, 112, 147, 44, 110, 121, 125, 96, 41, 51, 50, 140, 56, 47, 152, 540, 63, 807, 28, 42, 250, 138, 582, 98, 643, 32, 107, 140, 112, 26, 85, 138, 540, 53, 20, 125, 371, 38, 36, 10, 52, 118, 136, 102, 420, 150, 112, 71, 14, 20, 7, 24, 18, 12, 807, 37, 67, 110, 62, 33, 21, 95, 220, 511, 102, 811, 30, 83, 84, 305, 620, 15, 2, 108, 220, 106, 353, 105, 106, 60, 275, 72, 8, 50, 205, 185, 112, 125, 540, 65, 106, 807, 188, 96, 110, 16, 73, 33, 807, 150, 409, 400, 50, 154, 285, 96, 106, 316, 270, 205, 101, 811, 400, 8, 44, 37, 52, 40, 241, 34, 205, 38, 16, 46, 47, 85, 24, 44, 15, 64, 73, 138, 807, 85, 78, 110, 33, 420, 505, 53, 37, 38, 22, 31, 10, 110, 106, 101, 140, 15, 38, 3, 5, 44, 7, 98, 287, 135, 150, 96, 33, 84, 125, 807, 191, 96, 511, 118, 440, 370, 643, 466, 106, 41, 107, 603, 220, 275, 30, 150, 105, 49, 53, 287, 250, 208, 134, 7, 53, 12, 47, 85, 63, 138, 110, 21, 112, 140, 485, 486, 505, 14, 73, 84, 575, 1005, 150, 200, 16, 42, 5, 4, 25, 42, 8, 16, 811, 125, 160, 32, 205, 603, 807, 81, 96, 405, 41, 600, 136, 14, 20, 28, 26, 353, 302, 246, 8, 131, 160, 140, 84, 440, 42, 16, 811, 40, 67, 101, 102, 194, 138, 205, 51, 63, 241, 540, 122, 8, 10, 63, 140, 47, 48, 140, 288.

Figure 22 The second Beale cipher.

317, 8, 92, 73, 112, 89, 67, 318, 28, 96, 107, 41, 631, 78, 146, 397, 118, 98, 114, 246, 348, 116, 74, 88, 12, 65, 32, 14, 81, 19, 76, 121, 216, 85, 33, 66, 15, 108, 68, 77, 43, 24, 122, 96, 117, 36, 211, 301, 15, 44, 11, 46, 89, 18, 136, 68, 317, 28, 90, 82, 304, 71, 43, 221, 198, 176, 310, 319, 81, 99, 264, 380, 56, 37, 319, 2, 44, 53, 28, 44, 75, 98, 102, 37, 85, 107, 117, 64, 88, 136, 48, 154, 99, 175, 89, 315, 326, 78, 96, 214, 218, 311, 43, 89, 51, 90, 75, 128, 96, 33, 28, 103, 84, 65, 26, 41, 246, 84, 270, 98, 116, 32, 59, 74, 66, 69, 240, 15, 8, 121, 20, 77, 89, 31, 11, 106, 81, 191, 224, 328, 18, 75, 52, 82, 117, 201, 39, 23, 217, 27, 21, 84, 35, 54, 109, 128, 49, 77, 88, 1, 81, 217, 64, 55, 83, 116, 251, 269, 311, 96, 54, 32, 120, 18, 132, 102, 219, 211, 84, 150, 219, 275, 312, 64, 10, 106, 87, 75, 47, 21, 29, 37, 81, 44, 18, 126, 115, 132, 160, 181, 203, 76, 81, 299, 314, 337, 351, 96, 11, 28, 97, 318, 238, 106, 24, 93, 3, 19, 17, 26, 60, 73, 88, 14, 126, 138, 234, 286, 297, 321, 365, 264, 19, 22, 64, 56, 107, 90, 123, 111, 214, 136, 7, 33, 45, 40, 13, 28, 46, 42, 107, 196, 227, 344, 198, 203, 247, 116, 19, 8, 212, 230, 31, 6, 328, 65, 48, 52, 59, 41, 122, 33, 117, 11, 18, 25, 71, 36, 45, 83, 76, 89, 92, 31, 65, 70, 83, 96, 27, 33, 44, 50, 61, 24, 112, 136, 149, 176, 180, 194, 143, 171, 205, 296, 87, 12, 44, 51, 89, 98, 34, 41, 208, 173, 66, 9, 35, 16, 95, 8, 113, 175, 90, 56, 203, 19, 177, 183, 206, 157, 200, 218, 260, 291, 305, 618, 951, 320, 18, 124, 78, 65, 19, 32, 124, 48, 53, 57, 84, 96, 207, 244, 66, 82, 119, 71, 11, 86, 77, 213, 54, 82, 316, 245, 303, 86, 97, 106, 212, 18, 37, 15, 81, 89, 16, 7, 81, 39, 96, 14, 43, 216, 118, 29, 55, 109, 136, 172, 213, 64, 8, 227, 304, 611, 221, 364, 819, 375, 128, 296, 1, 18, 53, 76, 10, 15, 23, 19, 71, 84, 120, 134, 66, 73, 89, 96, 230, 48, 77, 26, 101, 127, 936, 218, 439, 178, 171, 61, 226, 313, 215, 102, 18, 167, 262, 114, 218, 66, 59, 48, 27, 19, 13, 82, 48, 162, 119, 34, 127, 139, 34, 128, 129, 74, 63, 120, 11, 54, 61, 73, 92, 180, 66, 75, 101, 124, 265, 89, 96, 126, 274, 896, 917, 434, 461, 235, 890, 312, 413, 328, 381, 96, 105, 217, 66, 118, 22, 77, 64, 42, 12, 7, 55, 24, 83, 67, 97, 109, 121, 135, 181, 203, 219, 228, 256, 21, 34, 77, 319, 374, 382, 675, 684, 717, 864, 203, 4, 18, 92, 16, 63, 82, 22, 46, 55, 69, 74, 112, 134, 186, 175, 119, 213, 416, 312, 343, 264, 119, 186, 218, 343, 417, 845, 951, 124, 209, 49, 617, 856, 924, 936, 72, 19, 28, 11, 35, 42, 40, 66, 85, 94, 112, 65, 82, 115, 119, 236, 244, 186, 172, 112, 85, 6, 56, 38, 44, 85, 72, 32, 47, 73, 96, 124, 217, 314, 319, 221, 644, 817, 821, 934, 922, 416, 975, 10, 22, 18, 46, 137, 181, 101, 39, 86, 103, 116, 138, 164, 212, 218, 296, 815, 380, 412, 460, 495, 675, 820, 952.

Figure 23 The third Beale cipher.

this criterion is the so-called *book cipher*, in which a book, or any other piece of text, is itself the key.

First, the cryptographer sequentially numbers every word in the keytext. Thereafter, each number acts as a substitute for the initial letter of its associated word. [1]For [2]example, [3]if [4]the [5]sender [6]and [7]receiver [8]agreed [9]that [10]this [11]sentence [12]were [13]to [14]be [15]the [16]keytext, [17]then [18]every [19]word [20]would [21]be [22]numerically [23]labelled, [24]each [25]number [26]providing [27]the [28]basis [29]for [30]encryption. Next, a list would be drawn up matching each number to the initial letter of its associated word:

1 = f	11 = s	21 = b
2 = e	12 = w	22 = n
3 = i	13 = t	23 = l
4 = t	14 = b	24 = e
5 = s	15 = t	25 = n
6 = a	16 = k	26 = p
7 = r	17 = t	27 = t
8 = a	18 = e	28 = b
9 = t	19 = w	29 = f
10 = t	20 = w	30 = e

A message can now be encrypted by substituting letters in the plaintext for numbers according to the list. In this list, the plaintext letter f would be substituted with 1, and the plaintext letter e could be substituted with either 2, 18, 24 or 30. Because our keytext is such a short sentence, we do not have numbers that could replace rare letters such as x and z, but we do have enough substitutes to encipher the word beale, which could be 14-2-8-23-18. If the intended receiver has a copy of the keytext, then deciphering the encrypted message is trivial. However, if a third party intercepts only the ciphertext, then cryptanalysis depends on somehow identifying the keytext. The author of the pamphlet wrote, 'With this idea, a test was made of every book I could procure, by numbering its letters and comparing the numbers with those of the manuscript; all to no purpose, however, until the Declaration of Independence afforded the clue to one of the papers, and revived all my hopes.'

The Declaration of Independence turned out to be the keytext for the second Beale cipher, and by numbering the words in the Declaration it is possible to unravel it. Figure 24 shows the start of the Declaration of Independence, with every tenth word numbered to help the reader see how the decipherment works. Figure 22 shows the ciphertext – the first number is **115**, and the 115th word in the Declaration is 'instituted', so the first number represents **i**. The second number in the ciphertext is **73**, and the 73rd word in the Declaration is 'hold', so the second number represents **h**. Here is the whole decipherment, as printed in the pamphlet:

> I have deposited in the county of Bedford, about four miles from Buford's, in an excavation or vault, six feet below the surface of the ground, the following articles, belonging jointly to the parties whose names are given in number "3," herewith:
>
> The first deposit consisted of one thousand and fourteen pounds of gold, and three thousand eight hundred and twelve pounds of silver, deposited November, 1819. The second was made December, 1821, and consisted of nineteen hundred and seven pounds of gold, and twelve hundred and eighty-eight pounds of silver; also jewels, obtained in St. Louis in exchange for silver to save transportation, and valued at $13,000.
>
> The above is securely packed in iron pots, with iron covers. The vault is roughly lined with stone, and the vessels rest on solid stone, and are covered with others. Paper number "1" describes the exact locality of the vault, so that no difficulty will be had in finding it.

It is worth noting that there are some errors in the ciphertext. For example, the decipherment includes the words 'four miles', which relies on the 95th word of the Declaration of Independence beginning with the letter *u*. However, the 95th word is '*in*alienable'. This could be the result of Beale's sloppy encryption, or it could be that Beale had a copy of the Declaration in which the 95th word was '*un*alienable', which does appear in some versions dating from the early nineteenth century. Either way, the successful decipherment clearly indicated the value of the treasure – at least $20 million at today's bullion prices.

Not surprisingly, once the author knew the value of the treasure, he spent increasing amounts of time analysing the other two cipher sheets, particularly the first Beale cipher, which describes the treasure's location. Despite strenuous efforts he failed, and the ciphers brought him nothing but sorrow:

When, in the course of human events, it becomes [10]necessary for one people to dissolve the political bands which [20]have connected them with another, and to assume among the [30]powers of the earth, the separate and equal station to [40]which the laws of nature and of nature's God entitle [50]them, a decent respect to the opinions of mankind requires [60]that they should declare the causes which impel them to [70]the separation.

We hold these truths to be self-evident, [80]that all men are created equal, that they are endowed [90]by their Creator with certain inalienable rights, that among these [100]are life, liberty and the pursuit of happiness; That to [110]secure these rights, governments are instituted among men, deriving their [120]just powers from the consent of the governed; That whenever [130]any form of government becomes destructive of these ends, it [140]is the right of the people to alter or to [150]abolish it, and to institute a new government, laying its [160]foundation on such principles and organizing its powers in such [170]form, as to them shall seem most likely to effect [180]their safety and happiness. Prudence, indeed, will dictate that governments [190]long established should not be changed for light and transient [200]causes; and accordingly all experience hath shewn, that mankind are [210]more disposed to suffer, while evils are sufferable, than to [220]right themselves by abolishing the forms to which they are [230]accustomed.

But when a long train of abuses and usurpations, [240]pursuing invariably the same object evinces a design to reduce them [250]under absolute despotism, it is their right, it is their [260]duty, to throw off such government, and to provide new [270]Guards for their future security. Such has been the patient [280]sufferance of these Colonies; and such is now the necessity [290]which constrains them to alter their former systems of government. [300]The history of the present King of Great Britain is [310]a history of repeated injuries and usurpations, all having in [320]direct object the establishment of an absolute tyranny over these [330]States. To prove this, let facts be submitted to a [340]candid world.

Figure 24 The first three paragraphs of the Declaration of Independence, with every tenth word numbered. This is the key for deciphering the second Beale cipher.

> In consequence of the time lost in the above investigation, I have been reduced from comparative affluence to absolute penury, entailing suffering upon those it was my duty to protect, and this, too, in spite of their remonstrations. My eyes were at last opened to their condition, and I resolved to sever at once, and forever, all connection with the affair, and retrieve, if possible, my errors. To do this, as the best means of placing temptation beyond my reach, I determined to make public the whole matter, and shift from my shoulders my responsibility to Mr. Morriss.

Thus the ciphers, along with everything else known by the author, were published in 1885. Although a warehouse fire destroyed most of the pamphlets, those that survived caused quite a stir in Lynchburg. Among the most ardent treasure-hunters attracted to the Beale ciphers were the Hart brothers, George and Clayton. For years they pored over the two remaining ciphers, mounting various forms of cryptanalytic attack, occasionally fooling themselves into believing that they had a solution. A false line of attack will sometimes generate a few tantalising words within a sea of gibberish, which then encourages the cryptanalyst to devise a series of caveats to excuse the gibberish. To an unbiased observer the decipherment is clearly nothing more than wishful thinking, but to the blinkered treasure-hunter it makes complete sense. One of the Harts' tentative decipherments encouraged them to use dynamite to excavate a particular site; unfortunately, the resulting crater yielded no gold. Although Clayton Hart gave up in 1912, George continued working on the Beale ciphers until 1952. An even more persistent Beale fanatic has been Hiram Herbert, Jr, who first became interested in 1923 and whose obsession continued right through to the 1970s. He, too, had nothing to show for his efforts.

Professional cryptanalysts have also embarked on the Beale treasure trail. Herbert O. Yardley, who founded the U.S. Cipher Bureau (known as the American Black Chamber) at the end of the First World War, was intrigued by the Beale ciphers, as was Colonel William Friedman, the dominant figure in American cryptanalysis during the first half of the twentieth century. While he was in charge of the Signal Intelligence Service, he made the Beale ciphers part of the training programme, presumably because, as his wife once said, he believed the ciphers to be of 'diabolical ingenuity, specifically designed to lure the unwary reader'. The

Friedman archive, established after his death in 1969 at the George C. Marshall Research Center, is frequently consulted by military historians, but the great majority of visitors are eager Beale devotees, hoping to follow up some of the great man's leads. More recently, one of the major figures in the hunt for the Beale treasure has been Carl Hammer, retired director of computer science at Sperry Univac and one of the pioneers of computer cryptanalysis. According to Hammer, 'the Beale ciphers have occupied at least 10% of the best cryptanalytic minds in the country. And not a dime of this effort should be begrudged. The work – even the lines that have led into blind alleys – has more than paid for itself in advancing and refining computer research.' Hammer has been a prominent member of the Beale Cypher and Treasure Association, founded in the 1960s to encourage interest in the Beale mystery. Initially, the Association required that any member who discovered the treasure should share it with the other members, but this obligation seemed to deter many Beale prospectors from joining, and so the Association soon dropped the condition.

Despite the combined efforts of the Association, amateur treasure-hunters and professional cryptanalysts, the first and third Beale ciphers have remained a mystery for over a century, and the gold, silver and jewels have yet to be found. Many attempts at decipherment have revolved around the Declaration of Independence, which was the key for the second Beale cipher. Although a straightforward numbering of the words of the Declaration yields nothing useful for the first and third ciphers, cryptanalysts have tried various other schemes, such as numbering it backwards or numbering alternate words, but so far nothing has worked. One problem is that the first cipher contains numbers as high as **2906**, whereas the Declaration contains only 1,322 words. Other texts and books have been considered as potential keys, and many cryptanalysts have looked into the possibility of an entirely different encryption system.

You might be surprised by the strength of the unbroken Beale ciphers, especially bearing in mind that when we left the ongoing battle between codemakers and codebreakers, it was the codebreakers who were on top. Babbage and Kasiski had invented a way of breaking the Vigenère cipher, and codemakers were struggling to find something to replace it. How did Beale come up with something that is so formidable? The answer is that the Beale ciphers were created under circumstances that gave the cryptographer a great advantage. The messages were a one-off, and, because they related to such a valuable treasure, Beale might have been prepared to create a special one-off keytext for the first and third ciphers. Indeed, if the keytext was penned by Beale himself, this would explain why searches of published material have not revealed it. We can imagine that Beale might have written a 2,000-word private essay on the subject of buffalo-hunting, of which there was only one copy. Only the holder of this essay, the unique keytext, would be able to decipher the first and third Beale ciphers. Beale mentioned that he had left the key in 'the hand of a friend' in St Louis, but if the friend lost or destroyed the key, then cryptanalysts might never be able to crack the Beale ciphers.

Creating a one-off keytext for a message is much more secure than using a key based on a published book, but it is practical only if the sender has the time to create the keytext and is able to convey it to the intended recipient, requirements that are not feasible for routine, day-to-day communications. In Beale's case, he could compose his keytext at leisure, deliver it to his friend in St Louis whenever he happened to be passing through, and then have it posted or collected at some arbitrary time in the future, whenever the treasure was to be reclaimed.

An alternative theory for explaining the indecipherability of the Beale ciphers is that the author of the pamphlet deliberately sabotaged them before having them published. Perhaps the author merely wanted to flush out the key, which was apparently in the hands of Beale's friend in St Louis. If he had accurately published the ciphers, then the friend would have been able to decipher them and collect the gold, and the author would have received no reward for his efforts. However, if the ciphers were corrupted in some way, then the friend would eventually realise that he needed the author's help, and would contact the publisher, Ward, who in turn would contact the author. The author could then hand over the accurate ciphers in exchange for a share of the treasure.

It is also possible that the treasure was found many years ago, and that the discoverer spirited it away without being spotted by local residents. Beale enthusiasts with a penchant for conspiracy theories have suggested that the National Security Agency (NSA) has already found the treasure. America's central government cipher facility has access to the most powerful computers and some of the most brilliant minds in the world, and they may have discovered something about the ciphers that has eluded everybody else. The lack of any announcement would be in keeping with the NSA's hush-hush reputation – it has been proposed that NSA does not stand for National Security Agency, but rather 'Never Say Anything' or 'No Such Agency'.

Finally, we cannot exclude the possibility that the Beale ciphers are an elaborate hoax, and that Beale never existed. Sceptics have suggested that the unknown author, inspired by Poe's 'The Gold Bug', fabricated the whole story and published the pamphlet as a way of profiting from the greed of others. Supporters of the hoax theory have searched for inconsistencies and flaws in the Beale story. For example, according to the pamphlet, Beale's letter, which was locked in the iron box and supposedly written in 1822, contains the word 'stampede', but this word was not seen in print until 1834. However, it is quite possible that the word was in common use in the Wild West at a much earlier date, and Beale could have learnt of it on his travels.

One of the foremost non-believers is the cryptographer Louis Kruh, who claims to have found evidence that the pamphlet's author also wrote Beale's letters, the one supposedly sent from St Louis and the one supposedly contained in the box. He performed a textual analysis on the words attributed to the author and the words attributed to Beale to see if there were any similarities. Kruh compared aspects such as the percentage of sentences beginning with 'The', 'Of' and 'And', the average number of commas and semicolons per sentence, and the writing style – the use of negatives, negative passives, infinitives, relative clauses, and so on. In addition to the author's words and Beale's letters, the analysis also took in the writing of three other nineteenth-century Virginians. Of the five sets of writing, those authored by Beale and the pamphlet's author bore the closest resemblance, suggesting that they may have been written by the same person. In other words, this suggests that the author faked the letters attributed to Beale and fabricated the whole story.

On the other hand, evidence for the integrity of the Beale ciphers is provided from various sources. First, if the undeciphered ciphers were hoaxes, we might expect the hoaxer to have chosen the numbers with little or no attention. However, the numbers give rise to various intricate patterns. One of the patterns can be found by using the Declaration of

Independence as a key for the first cipher. This yields no discernible words, but it does give sequences such as **abfdefghiijklmmnohpp**. Although this is not a perfect alphabetical list, it is certainly not random. James Gillogly, President of the American Cryptogram Association, estimated that the chances of this and other sequences appearing by chance are less that one in a hundred million million, suggesting that there is a cryptographic principle underlying the first cipher. One theory is that the Declaration is indeed the key, but the resulting text requires a second stage of decipherment; in other words, the first Beale cipher was enciphered by a two-stage process, so-called superencipherment. If this is so, then the alphabetical sequence might have been put there as a sign of encouragement, a hint that the first stage of decipherment has been successfully completed.

Further evidence favouring the probity of the ciphers comes from historical research, which can be used to verify the story of Thomas Beale. Peter Viemeister, a local historian, has gathered much of the research in his book *The Beale Treasure – History of a Mystery*. Viemeister began by asking if there was any evidence that Thomas Beale actually existed. Using the census of 1790 and other documents, Viemeister has identified several Thomas Beales who were born in Virginia and whose backgrounds fit the few known details. Viemeister has also attempted to corroborate the other details in the pamphlet, such as Beale's trip to Santa Fé and his discovery of gold. For example, there is a Cheyenne legend dating from around 1820 which tells of gold and silver being taken from the West and buried in Eastern Mountains. Also, the 1820 postmaster's list in St Louis contains a 'Thomas Beall', which fits in with the pamphlet's claim that Beale passed through the city in 1820 on his journey westward after leaving Lynchburg. The pamphlet also says that Beale sent a letter from St Louis in 1822.

So there does seem to be a basis for the tale of the Beale ciphers, and consequently it continues to enthral cryptanalysts and treasure-hunters, such as Joseph Jancik, Marilyn Parsons and their dog Muffin. In February 1983 they were charged with 'violation of a sepulchre', after being caught digging in the cemetery of Mountain View Church in the middle of the night. Having discovered nothing other than a coffin, they spent the rest of the weekend in the county jail and were eventually fined $500. These amateur gravediggers can console themselves with the knowledge that they were hardly any less successful than Mel Fisher, the professional treasure-hunter who salvaged $40 million worth of gold from the sunken Spanish galleon *Nuestra Señora de Atocha*, which he discovered off Key West, Florida, in 1985. In November 1989, Fisher received a tip-off from a Beale expert in Florida, who believed that Beale's hoard was buried at Graham's Mill in Bedford County, Virginia. Supported by a team of wealthy investors, Fisher bought the site under the name of Mr Voda, in order to avoid arousing any suspicion. Despite a lengthy excavation, he discovered nothing.

Some treasure-hunters have abandoned hope of cracking the two undeciphered sheets, and have concentrated instead on gleaning clues from the one cipher that has been deciphered. For example, as well as describing the contents of the buried treasure, the solved cipher states that it is deposited 'about four miles from Buford's', which probably refers to the community of Buford or, more specifically, to Buford's Tavern, located at the centre of Figure 25. The cipher also mentions that 'the vault is roughly lined with stone,' so many treasure-hunters have searched along Goose Creek, a rich source of large stones. Each summer the region attracts hopefuls, some armed with metal detectors, others accompanied by psychics or diviners. The nearby town of Bedford has a number of businesses which gladly hire out equipment, including industrial diggers. Local farmers tend to be less welcoming to the strangers, who often trespass on their land, damage their fences and dig giant holes.

Having read the tale of the Beale ciphers, you might be encouraged to take up the challenge yourself. The lure of an unbroken nineteenth-century cipher, together with a treasure worth $20 million, might prove irresistible. However, before you set off on the treasure trail, take heed of the advice given by the author of the pamphlet:

> Before giving the papers to the public, I would say a word to those who may take an interest in them, and give them a little advice, acquired by bitter experience. It is, to devote only such time as can be spared from your legitimate business to the task, and if you can spare no time, let the matter alone . . . Again, never, as I have done, sacrifice your own and your family's interests to what may prove an illusion; but, as I have already said, when your day's work is done, and you are comfortably seated by your good fire, a short time devoted to the subject can injure no one, and may bring its reward.

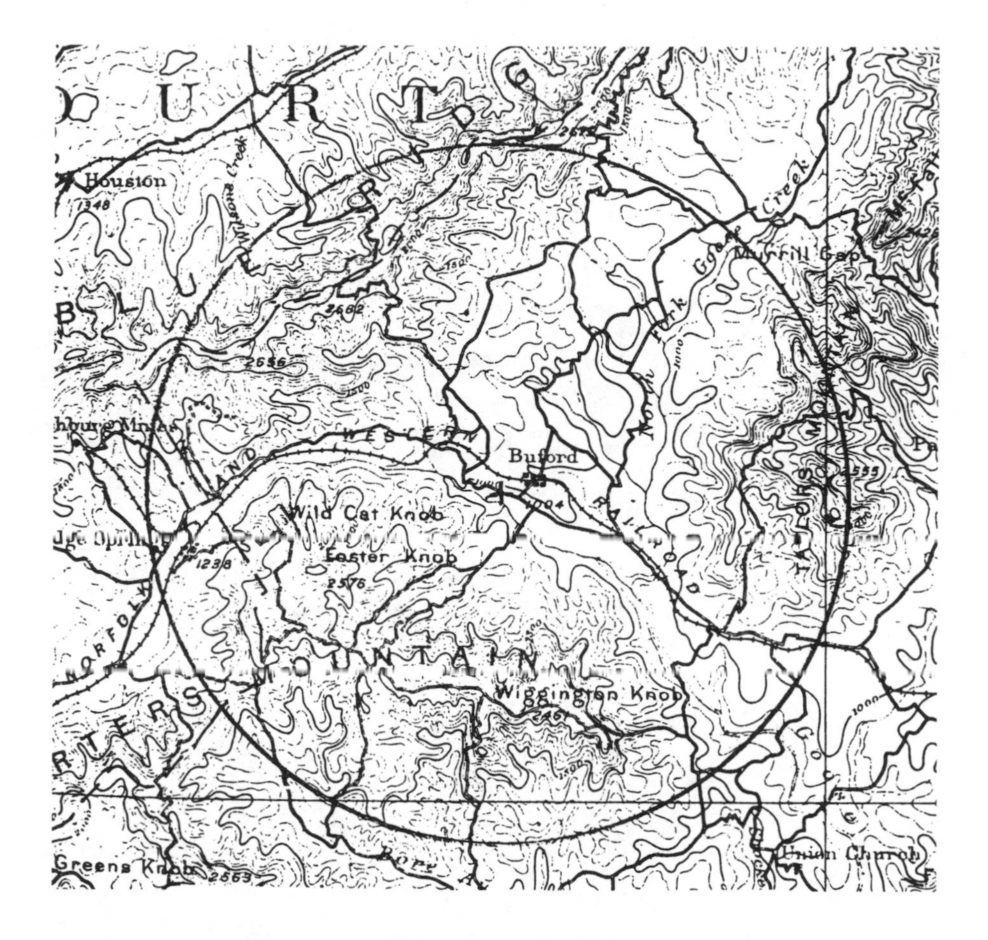

Figure 25 Part of a U.S. Geological Survey map of 1891. The circle has a radius of four miles, and is centred on Buford's Tavern, a location alluded to in the second cipher.

3 The Mechanisation of Secrecy

At the end of the nineteenth century, cryptography was in disarray. Ever since Babbage and Kasiski had destroyed the security of the Vigenère cipher, cryptographers had been searching for a new cipher, something that would re-establish secret communication, thereby allowing businessmen and the military to exploit the immediacy of the telegraph without their communications being stolen and deciphered. Furthermore, at the turn of the century, the Italian physicist Guglielmo Marconi invented an even more powerful form of telecommunication, which made the need for secure encryption even more pressing.

In 1894, Marconi began experimenting with a curious property of electrical circuits. Under certain conditions, if one circuit carried an electric current, this could induce a current in another isolated circuit some distance away. By enhancing the design of the two circuits, increasing the power and adding aerials, Marconi could soon transmit and receive pulses of information across distances of up to 2.5 km. He had invented radio. The telegraph had already been established for half a century, but it required a wire to transport a message between sender and receiver. Marconi's system had the great advantage of being wireless – the signal travelled, as if by magic, through the air.

In 1896, in search of financial backing for his idea, Marconi emigrated to Britain, where he filed his first patent. Continuing his experiments, he increased the range of his radio communications, first transmitting a message 15 km across the Bristol Channel, and then 53 km across the English Channel to France. At the same time he began to look for commercial applications for his invention, pointing out to potential backers the two main advantages of radio: it did not require the construction of expensive telegraph lines, and it had the potential to send messages between otherwise isolated locations. He pulled off a magnificent publicity stunt in 1899, when he equipped two ships with radios so that journalists covering the America's Cup, the world's most important yacht race, could send reports back to New York for the following day's newspapers.

Interest increased still further when Marconi shattered the myth that radio communication was limited by the horizon. Critics had argued that because radio waves could not bend and follow the curvature of the Earth, radio communication would be limited to a hundred kilometres or so. Marconi attempted to prove them wrong by sending a message from Poldhu in Cornwall to St John's in Newfoundland, a distance of 3,500 km. In December 1901, for three hours each day, the Poldhu transmitter sent the letter S (dot-dot-dot) over and over again, while Marconi stood on the windy cliffs of Newfoundland trying to detect the radio waves. Day after day, he wrestled to raise aloft a giant kite, which in turn hoisted his antenna high into the air. A little after midday on 12 December, Marconi detected three faint dots, the first transatlantic radio message. The explanation of Marconi's achievement remained a mystery until 1924, when physicists discovered the ionosphere, a layer of the atmosphere whose lower boundary is about 60 km above the Earth. The ionosphere acts as a mirror, allowing radio waves to bounce off it. Radio waves also bounce off the Earth's surface, so radio messages could effectively reach anywhere in the world after a series of reflections between the ionosphere and the Earth.

Marconi's invention tantalised the military, who viewed it with a mixture of desire and trepidation. The tactical advantages of radio are obvious: it allows direct communication between any two points without the need for a wire between the locations. Laying such a wire is often impractical, sometimes impossible. Previously, a naval commander based in port had no way of communicating with his ships, which might disappear for months on end, but radio would enable him to coordinate a fleet wherever the ships might be. Similarly, radio would allow generals to direct their campaigns, keeping them in continual contact with battalions, regardless of their movements. All this is made possible by the nature of radio waves, which emanate in all directions, and reach receivers wherever they may be. However, this all-pervasive property of radio is also its greatest military weakness, because messages will inevitably reach the enemy as well as the intended recipient. Consequently, reliable encryption became a necessity. If the enemy were going to be able to intercept every radio message, then cryptographers had to find a way of preventing them from deciphering these messages.

The mixed blessings of radio – ease of communication and ease of interception – were brought into sharp focus at the outbreak of the First World War. All sides were keen to exploit the power of radio, but were also unsure of how to guarantee security. Together, the advent of radio and the Great War intensified the need for effective encryption. The hope was that there would be a breakthrough, some new cipher that would re-establish secrecy for military commanders. However, between 1914 and 1918 there was to be no great discovery, merely a catalogue of cryptographic failures. Codemakers conjured up several new ciphers, but one by one they were broken.

One of the most famous wartime ciphers was the German *ADFGVX cipher*, introduced on 5 March 1918, just before the major German offensive that began on 21 March. Like any attack, the German thrust would benefit from the element of surprise, and a committee of cryptographers had selected the ADFGVX cipher from a variety of candidates, believing that it offered the best security. In fact, they were confident that it was unbreakable. The cipher's strength lay in its convoluted nature, a mixture of a substitution and transposition (see Appendix F).

By the beginning of June 1918, the German artillery was only 100 km from Paris, and was preparing for one final push. The only hope for the Allies was to break the ADFGVX cipher to find just where the Germans were planning to punch through their defences. Fortunately, they had a secret weapon, a cryptanalyst by the name of Georges Painvin. This dark, slender Frenchman with a penetrating mind had recognised his talent for cryptographic conundrums only after a chance meeting with a member of the Bureau du Chiffre soon after the outbreak of war. Thereafter, his priceless skill was devoted to pinpointing the weaknesses in German ciphers. He grappled day and night with the ADFGVX cipher, in the process losing 15 kg in weight.

Eventually, on the night of 2 June, he cracked an ADFGVX message. Painvin's breakthrough led to a spate of other decipherments, including a message that contained the order 'Rush munitions. Even by day if not seen.' The preamble to the message indicated that it was sent from somewhere between Montdidier and Compiègne, some 80 km to the north of Paris. The urgent need for munitions implied that this was to be the location of the imminent German thrust. Aerial reconnaissance confirmed that this was the case. Allied soldiers were sent to reinforce this stretch of the front line, and a week later the German onslaught began. Having lost the element of surprise, the German army was beaten back in a hellish battle that lasted five days.

The breaking of the ADFGVX cipher typified cryptography during the First World War. Although there was a flurry of new ciphers, they were all variations or combinations of nineteenth-century ciphers that had already been broken. While some of them initially offered security, it was never long before cryptanalysts got the better of them. The biggest problem for cryptanalysts was dealing with the sheer volume of traffic. Before the advent of radio, intercepted messages were rare and precious items, and cryptanalysts cherished each one. However, in the First World War, the amount of radio traffic was enormous, and every single message could be intercepted, generating a steady flow of ciphertexts to occupy the minds of the cryptanalysts. It is estimated that the French intercepted a hundred million words of German communications during the course of the Great War.

Of all the wartime cryptanalysts, the French were the most effective. When they entered the war, they already had the strongest team of codebreakers in Europe, a consequence of the humiliating French defeat in the Franco-Prussian War. Napoleon III, keen to restore his declining popularity, had invaded Prussia in 1870, but he had not anticipated the alliance between Prussia in the north and the southern German states. Led by Otto von Bismarck, the Prussians steamrollered the French army, annexing the provinces of Alsace and Lorraine and bringing an end to French domination of Europe. Thereafter, the continued threat of the newly united Germany seems to have been the spur for French cryptanalysts to master the skills necessary to provide France with detailed intelligence about the plans of its enemy.

It was in this climate that Auguste Kerckhoffs wrote his treatise *La Cryptographie militaire*. Although Kerckhoffs was Dutch, he spent most of his life in France, and his writings provided the French with an exceptional guide to the principles of cryptanalysis. By the time the First World

Figure 26 Lieutenant Georges Painvin.

War had begun, three decades later, the French military had implemented Kerckhoffs' ideas on an industrial scale. While lone geniuses like Painvin sought to break new ciphers, teams of experts, each with specially developed skills for tackling a particular cipher, concentrated on the day-to-day decipherments. Time was of the essence, and conveyor-belt cryptanalysis could provide intelligence quickly and efficiently.

Sun-Tzu, author of the *Art of War*, a text on military strategy dating from the fourth century BC, stated that: 'Nothing should be as favourably regarded as intelligence; nothing should be as generously rewarded as intelligence; nothing should be as confidential as the work of intelligence.' The French were fervent believers in the words of Sun-Tzu, and in addition to honing their cryptanalytic skills they also developed several ancillary techniques for gathering radio intelligence, methods that did not involve decipherment. For example, the French listening posts learnt to recognise a radio operator's *fist*. Once encrypted, a message is sent in Morse code, as a series of dots and dashes, and each operator can be identified by his pauses, the speed of transmission, and the relative lengths of dots and dashes. A fist is the equivalent of a recognisable style of handwriting. As well as operating listening posts, the French established six direction-finding stations which were able to detect where each message was coming from. Each station moved its antenna until the incoming signal was strongest, which identified a direction for the source of a message. By combining the directional information from two or more stations it was possible to locate the exact source of the enemy transmission. By combining fist information with direction-finding, it was possible to establish both the identity and the location of, say, a particular battalion. French intelligence could then track its path over the course of several days, and potentially deduce its destination and objective. This form of intelligence gathering, known as traffic analysis, was particularly valuable after the introduction of a new cipher. Each new cipher would make cryptanalysts temporarily impotent, but even if a message was indecipherable it could still yield information via traffic analysis.

The vigilance of the French was in sharp contrast to the attitude of the Germans, who entered the war with no military cryptanalytic bureau. Not until 1916 did they set up the Abhorchdienst, an organisation devoted to intercepting Allied messages. Part of the reason for their tardiness in establishing the Abhorchdienst was that the German army had advanced into French territory in the early phase of the war. The French, as they retreated, destroyed the landlines, forcing the advancing Germans to rely on radios for communication. While this gave the French a continuous supply of German intercepts, the opposite was not true. As the French were retreating back into their own territory, they still had access to their own landlines, and had no need to communicate by radio. With a lack of French radio communication, the Germans could not make many interceptions, and hence they did not bother to develop their cryptanalytic department until two years into the war.

The British and the Americans also made important contributions to Allied cryptanalysis. The supremacy of the Allied codebreakers and their influence on the Great War are best illustrated by the decipherment of a German telegram that was intercepted by the British on 17 January 1917. The story of this decipherment shows how cryptanalysis can affect the course of war at the very highest level, and demonstrates the potentially devastating repercussions of employing inadequate encryption. Within a matter of weeks, the deciphered telegram would force America to rethink its policy of neutrality, thereby shifting the balance of the war.

Despite calls from politicians in Britain and America, President Woodrow Wilson had spent the first two years of the war steadfastly refusing to send American troops to support the Allies. Besides not wanting to sacrifice his nation's youth on the bloody battlefields of Europe, he was convinced that the war could be ended only by a negotiated settlement, and he believed that he could best serve the world if he remained neutral and acted as a mediator. In November 1916, Wilson saw hope for a settlement when Germany appointed a new Foreign Minister, Arthur Zimmermann, a jovial giant of a man who appeared to herald a new era of enlightened German diplomacy. American newspapers ran headlines such as OUR FRIEND ZIMMERMANN and LIBERALIZATION OF GERMANY, and one article proclaimed him as 'one of the most auspicious omens for the future of German–American relations'. However, unknown to the Americans, Zimmermann had no intention of pursuing peace. Instead, he was plotting to extend Germany's military aggression.

Back in 1915, a submerged German U-boat had been responsible for sinking the ocean liner *Lusitania*, drowning 1,198 passengers, including 128 US civilians. The loss of the *Lusitania* would have drawn America into the war, were it not for Germany's reassurances that henceforth U-boats would surface before attacking, a restriction that was intended to avoid accidental attacks on civilian ships. However, on 9 January 1917, Zimmermann attended a momentous meeting at the German castle of Pless, where the Supreme High Command was trying to persuade the Kaiser that it was time to renege on their promise, and embark on a course of unrestricted submarine warfare. German commanders knew that their U-boats were almost invulnerable if they launched their torpedoes while remaining submerged, and they believed that this would prove to be the decisive factor in determining the outcome of the war. Germany had been constructing a fleet of two hundred U-boats, and the Supreme High Command argued that unrestricted U-boat aggression would cut off Britain's supply lines and starve it into submission within six months.

A swift victory was essential. Unrestricted submarine warfare and the inevitable sinking of US civilian ships would almost certainly provoke America into declaring war on Germany. Bearing this in mind, Germany needed to force an Allied surrender before America could mobilise its troops and make an impact in the European arena. By the end of the meeting at Pless, the Kaiser was convinced that a swift victory could be achieved, and he signed an order to proceed with unrestricted U-boat warfare, which would take effect on 1 February.

In the three weeks that remained, Zimmermann devised an insurance policy. If unrestricted U-boat warfare increased the likelihood of America entering the war, then Zimmermann had a plan that would delay and weaken American involvement in Europe, and which might even discourage it completely. Zimmermann's idea was to propose an alliance with Mexico, and persuade the President of Mexico to invade America and reclaim territories such as Texas, New Mexico and Arizona. Germany would support Mexico in its battle with their common enemy, aiding it financially and militarily.

Furthermore, Zimmermann wanted the Mexican president to act as a mediator and persuade Japan that it too should attack America. This way, Germany would pose a threat to America's east coast, Japan would attack from the west, while Mexico invaded from the south. Zimmermann's main motive was to pose America such problems at home that it could not afford to send troops to Europe. Thus Germany could win the battle

Figure 27 Arthur Zimmermann.

at sea, win the war in Europe and then withdraw from the American campaign. On 16 January, Zimmermann encapsulated his proposal in a telegram to the German Ambassador in Washington, who would then retransmit it to the German Ambassador in Mexico, who would finally deliver it to the Mexican President. Figure 28 shows the encrypted telegraph; the actual message is as follows:

> We intend to begin unrestricted submarine warfare on the first of February. We shall endeavour in spite of this to keep the United States neutral. In the event of this not succeeding, we make Mexico a proposal of alliance on the following basis: make war together, make peace together, generous financial support, and an understanding on our part that Mexico is to reconquer the lost territory in Texas, New Mexico and Arizona. The settlement in detail is left to you.
>
> You will inform the President [of Mexico] of the above most secretly, as soon as the outbreak of war with the United States is certain, and add the suggestion that he should, on his own initiative, invite Japan to immediate adherence and at the same time mediate between Japan and ourselves.
>
> Please call the President's attention to the fact that the unrestricted employment of our submarines now offers the prospect of compelling England to make peace within a few months. Acknowledge receipt.
>
> Zimmermann

Zimmermann had to encrypt his telegram because Germany was aware that the Allies were intercepting all its transatlantic communications, a consequence of Britain's first offensive action of the war. Before dawn on the first day of the First World War, the British ship *Telconia* approached the German coast under cover of darkness, dropped anchor, and hauled up a clutch of undersea cables. These were Germany's transatlantic cables – its communication links to the rest of the world. By the time the sun had risen, they had been severed. This act of sabotage was aimed at destroying Germany's most secure means of communication, thereby forcing German messages to be sent via insecure radio links or via cables owned by other countries. Zimmermann was forced to send his encrypted telegram via Sweden and, as a back-up, via the more direct American-owned cable. Both routes touched England, which meant that the text of the Zimmermann telegram, as it would become known, soon fell into British hands.

The intercepted telegram was immediately sent to Room 40, the Admiralty's cipher bureau, named after the office in which it was initially housed. Room 40 was a strange mixture of linguists, classical scholars and crossword addicts, capable of the most ingenious feats of cryptanalysis. For example, the Reverend Montgomery, a gifted translator of German theological works, had deciphered a secret message hidden in a postcard addressed to Sir Henry Jones, 184 King's Road, Tighnabruaich, Scotland.

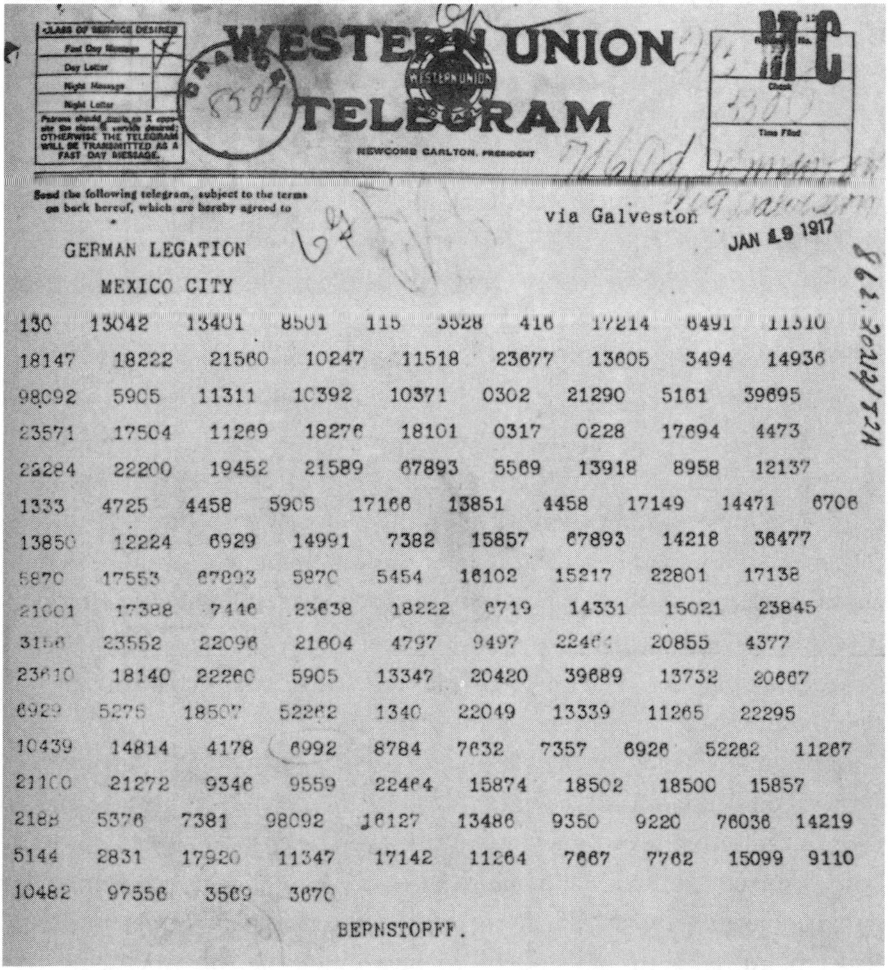

WESTERN UNION TELEGRAM

NEWCOMB CARLTON, PRESIDENT

Send the following telegram, subject to the terms on back hereof, which are hereby agreed to

via Galveston

JAN 19 1917

GERMAN LEGATION

MEXICO CITY

130 13042 13401 8501 115 3528 416 17214 6491 11310
18147 18222 21560 10247 11518 23677 13605 3494 14936
98092 5905 11311 10392 10371 0302 21290 5161 39695
23571 17504 11269 18276 18101 0317 0228 17694 4473
22284 22200 19452 21589 67893 5569 13918 8958 12137
1333 4725 4458 5905 17166 13851 4458 17149 14471 6706
13850 12224 6929 14991 7382 15857 67893 14218 36477
5870 17553 67893 5870 5454 16102 15217 22801 17138
21001 17388 7446 23638 18222 6719 14331 15021 23845
3156 23552 22096 21604 4797 9497 22464 20855 4377
23610 18140 22260 5905 13347 20420 39689 13732 20667
6929 5275 18507 52262 1340 22049 13339 11265 22295
10439 14814 4178 6992 8784 7632 7357 6926 52262 11267
21100 21272 9346 9559 22464 15874 18502 18500 15857
2188 5376 7381 98092 16127 13486 9350 9220 76036 14219
5144 2831 17920 11347 17142 11264 7667 7762 15099 9110
10482 97556 3569 3670

BEPNSTOPFF.

Figure 28 The Zimmermann telegram, as forwarded by von Bernstorff, the German Ambassador in Washington, to Eckhardt, the German Ambassador in Mexico City.

The postcard had been sent from Turkey, so Sir Henry had assumed that it was from his son, a prisoner of the Turks. However, he was puzzled because the postcard was blank, and the address was peculiar – the village of Tighnabruaich was so tiny that none of the houses had numbers and there was no King's Road. Eventually, the Reverend Montgomery spotted the postcard's cryptic message. The address alluded to the Bible, First Book of Kings, Chapter 18, Verse 4: 'Obadiah took a hundred prophets, and hid them fifty in a cave, and fed them with bread and water.' Sir Henry's son was simply reassuring his family that he was being well looked after by his captors.

When the encrypted Zimmermann telegram arrived in Room 40, it was Montgomery who was made responsible for deciphering it, along with Nigel de Grey, a publisher seconded from the firm of William Heinemann. They saw immediately that they were dealing with a form of encryption used only for high-level diplomatic communications, and tackled the telegram with some urgency. The decipherment was far from trivial, but they were able to draw upon previous analyses of other similarly encrypted telegrams. Within a few hours the codebreaking duo had been able to recover a few chunks of text, enough to see that they were uncovering a message of the utmost importance. Montgomery and de Grey persevered with their task, and by the end of the day they could discern the outline of Zimmermann's terrible plans. They realised the dreadful implications of unrestricted U-boat warfare, but at the same time they could see that the German Foreign Minister was encouraging an attack on America, which was likely to provoke President Wilson into abandoning America's neutrality. The telegram contained the deadliest of threats, but also the possibility of America joining the Allies.

Montgomery and de Grey took the partially deciphered telegram to Admiral Sir William Hall, Director of Naval Intelligence, expecting him to pass the information to the Americans, thereby drawing them into the war. However, Admiral Hall merely placed the partial decipherment in his safe, encouraging his cryptanalysts to continue filling in the gaps. He was reluctant to hand the Americans an incomplete decipherment, in case there was a vital caveat that had not yet been deciphered. He also had another concern lurking in the back of his mind. If the British gave the Americans the deciphered Zimmermann telegram, and the Americans reacted by publicly condemning Germany's proposed aggression, then the Germans would conclude that their method of encryption had been broken. This would goad them into developing a new and stronger encryption system, thus choking a vital channel of intelligence. In any case, Hall was aware that the all-out U-boat onslaught would begin in just two weeks, which in itself might be enough to incite President Wilson into declaring war on Germany. There was no point jeopardising a valuable source of intelligence when the desired outcome might happen anyway.

On 1 February, as ordered by the Kaiser, Germany instigated unrestricted naval warfare. On 2 February Woodrow Wilson held a cabinet meeting to decide the American response. On 3 February he spoke to Congress and announced that America would continue to remain neutral, acting as a peacemaker, not a combatant. This was contrary to Allied and German expectations. American reluctance to join the Allies left Admiral Hall with no choice but to exploit the Zimmermann telegram.

In the fortnight since Montgomery and de Grey had first contacted Hall, they had completed the decipherment. Furthermore, Hall had found a way of keeping Germany from suspecting that their security had been breached. He realised that von Bernstorff, the German Ambassador in Washington, would have forwarded the message to von Eckhardt, the German Ambassador in Mexico, having first made some minor changes. For example, von Bernstorff would have removed the instructions aimed at himself, and would also have changed the address. Von Eckhardt would then have delivered this revised version of the telegram, unencrypted, to the Mexican President. If Hall could somehow obtain this Mexican version of the Zimmermann telegram, then it could be published in the newspapers and the Germans would assume that it had been stolen from the Mexican Government, not intercepted and cracked by the British on its way to America. Hall contacted a British agent in Mexico, known only as Mr H., who in turn infiltrated the Mexican Telegraph Office. Mr H. was able to obtain exactly what he needed – the Mexican version of the Zimmermann telegram.

It was this version of the telegram that Hall handed to Arthur Balfour, the British Secretary of State for Foreign Affairs. On 23 February, Balfour summoned the American Ambassador, Walter Page, and presented him with the Zimmermann telegram, later calling this 'the most dramatic moment in all my life'. Four days later, President Wilson saw for himself the 'eloquent evidence', as he called it, proof that Germany was encouraging direct aggression against America.

The telegram was released to the press and, at last, the American nation was confronted with the reality of Germany's intentions. Although there was little doubt among the American people that they should retaliate, there was some concern within the US administration that the telegram might be a hoax, manufactured by the British to guarantee American involvement in the war. However, the question of authenticity soon vanished when Zimmermann publicly admitted his authorship. At a press

Figure 29 'Exploding in his Hands', a cartoon by Rollin Kirby published on 3 March 1917 in *The World*.

conference in Berlin, without being pressured, he simply stated, 'I cannot deny it. It is true.'

In Germany, the Foreign Office began an investigation into how the Americans had obtained the Zimmermann telegram. They fell for Admiral Hall's ploy, and came to the conclusion that 'various indications suggest that the treachery was committed in Mexico'. Meanwhile, Hall continued to distract attention from the work of British cryptanalysts. He planted a story in the British press criticising his own organisation for not intercepting the Zimmermann telegram, which in turn led to a spate of articles attacking the British secret service and praising the Americans.

At the beginning of the year, Wilson had said that it would be a 'crime against civilisation' to lead his nation to war, but by 2 April 1917 he had changed his mind: 'I advise that the Congress declare the recent course of the Imperial Government to be in fact nothing less than war against the government and people of the United States, and that it formally accept the status of belligerent which has thus been thrust upon it.' A single breakthrough by Room 40 cryptanalysts had succeeded where three years of intensive diplomacy had failed. Barbara Tuchman, American historian and author of *The Zimmermann Telegram*, offered the following analysis:

> Had the telegram never been intercepted or never been published, inevitably the Germans would have done something else that would have brought us in eventually. But the time was already late and, had we delayed much longer, the Allies might have been forced to negotiate. To that extent the Zimmermann telegram altered the course of history . . . In itself the Zimmermann telegram was only a pebble on the long road of history. But a pebble can kill a Goliath, and this one killed the American illusion that we could go about our business happily separate from other nations. In world affairs it was a German Minister's minor plot. In the lives of the American people it was the end of innocence.

The Holy Grail of Cryptography

The First World War saw a series of victories for cryptanalysts, culminating in the decipherment of the Zimmermann telegram. Ever since the cracking of the Vigenère cipher in the nineteenth century, codebreakers had maintained the upper hand over the codemakers. Then, towards the end of the war, when cryptographers were in a state of utter despair, scientists in America made an astounding breakthrough. They discovered that the Vigenère cipher could be used as the basis for a new, more formidable form of encryption. In fact, this new cipher could offer perfect security.

The fundamental weakness of the Vigenère cipher is its cyclical nature. If the keyword is five letters long, then every fifth letter of the plaintext is encrypted according to the same cipher alphabet. If the cryptanalyst can identify the length of the keyword, the ciphertext can be treated as a series of five monoalphabetic ciphers, and each one can be broken by frequency analysis. However, consider what happens as the keyword gets longer.

Imagine a plaintext of 1,000 letters encrypted according to the Vigenère cipher, and imagine that we are trying to cryptanalyse the resulting ciphertext. If the keyword used to encipher the plaintext were only 5 letters long, the final stage of cryptanalysis would require applying frequency analysis to 5 sets of 200 letters, which is easy. But if the keyword had been 20 letters long, the final stage would be a frequency analysis of 20 sets of 50 letters, which is considerably harder. And if the keyword had been 1,000 letters long, you would be faced with frequency analysis of 1,000 sets of 1 letter each, which is completely impossible. In other words, if the keyword (or keyphrase) is as long as the message, then the cryptanalytic technique developed by Babbage and Kasiski will not work.

Using a key as long as the message is all well and good, but this requires the cryptographer to create a lengthy key. If the message is hundreds of letters long, the key also needs to be hundreds of letters long. Rather than inventing a long key from scratch, it might be tempting to base it on, say, the lyrics of a song. Alternatively, the cryptographer could pick up a book on birdwatching and base the key on a series of randomly chosen bird names. However, such shortcut keys are fundamentally flawed.

In the following example, I have enciphered a piece of ciphertext using the Vigenère cipher, using a keyphrase that is as long as the message. All the cryptanalytic techniques that I have previously described will fail. None the less, the message can be deciphered.

Key	? ?
Plaintext	? ?
Ciphertext	V H R M H E U Z N F Q D E Z R W X F I D K

This new system of cryptanalysis begins with the assumption that the ciphertext contains some common words, such as **the**. Next, we randomly place **the** at various points in the plaintext, as shown below, and deduce what sort of keyletters would be required to turn **the** into the appropriate ciphertext. For example, if we pretend that **the** is the first word of the plaintext, then what would this imply for the first three letters of the key? The first letter of the key would encrypt **t** into **V**. To work out the first letter of the key, we take a Vigenère square, look down the column headed by **t** until we reach **V**, and find that the letter that begins that row is **C**. This process is repeated with **h** and **e**, which would be encrypted as **H** and **R** respectively, and eventually we have candidates for the first three letters of the key, **CAN**. All of this comes from the assumption that **the** is the first word of the plaintext. We place **the** in a few other positions, and, once again, deduce the corresponding keyletters. (You can check the relationship between each plaintext letter and ciphertext letter by referring to the Vigenère square in Table 9.)

Key	**C**	**A**	**N**	?	?	?	**B**	**S**	**J**	?	?	?	?	?	**Y**	**P**	**T**	?	?	?	?
Plaintext	**t**	**h**	**e**	?	?	?	**t**	**h**	**e**	?	?	?	?	?	**t**	**h**	**e**	?	?	?	?
Ciphertext	**V**	**H**	**R**	**M**	**H**	**E**	**U**	**Z**	**N**	**F**	**Q**	**D**	**E**	**Z**	**R**	**W**	**X**	**F**	**I**	**D**	**K**

We have tested three **the**'s against three arbitrary fragments of the ciphertext, and generated three guesses as to the elements of certain parts of the key. How can we tell whether any of the **the**'s are in the right position? We suspect that the key consists of sensible words, and we can use this to our advantage. If a **the** is in a wrong position, it will probably result in a random selection of keyletters. However, if it is in a correct position, the keyletters should make some sense. For example, the first **the** yields the keyletters **CAN**, which is encouraging because this is a perfectly reasonable English syllable. It is possible that this **the** is in the correct position. The second **the** yields **BSJ**, which is a very peculiar combination of consonants, suggesting that the second **the** is probably a mistake. The third **the** yields **YPT**, an unusual syllable but one which is worth further investigation. If **YPT** really were part of the key, it would be within a larger word, the only possibilities being **APOCALYPTIC**, **CRYPT** and **EGYPT**, and derivatives of these words. How can we find out if one of these words is part of the key? We can test each hypothesis by inserting the three candidate words in the key, above the appropriate section of the ciphertext, and working out the corresponding plaintext:

```
Key         C A N ? ? ? ? ? A P O C A L Y P T I C ? ?
Plaintext   t h e ? ? ? ? ? n q c b e o t h e x g ? ?
Ciphertext  V H R M H E U Z N F Q D E Z R W X F I D K
```

```
Key         C A N ? ? ? ? ? ? ? ? ? C R Y P T ? ? ? ?
Plaintext   t h e ? ? ? ? ? ? ? ? ? c i t h e ? ? ? ?
Ciphertext  V H R M H E U Z N F Q D E Z R W X F I D K
```

```
Key         C A N ? ? ? ? ? ? ? ? ? E G Y P T ? ? ? ?
Plaintext   t h e ? ? ? ? ? ? ? ? ? a t t h e ? ? ? ?
Ciphertext  V H R M H E U Z N F Q D E Z R W X F I D K
```

If the candidate word is not part of the key, it will probably result in a random piece of plaintext, but if it is part of the key the resulting plaintext should make some sense. With **APOCALYPTIC** as part of the key the resulting plaintext is gibberish of the highest quality. With **CRYPT**, the resulting plaintext is **cithe**, which is not an inconceivable piece of plaintext. However, if **EGYPT** were part of the key it would generate **atthe**, a more promising combination of letters, probably representing the words **at the**.

For the time being let us assume that the most likely possibility is that **EGYPT** is part of the key. Perhaps the key is a list of countries. This would suggest that **CAN**, the piece of the key that corresponds to the first **the**, is the start of **CANADA**. We can test this hypothesis by working out more of the plaintext, based on the assumption that **CANADA**, as well as **EGYPT**, is part of the key:

```
Key         C A N A D A ? ? ? ? ? ? E G Y P T ? ? ? ?
Plaintext   t h e m e e ? ? ? ? ? ? a t t h e ? ? ? ?
Ciphertext  V H R M H E U Z N F Q D E Z R W X F I D K
```

Our assumption seems to be making sense. **CANADA** implies that the plaintext begins with **themee** which perhaps is the start of **the meeting**. Now that we have deduced some more letters of the plaintext, **ting**, we can deduce the corresponding part of the key, which turns out to be

BRAZ. Surely this is the beginning of **BRAZIL**. Using the combination of **CANADABRAZILEGYPT** as the bulk of the key, we get the following decipherment: **the meeting is at the ????.**

In order to find the final word of the plaintext, the location of the meeting, the best strategy would be to complete the key by testing one by one the names of all possible countries, and deducing the resulting plaintext. The only sensible plaintext is derived if the final piece of the key is **CUBA**:

Key	**C A N A D A B R A Z I L E G Y P T C U B A**
Plaintext	**t h e m e e t i n g i s a t t h e d o c k**
Ciphertext	**V H R M H E U Z N F Q D E Z R W X F I D K**

Table 9 Vigenère square

Plain	a	b	c	d	e	f	g	h	i	j	k	l	m	n	o	p	q	r	s	t	u	v	w	x	y	z
1	B	C	D	E	F	G	H	I	J	K	L	M	N	O	P	Q	R	S	T	U	V	W	X	Y	Z	A
2	C	D	E	F	G	H	I	J	K	L	M	N	O	P	Q	R	S	T	U	V	W	X	Y	Z	A	B
3	D	E	F	G	H	I	J	K	L	M	N	O	P	Q	R	S	T	U	V	W	X	Y	Z	A	B	C
4	E	F	G	H	I	J	K	L	M	N	O	P	Q	R	S	T	U	V	W	X	Y	Z	A	B	C	D
5	F	G	H	I	J	K	L	M	N	O	P	Q	R	S	T	U	V	W	X	Y	Z	A	B	C	D	E
6	G	H	I	J	K	L	M	N	O	P	Q	R	S	T	U	V	W	X	Y	Z	A	B	C	D	E	F
7	H	I	J	K	L	M	N	O	P	Q	R	S	T	U	V	W	X	Y	Z	A	B	C	D	E	F	G
8	I	J	K	L	M	N	O	P	Q	R	S	T	U	V	W	X	Y	Z	A	B	C	D	E	F	G	H
9	J	K	L	M	N	O	P	Q	R	S	T	U	V	W	X	Y	Z	A	B	C	D	E	F	G	H	I
10	K	L	M	N	O	P	Q	R	S	T	U	V	W	X	Y	Z	A	B	C	D	E	F	G	H	I	J
11	L	M	N	O	P	Q	R	S	T	U	V	W	X	Y	Z	A	B	C	D	E	F	G	H	I	J	K
12	M	N	O	P	Q	R	S	T	U	V	W	X	Y	Z	A	B	C	D	E	F	G	H	I	J	K	L
13	N	O	P	Q	R	S	T	U	V	W	X	Y	Z	A	B	C	D	E	F	G	H	I	J	K	L	M
14	O	P	Q	R	S	T	U	V	W	X	Y	Z	A	B	C	D	E	F	G	H	I	J	K	L	M	N
15	P	Q	R	S	T	U	V	W	X	Y	Z	A	B	C	D	E	F	G	H	I	J	K	L	M	N	O
16	Q	R	S	T	U	V	W	X	Y	Z	A	B	C	D	E	F	G	H	I	J	K	L	M	N	O	P
17	R	S	T	U	V	W	X	Y	Z	A	B	C	D	E	F	G	H	I	J	K	L	M	N	O	P	Q
18	S	T	U	V	W	X	Y	Z	A	B	C	D	E	F	G	H	I	J	K	L	M	N	O	P	Q	R
19	T	U	V	W	X	Y	Z	A	B	C	D	E	F	G	H	I	J	K	L	M	N	O	P	Q	R	S
20	U	V	W	X	Y	Z	A	B	C	D	E	F	G	H	I	J	K	L	M	N	O	P	Q	R	S	T
21	V	W	X	Y	Z	A	B	C	D	E	F	G	H	I	J	K	L	M	N	O	P	Q	R	S	T	U
22	W	X	Y	Z	A	B	C	D	E	F	G	H	I	J	K	L	M	N	O	P	Q	R	S	T	U	V
23	X	Y	Z	A	B	C	D	E	F	G	H	I	J	K	L	M	N	O	P	Q	R	S	T	U	V	W
24	Y	Z	A	B	C	D	E	F	G	H	I	J	K	L	M	N	O	P	Q	R	S	T	U	V	W	X
25	Z	A	B	C	D	E	F	G	H	I	J	K	L	M	N	O	P	Q	R	S	T	U	V	W	X	Y
26	A	B	C	D	E	F	G	H	I	J	K	L	M	N	O	P	Q	R	S	T	U	V	W	X	Y	Z

So, a key that is as long as the message is not sufficient to guarantee security. The insecurity in the example above arises because the key was constructed from meaningful words. We began by randomly inserting **the** throughout the plaintext, and working out the corresponding keyletters. We could tell when we had put a **the** in the correct place, because the keyletters looked as if they might be part of meaningful words. Thereafter, we used these snippets in the key to deduce whole words in the key. In turn this gave us more snippets in the message, which we could expand into whole words, and so on. This entire process of toing and froing between the message and the key was only possible because the key had an inherent structure and consisted of recognisable words. However, in 1918 cryptographers began experimenting with keys that were devoid of structure. The result was an unbreakable cipher.

As the Great War drew to a close, Major Joseph Mauborgne, head of cryptographic research for the US Army, introduced the concept of a random key – one that consisted not of a recognisable series of words, but rather a random series of letters. He advocated employing these random keys as part of a Vigenère cipher to give an unprecedented level of security. The first stage of Mauborgne's system was to compile a thick pad consisting of hundreds of sheets of paper, each sheet bearing a unique key in the form of lines of randomly sequenced letters. There would be two copies of the pad, one for the sender and one for the receiver. To encrypt a message, the sender would apply the Vigenère cipher using the first sheet of the pad as the key. Figure 30 shows three sheets from such a pad (in reality each sheet would contain hundreds of letters), followed by a message encrypted using the random key on the first sheet. The receiver can easily decipher the ciphertext by using the identical key and reversing the Vigenère cipher. Once that message has been successfully sent, received and deciphered, both the sender and the receiver destroy the sheet that acted as the key, so that it is never used again. When the next message is encrypted, the next random key in the pad is employed, which is also subsequently destroyed, and so on. Because each key is used once, and only once, this system is known as a *one-time pad cipher*.

The one-time pad cipher overcomes all previous weaknesses. Imagine that the message **attack the valley at dawn** has been enciphered as in Figure 30, sent via a radio transmitter and intercepted by the enemy. The ciphertext is handed to an enemy cryptanalyst, who then attempts to decipher it. The first hurdle is that, by definition, there is no repetition in a random key, so the method of Babbage and Kasiski cannot break the one-time pad cipher. As an alternative, the enemy cryptanalyst might try placing the word the in various places, and deduce the corresponding piece of the key, just as we did when we attempted to decipher the previous message. If the cryptanalyst tries putting the at the beginning of the message, which is incorrect, then the corresponding segment of key would be revealed as WXB, which is a random series of letters. If the cryptanalyst tries placing the so that it begins at the seventh letter of the message, which happens to be correct, then the corresponding segment of key would be revealed as QKJ, which is also a random series of letters. In other words, the cryptanalyst cannot tell whether the trial word is, or is not, in the correct place.

In desperation, the cryptanalyst might consider an exhaustive search of all possible keys. The ciphertext consists of 21 letters, so the cryptanalyst knows that the key consists of 21 letters. This means that there are roughly 500,000,000,000,000,000,000,000,000,000 possible keys to test, which is completely beyond what is humanly or mechanically feasible. However, even if the cryptanalyst could test all these keys, there is an even greater obstacle to be overcome. By checking every possible key the

Sheet 1	**Sheet 2**	**Sheet 3**
P L M O E	O I W V H	J A B P R
Z Q K J Z	P I Q Z E	M F E C F
L R T E A	T S E B L	L G U X D
V C R C B	C Y R U P	D A G M R
Y N N R B	D U V N M	Z K W Y I

Key	P L M O E Z Q K J Z L R T E A V C R C B Y
Plaintext	a t t a c k t h e v a l l e y a t d a w n
Ciphertext	P E F O G J J R N U L C E I Y V V U C X L

Figure 30 Three sheets, each a potential key for a one-time pad cipher. The message is enciphered using Sheet 1.

cryptanalyst will certainly find the right message – but every wrong message will also be revealed. For example, the following key applied to the same ciphertext generates a completely different message:

Key	M A A K T G Q K J N D R T I F D B H K T S
Plaintext	d e f e n d t h e h i l l a t s u n s e t
Ciphertext	P E F O G J J R N U L C E I Y V V U C X L

If all the different keys could be tested, every conceivable 21-letter message would be generated, and the cryptanalyst would be unable to distinguish between the right one and all the others. This difficulty would not have arisen had the key been a series of words or a phrase, because the incorrect messages would almost certainly have been associated with a meaningless key, whereas the correct message would be associated with a sensible key.

The security of the one-time pad cipher is wholly due to the randomness of the key. The key injects randomness into the ciphertext, and if the ciphertext is random then it has no patterns, no structure, nothing the cryptanalyst can latch onto. In fact, it can be mathematically proved that it is impossible for a cryptanalyst to crack a message encrypted with a one-time pad cipher. In other words, the one-time pad cipher is not merely believed to be unbreakable, just as the Vigenère cipher was in the nineteenth century, *it really is absolutely secure.* The one-time pad offers a guarantee of secrecy: the Holy Grail of cryptography.

At last, cryptographers had found an unbreakable system of encryption. However, the perfection of the one-time pad cipher did not end the quest for secrecy: the truth of the matter is that it was hardly ever used. Although it is perfect in theory, it is flawed in practice because the cipher suffers from two fundamental difficulties. First, there is the practical problem of making large quantities of random keys. In a single day an army might exchange hundreds of messages, each containing thousands of characters, so radio operators would require a daily supply of keys equivalent to millions of randomly arranged letters. Supplying so many random sequences of letters is an immense task.

Some early cryptographers assumed that they could generate huge amounts of random keys by haphazardly tapping away at a typewriter. However, whenever this was tried, the typist would tend to get into the habit of typing a character using the left hand, and then a character using the right hand, and thereafter alternate between the two sides. This might be a quick way of generating a key, but the resulting sequence has structure, and is no longer random – if the typist hits the letter D, from the left side of the keyboard, then the next letter is predictable in as much as it is probably from the right side of the keyboard. If a one-time pad key was to be truly random, a letter from the left side of the keyboard should be followed by another letter from the left side of the keyboard on roughly half the occasions.

Cryptographers have come to realise that it requires a great deal of time, effort and money to create a random key. The best random keys are created by harnessing natural physical processes, such as radioactivity, which is known to exhibit truly random behaviour. The cryptographer could place a lump of radioactive material on a bench, and detect its emissions with a Geiger counter. Sometimes the emissions follow each other in rapid succession, sometimes there are long delays – the time between emissions is unpredictable and random. The cryptographer could then connect a display to the Geiger counter, which rapidly cycles through the alphabet at a fixed rate, but which freezes momentarily as soon as an emission is detected. Whatever letter is on the display could be used as the next letter of the random key. The display restarts and once again cycles through the alphabet until it is stopped at random by the next emission, the letter frozen on the display is added to the key, and so on. This arrangement would be guaranteed to generate a truly random key, but it is impractical for day-to-day cryptography.

Even if you could fabricate enough random keys, there is a second problem, namely the difficulty of distributing them. Imagine a battlefield scenario in which hundreds of radio operators are part of the same communications network. To start with, every single person must have identical copies of the one-time pad. Next, when new pads are issued, they must be distributed to everybody simultaneously. Finally, everybody must remain in step, making sure that they are using the right sheet of the one-time pad at the right time. Widespread use of the one-time pad would fill the battlefield with couriers and bookkeepers. Furthermore, if the enemy captures just one set of keys, then the whole communication system is compromised.

It might be tempting to cut down on the manufacture and distribution of keys by reusing one-time pads, but this is a cryptographic cardinal sin.

Reusing a one-time pad would allow an enemy cryptanalyst to decipher messages with relative ease. The technique used to prise open two pieces of ciphertext encrypted with the same one-time pad key is explained in Appendix G, but for the time being the important point is that there can be no shortcuts in using the one-time pad cipher. The sender and receiver must use a new key for every message.

A one-time pad is practicable only for people who need ultra-secure communication, and who can afford to meet the enormous costs of manufacturing and securely distributing the keys. For example, the hotline between the presidents of Russia and America is secured via a one-time pad cipher.

The practical flaws of the theoretically perfect one-time pad meant that Mauborgne's idea could never be used in the heat of battle. In the aftermath of the First World War and all its cryptographic failures, the search continued for a practical system that could be employed in the next conflict. Fortunately for cryptographers, it would not be long before they made a breakthrough, something that would re-establish secret communication on the battlefield. In order to strengthen their ciphers, cryptographers were forced to abandon their pencil-and-paper approach to secrecy, and exploit the very latest technology to scramble messages.

The Development of Cipher Machines – from Cipher Discs to the Enigma

The earliest cryptographic machine is the cipher disc, invented in the fifteenth century by the Italian architect Leon Alberti, one of the fathers of the polyalphabetic cipher. He took two copper discs, one slightly larger than the other, and inscribed the alphabet around the edge of both. By placing the smaller disk on top of the larger one and fixing them with a needle to act as an axis, he constructed something similar to the cipher disc shown in Figure 31. The two discs can be independently rotated so that the two alphabets can have different relative positions, and can thus be used to encrypt a message with a simple Caesar shift. For example, to encrypt a message with a Caesar shift of one place, position the outer A next to the inner B – the outer disc is the plain alphabet, and the inner disc represents the cipher alphabet. Each letter in the plaintext message is looked up on the outer disc, and the corresponding letter on the inner disc is written down as part of the ciphertext. To send a message with a Caesar shift of five places, simply rotate the discs so that the outer A is next to the inner F, and then use the cipher disc in its new setting.

Even though the cipher disc is a very basic device, it does ease encipherment, and it endured for five centuries. The version shown in Figure 31 was used in the American Civil War. Figure 32 shows a Code-o-Graph, a cipher disc used by the eponymous hero of *Captain Midnight*, one of the early American radio dramas. Listeners could obtain their own Code-o-Graph by writing to the programme sponsors, Ovaltine, and enclosing a label from one of their containers. Occasionally the programme would end with a secret message from Captain Midnight, which could be deciphered by loyal listeners using the Code-o-Graph.

The cipher disc can be thought of as a 'scrambler', taking each plaintext letter and transforming it into something else. The mode of operation described so far is straightforward, and the resulting cipher is relatively

Figure 31 A U.S. Confederate cipher disc used in the American Civil War.

trivial to break, but the cipher disc can be used in a more complicated way. Its inventor, Alberti, suggested changing the setting of the disc - during the message, which in effect generates a polyalphabetic cipher instead of a monoalphabetic cipher. For example, Alberti could have used his disc to encipher the word **goodbye**, using the keyword **LEON**. He would begin by setting his disc according to the first letter of the keyword, moving the outer A next to the inner L. Then he would encipher the first letter of the message, g, by finding it on the outer disc and noting the corresponding letter on the inner disc, which is R. To encipher the second letter of the message, he would reset his disc according to the second letter of the keyword, moving the outer A next to the inner E. Then he would encipher o by finding it on the outer disc and noting the corresponding letter on the inner disc, which is S. The encryption process continues with the cipher disc being set according to the keyletter O, then N, then back to L, and so on. Alberti has effectively encrypted a message

Figure 32 Captain Midnight's Code-o-Graph, which enciphers each plaintext letter (outer disc) as a number (inner disc), rather than a letter.

using the Vigenère cipher with his first name acting as the keyword. The cipher disc speeds up encryption and reduces errors compared with performing the encryption via a Vigenère square.

The important feature of using the cipher disc in this way is the fact that the disc is changing its mode of scrambling during encryption. Although this extra level of complication makes the cipher harder to break, it does not make it unbreakable, because we are simply dealing with a mechanised version of the Vigenère cipher, and the Vigenère cipher was broken by Babbage and Kasiski. However, five hundred years after Alberti, a more complex reincarnation of his cipher disc would lead to a new generation of ciphers, an order of magnitude more difficult to crack than anything previously used.

In 1918, the German inventor Arthur Scherbius and his close friend Richard Ritter founded the company of Scherbius & Ritter, an innovative engineering firm that dabbled in everything from turbines to heated pillows. Scherbius was in charge of research and development, and was constantly looking for new opportunities. One of his pet projects was to replace the inadequate systems of cryptography used in the First World War by swapping pencil-and-paper ciphers with a form of encryption that exploited twentieth-century technology. Having studied electrical engineering in Hanover and Munich, he developed a piece of cryptographic machinery that was essentially an electrical version of Alberti's cipher disc. Called Enigma, Scherbius's invention would become the most fearsome system of encryption in history.

Scherbius's Enigma machine consisted of a number of ingenious components, which he combined into a formidable and intricate cipher machine. However, if we break the machine down into its constituent parts and rebuild it in stages, then its underlying principles will become apparent. The basic form of Scherbius's invention consists of three elements connected by wires: a keyboard for inputting each plaintext letter, a scrambling unit that encrypts each plaintext letter into a corresponding ciphertext letter, and a display board consisting of various lamps for indicating the ciphertext letter. Figure 33 shows a stylised layout of the machine, limited to a six-letter alphabet for simplicity. In order to encrypt a plaintext letter, the operator presses the appropriate plaintext letter on the keyboard, which sends an electric pulse through the central scrambling unit and out the other side, where it illuminates the corresponding ciphertext letter on the lampboard.

The scrambler, a thick rubber disc riddled with wires, is the most important part of the machine. From the keyboard, the wires enter the scrambler at six points, and then make a series of twists and turns within the scrambler before emerging at six points on the other side. The internal wirings of the scrambler determine how the plaintext letters will be encrypted. For example, in Figure 33 the wirings dictate that:

typing in a will illuminate the letter B, which means that a is encrypted as B,
typing in b will illuminate the letter A, which means that b is encrypted as A,
typing in c will illuminate the letter D, which means that c is encrypted as D,
typing in d will illuminate the letter F, which means that d is encrypted as F,
typing in e will illuminate the letter E, which means that e is encrypted as E,
typing in f will illuminate the letter C, which means that f is encrypted as C.

The message cafe would be encrypted as DBCE. With this basic set-up, the scrambler essentially defines a cipher alphabet, and the machine can be used to implement a simple monoalphabetic substitution cipher.

However, Scherbius's idea was for the scrambler disc to automatically rotate by one-sixth of a revolution each time a letter is encrypted (or one-twenty-sixth of a revolution for a complete alphabet of 26 letters). Figure 34(a) shows the same arrangement as in Figure 33; once again, typing in the letter b will illuminate the letter A. However, this time, immediately after typing a letter and illuminating the lampboard, the scrambler revolves by one-sixth of a revolution to the position shown in Figure 34(b). Typing in the letter b again will now illuminate a different letter, namely C. Immediately afterwards, the scrambler rotates once more, to the position shown in Figure 34(c). This time, typing in the letter b will illuminate E. Typing the letter b six times in a row would generate the ciphertext ACEBDC. In other words, the cipher alphabet changes after each encryption, and the encryption of the letter b is constantly changing. With this rotating set-up, the scrambler essentially defines six cipher alphabets, and the machine can be used to implement a polyalphabetic cipher.

The rotation of the scrambler is the most important feature of Scherbius's design. However, as it stands the machine suffers from one obvious weakness. Typing b six times will return the scrambler to its original

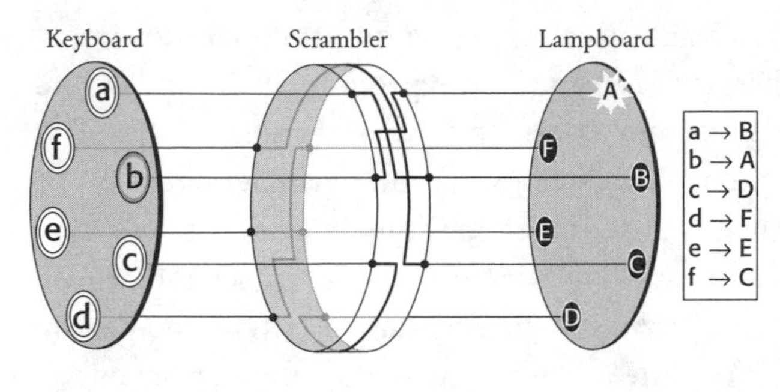

Figure 33 A simplified version of the Enigma machine with an alphabet of just six letters. The most important element of the machine is the scrambler. By typing in **b** on the keyboard, a current passes into the scrambler, follows the path of the internal wiring, and then emerges so as illuminate the **A** lamp. In short, **b** is encrypted as **A**. The box to the right indicates how each of the six letters is encrypted.

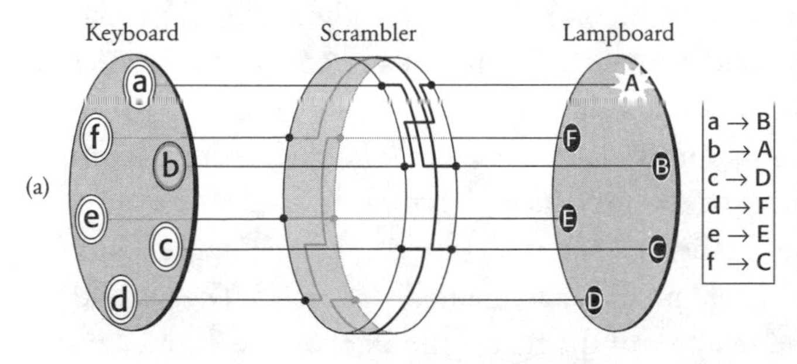

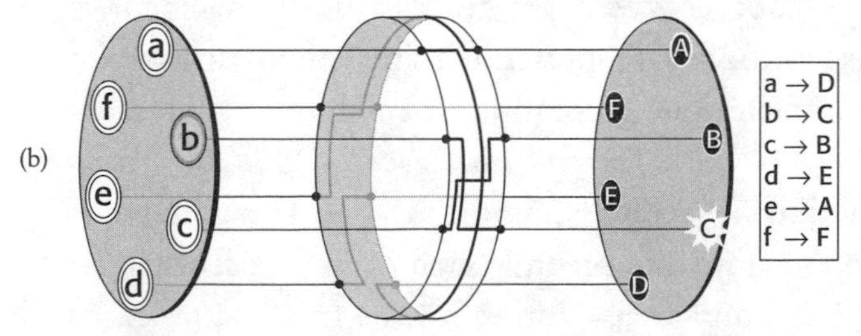

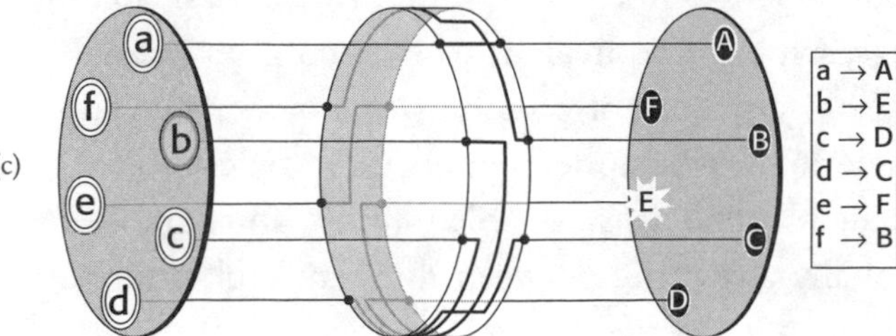

Figure 34 Every time a letter is typed into the keyboard and encrypted, the scrambler rotates by one place, thus changing how each letter is potentially encrypted. In (a) the scrambler encrypts **b** as **A**, but in (b) the new scrambler orientation encrypts **b** as **C**. In (c), after rotating one more place, the scrambler encrypts **b** as **E**. After encrypting four more letters, and rotating four more places, the scrambler returns to its original orientation.

position, and typing b again and again will repeat the pattern of encryption. In general, cryptographers are keen to avoid repetition because it leads to regularity and structure in the ciphertext, symptoms of a weak cipher. This problem can be alleviated by introducing a second scrambler disc.

Figure 35 is a schematic of a cipher machine with two scramblers. Because of the difficulty of drawing a three-dimensional scrambler with three-dimensional internal wirings, Figure 35 shows only a two-dimensional representation. Each time a letter is encrypted, the first scrambler rotates by one space, or in terms of the two-dimensional diagram, each wiring shifts down one place. In contrast, the second scrambler disc remains stationary for most of the time. It moves only after the first scrambler has made a complete revolution. The first scrambler is fitted with a tooth, and it is only when this tooth reaches a certain point that it knocks the second scrambler on one place.

In Figure 35(a), the first scrambler is in a position where it is just about to knock forward the second scrambler. Typing in and encrypting a letter moves the mechanism to the configuration shown in Figure 35(b), in which the first scrambler has moved on one place, and the second scrambler has also been knocked on one place. Typing in and encrypting another letter again moves the first scrambler on one place, Figure 35(c), but this time the second scrambler has remained stationary. The second scrambler will not move again until the first scrambler completes one revolution, which will take another five encryptions. This arrangement is similar to a car milometer – the rotor representing single miles turns quite quickly, and when it completes one revolution by reaching '9', it knocks the rotor representing tens of miles forward one place.

The advantage of adding a second scrambler is that the pattern of encryption is not repeated until the second scrambler is back where it started, which requires six complete revolutions of the first scrambler, or the encryption of 6 × 6, or 36 letters in total. In other words, there are 36 distinct scrambler settings, which is equivalent to switching between 36 cipher alphabets. With a full alphabet of 26 letters, the cipher machine would switch between 26 × 26, or 676 cipher alphabets. So by combining scramblers (sometimes called rotors), it is possible to build an encryption machine which is continually switching between different cipher alphabets. The operator types in a particular letter and, depending on the

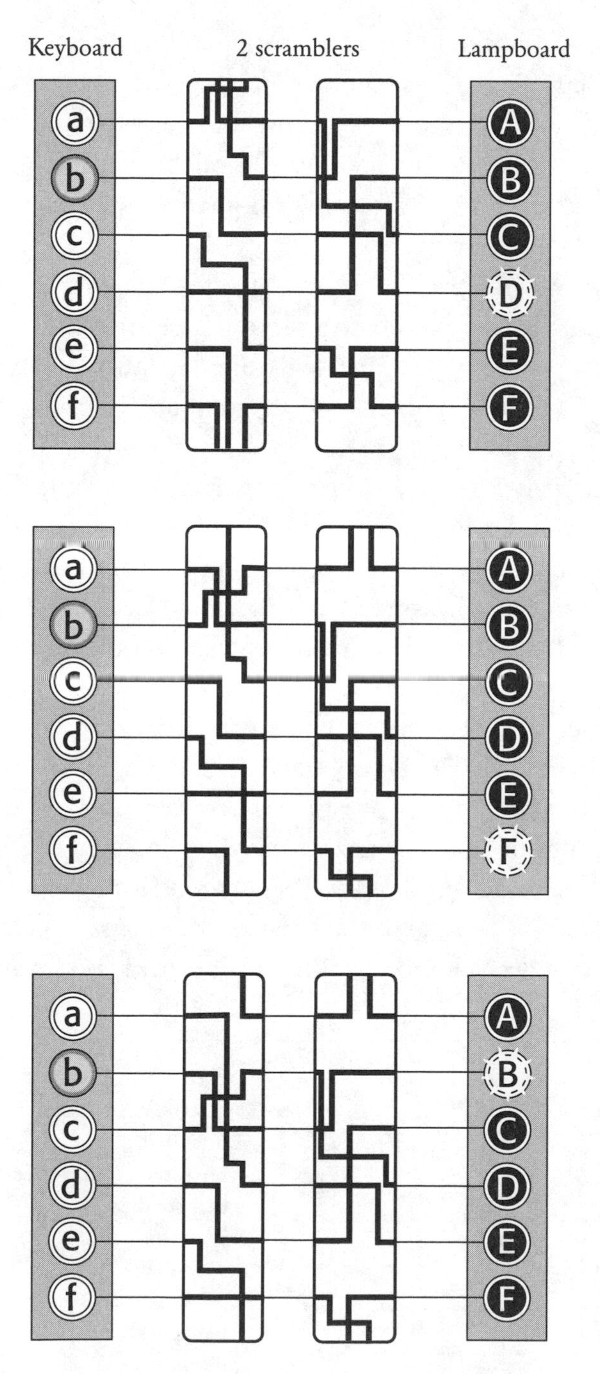

Figure 35 On adding a second scrambler, the pattern of encryption does not repeat until 36 letters have been enciphered, at which point both scramblers have returned to their original positions. To simplify the diagram, the scramblers are represented in just two dimensions; instead of rotating one place, the wirings move down one place. If a wire appears to leave the top or bottom of a scrambler, its path can be followed by continuing from the corresponding wire at the bottom or top of the same scrambler. In (a), **b** is encrypted as **D**. After encryption, the first scrambler rotates by one place, also nudging the second scrambler round one place – this happens only once during each complete revolution of the first wheel. This new setting is shown in (b), in which **b** is encrypted as **F**. After encryption, the first scrambler rotates by one place, but this time the second scrambler remains fixed. This new setting is shown in (c), in which **b** is encrypted as **B**.

scrambler arrangement, it can be encrypted according to any one of hundreds of cipher alphabets. Then the scrambler arrangement changes, so that when the next letter is typed into the machine it is encrypted according to a different cipher alphabet. Furthermore, all of this is done with great efficiency and accuracy, thanks to the automatic movement of scramblers and the speed of electricity.

Before explaining in detail how Scherbius intended his encryption machine to be used, it is necessary to describe two more elements of the Enigma, which are shown in Figure 36. First, Scherbius's standard encryption machine employed a third scrambler for extra complexity – for a full alphabet these three scramblers would provide 26 × 26 × 26, or 17,576 distinct scrambler arrangements. Second, Scherbius added a *reflector*. The reflector is a bit like a scrambler, inasmuch as it is a rubber disc with internal wirings, but it differs because it does not rotate, and the wires enter on one side and then re-emerge on the same side. With the reflector in place, the operator types in a letter, which sends an electrical signal through the three scramblers. When the reflector receives the incoming signal it sends it back through the same three scramblers, but along a different route. For example, with the set-up in Figure 36, typing the letter **b** would send a signal through the three scramblers and into the reflector, whereupon the signal would return back through the wirings to arrive at the letter **D**. The signal does not actually emerge through the keyboard, as it might seem from Figure 36, but instead is diverted to the lampboard. At first sight the reflector seems to be a pointless addition to the machine, because its static

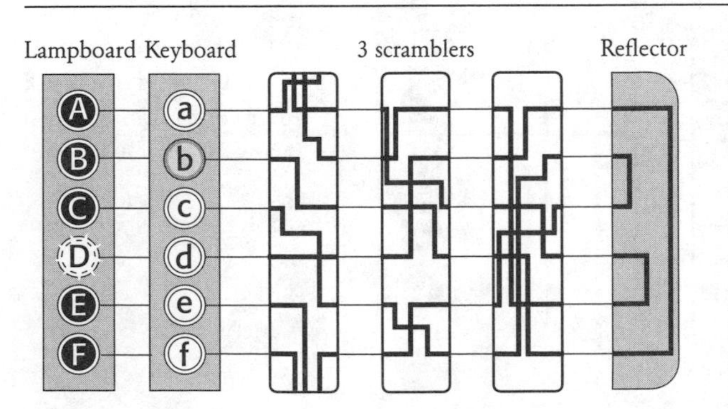

Figure 36 Scherbius's design of the Enigma included a third scrambler and a reflector that sends the current back through the scramblers. In this particular setting, typing in **b** eventually illuminates **D** on the lampboard, shown here adjacent to the keyboard.

nature means that it does not add to the number of cipher alphabets. However, its benefits become clear when we see how the machine was actually used to encrypt and decrypt a message.

An operator wishes to send a secret message. Before encryption begins, the operator must first rotate the scramblers to a particular starting position. There are 17,576 possible arrangements, and therefore 17,576 possible starting positions. The initial setting of the scramblers will determine how the message is encrypted. We can think of the Enigma machine in terms of a general cipher system, and the initial settings are what determine the exact details of the encryption. In other words, the initial settings provide the key. The initial settings are usually dictated by a codebook, which lists the key for each day, and which is available to everybody within the communications network. Distributing the codebook requires time and effort, but because only one key per day is required, it could be arranged for a codebook containing 28 keys to be sent out just once every four weeks. By comparison, if an army were to use a one-time pad cipher, it would require a new key for every message, and key distribution would be a much greater task. Once the scramblers have been set according to the codebook's daily requirement, the sender can begin encrypting. He types in the first letter of the message, sees which letter is illuminated on the lampboard, and notes it down as the first letter of the ciphertext. Then, the first scrambler having automatically stepped on by one place, the sender inputs the second letter of the message, and so on. Once he has generated the complete ciphertext, he hands it to a radio operator who transmits it to the intended receiver.

In order to decipher the message, the receiver needs to have another Enigma machine and a copy of the codebook that contains the initial scrambler settings for that day. He sets up the machine according to the book, types in the ciphertext letter by letter, and the lampboard indicates the plaintext. In other words, the sender typed in the plaintext to generate the ciphertext, and now the receiver types in the ciphertext to generate the plaintext – encipherment and decipherment are mirror processes. The ease of decipherment is a consequence of the reflector. From Figure 36 we can see that if we type in b and follow the electrical path, we come back to D. Similarly, if we type in d and follow the path, then we come back to B. The machine encrypts a plaintext letter into a ciphertext letter, and, as long as the machine is in the same setting, it will decrypt the same ciphertext letter back into the same plaintext letter.

It is clear that the key, and the codebook that contains it, must never be allowed to fall into enemy hands. It is quite possible that the enemy might capture an Enigma machine, but without knowing the initial settings used for encryption, they cannot easily decrypt an intercepted message. Without the codebook, the enemy cryptanalyst must resort to checking all the possible keys, which means trying all the 17,576 possible initial scrambler settings. The desperate cryptanalyst would set up the captured Enigma machine with a particular scrambler arrangement, input a short piece of the ciphertext, and see if the output makes any sense. If not, he would change to a different scrambler arrangement and try again. If he can check one scrambler arrangement each minute and works night and day, it would take almost two weeks to check all the settings. This is a moderate level of security, but if the enemy set a dozen people on the task, then all the settings could be checked within a day. Scherbius therefore decided to improve the security of his invention by increasing the number of initial settings and thus the number of possible keys.

He could have increased security by adding more scramblers (each new scrambler increases the number of keys by a factor of 26), but this would have increased the size of the Enigma machine. Instead, he added two other features. First, he simply made the scramblers removable and interchangeable. So, for example, the first scrambler disc could be moved to the third position, and the third scrambler disc to the first position. The arrangement of the scramblers affects the encryption, so the exact arrangement is crucial to encipherment and decipherment. There are six different ways to arrange the three scramblers, so this feature increases the number of keys, or the number of possible initial settings, by a factor of six.

The second new feature was the insertion of a *plugboard* between the keyboard and the first scrambler. The plugboard allows the sender to insert cables which have the effect of swapping some of the letters before they enter the scrambler. For example, a cable could be used to connect the a and b sockets of the plugboard, so that when the cryptographer wants to encrypt the letter b, the electrical signal actually follows the path through the scramblers that previously would have been the path for the letter a, and vice versa. The Enigma operator had six cables, which meant that six pairs of letters could be swapped, leaving fourteen letters unplugged and unswapped. The letters swapped by the plugboard are part of the machine's setting, and so must be specified in the codebook. Figure 37 shows the layout of the machine with the plugboard in place. Because the diagram deals only with a six-letter alphabet, only one pair of letters, a and b, have been swapped.

There is one more feature of Scherbius's design, known as the *ring*, which has not yet been mentioned. Although the ring does have some effect on encryption, it is the least significant part of the whole Enigma machine, and I have decided to ignore it for the purposes of this discussion. (Readers who would like to know about the exact role of the ring should refer to some of the books in the list of further reading, such as *Seizing the Enigma* by David Kahn. This list also includes two websites containing excellent Enigma emulators, which allow you to operate a virtual Enigma machine.)

Now that we know all the main elements of Scherbius's Enigma machine, we can work out the number of keys, by combining the number of possible plugboard cablings with the number of possible scrambler

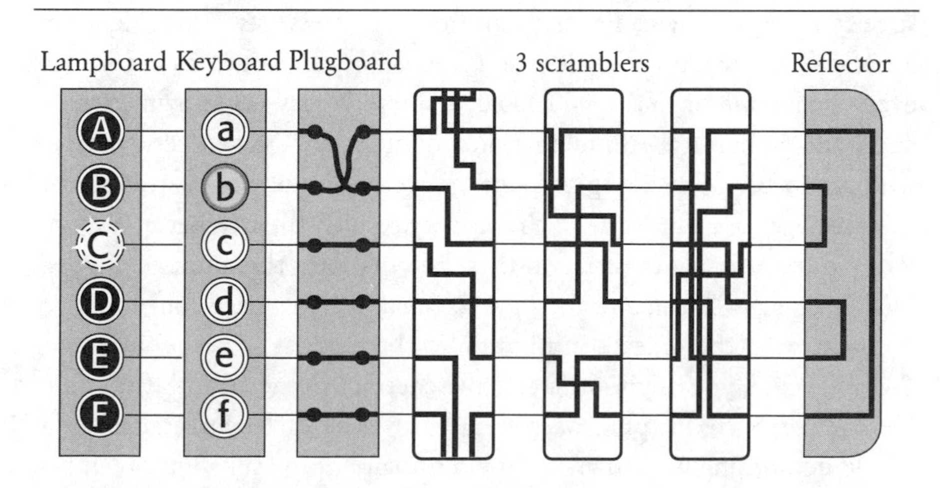

Figure 37 The plugboard sits between the keyboard and the first scrambler. By inserting cables it is possible to swap pairs of letters, so that, in this case, b is swapped with a. Now, b is encrypted by following the path previously associated with the encryption of a. In the real 26-letter Enigma, the user would have six cables for swapping six pairs of letters.

arrangements and orientations. The following list shows each variable of the machine and the corresponding number of possibilities for each one:

Scrambler orientations. Each of the 3 scramblers can be set in one of 26 orientations. There are therefore $26 \times 26 \times 26$ settings: 17,576

Scrambler arrangements. The three scramblers (1, 2 and 3) can be positioned in any of the following six orders: 123, 132, 213, 231, 312, 321. 6

Plugboard. The number of ways of connecting, thereby swapping, six pairs of letters out of 26 is enormous: 100,391,791,500

Total. The total number of keys is the multiple of these three numbers: $17{,}576 \times 6 \times 100{,}391{,}791{,}500$

$\approx 10{,}000{,}000{,}000{,}000{,}000$

As long as sender and receiver have agreed on the plugboard cablings, the order of the scramblers and their respective orientations, all of which specify the key, they can encrypt and decrypt messages easily. However, an enemy interceptor who does not know the key would have to check every single one of the 10,000,000,000,000,000 possible keys in order to crack the ciphertext. To put this into context, a persistent cryptanalyst who is capable of checking one setting every minute would need longer than the age of the universe to check every setting. (In fact, because I have ignored the effect of the rings in these calculations, the number of possible keys is even larger, and the time to break Enigma even longer.)

Since by far the largest contribution to the number of keys comes from the plugboard, you might wonder why Scherbius bothered with the scramblers. On its own, the plugboard would provide a trivial cipher, because it would do nothing more than act as a monoalphabetic substitution cipher, swapping around just 12 letters. The problem with the plugboard is that the swaps do not change once encryption begins, so on its own it would generate a ciphertext that could be broken by frequency analysis. The scramblers contribute a smaller number of keys, but their set-up is continually changing, which means that the resulting ciphertext cannot be broken by frequency analysis. By combining the scramblers with the plugboard, Scherbius protected his machine against frequency analysis, and at the same time gave it an enormous number of possible keys.

Scherbius took out his first patent in 1918. His cipher machine was contained in a compact box measuring only 34 × 28 × 15 cm, but it weighed a hefty 12 kg. Figure 39 shows an Enigma machine with the outer lid open, ready for use. It is possible to see the keyboard where the plaintext letters are typed in, and, above it, the lampboard which displays the resulting ciphertext letter. Below the keyboard is the plugboard; there are more than six pairs of letters swapped by the plugboard, because this particular Enigma machine is a slightly later modification of the original model, which is the version that has been described so far. Figure 40 shows an Enigma with the cover-plate removed to reveal more features, in particular the three scramblers.

Scherbius believed that Enigma was impregnable, and that its cryptographic strength would create a great demand for it. He tried to market the cipher machine to both the military and the business community,

Figure 38 Arthur Scherbius.

offering different versions to each. For example, he offered a basic version of Enigma to businesses, and a luxury diplomatic version with a printer rather than a lampboard to the Foreign Office. The price of an individual unit was as much as £20,000 in today's prices.

Unfortunately, the high cost of the machine discouraged potential buyers. Businesses said that they could not afford Enigma's security, but Scherbius believed that they could not afford to be without it. He argued that a vital message intercepted by a business rival could cost a company a fortune, but few businessmen took any notice of him. The German military were equally unenthusiastic, because they were oblivious to the damage caused by their insecure ciphers during the Great War. For example, they had been led to believe that the Zimmermann telegram had been stolen by American spies in Mexico, and so they blamed that failure on Mexican security. They still did not realise that the telegram had in fact been intercepted and deciphered by the British, and that the Zimmermann debacle was actually a failure of German cryptography.

Scherbius was not alone in his growing frustration. Three other inventors in three other countries had independently and almost simultaneously hit upon the idea of a cipher machine based on rotating scramblers. In the Netherlands in 1919, Alexander Koch took out patent No. 10,700, but he failed to turn his rotor machine into a commercial success and eventually sold the patent rights in 1927. In Sweden, Arvid Damm took out a similar patent, but by the time he died in 1927 he had also failed to find a market. In America, inventor Edward Hebern had complete faith in his invention, the so-called Sphinx of the Wireless, but his failure was the greatest of all.

In the mid-1920s, Hebern began building a $380,000 factory, but unfortunately this was a period when the mood in America was changing from paranoia to openness. The previous decade, in the aftermath of the First World War, the U.S. Government had established the American Black Chamber, a highly effective cipher bureau staffed by a team of twenty cryptanalysts, led by the flamboyant and brilliant Herbert Yardley. Later, Yardley wrote that 'The Black Chamber, bolted, hidden, guarded, sees all, hears all. Though the blinds are drawn and the windows heavily curtained, its far-seeking eyes penetrate the secret conference chambers at Washington, Tokyo, London, Paris, Geneva, Rome. Its sensitive ears catch the faintest whisperings in the foreign capitals of the world.' The

Figure 39 An army Enigma machine ready for use.

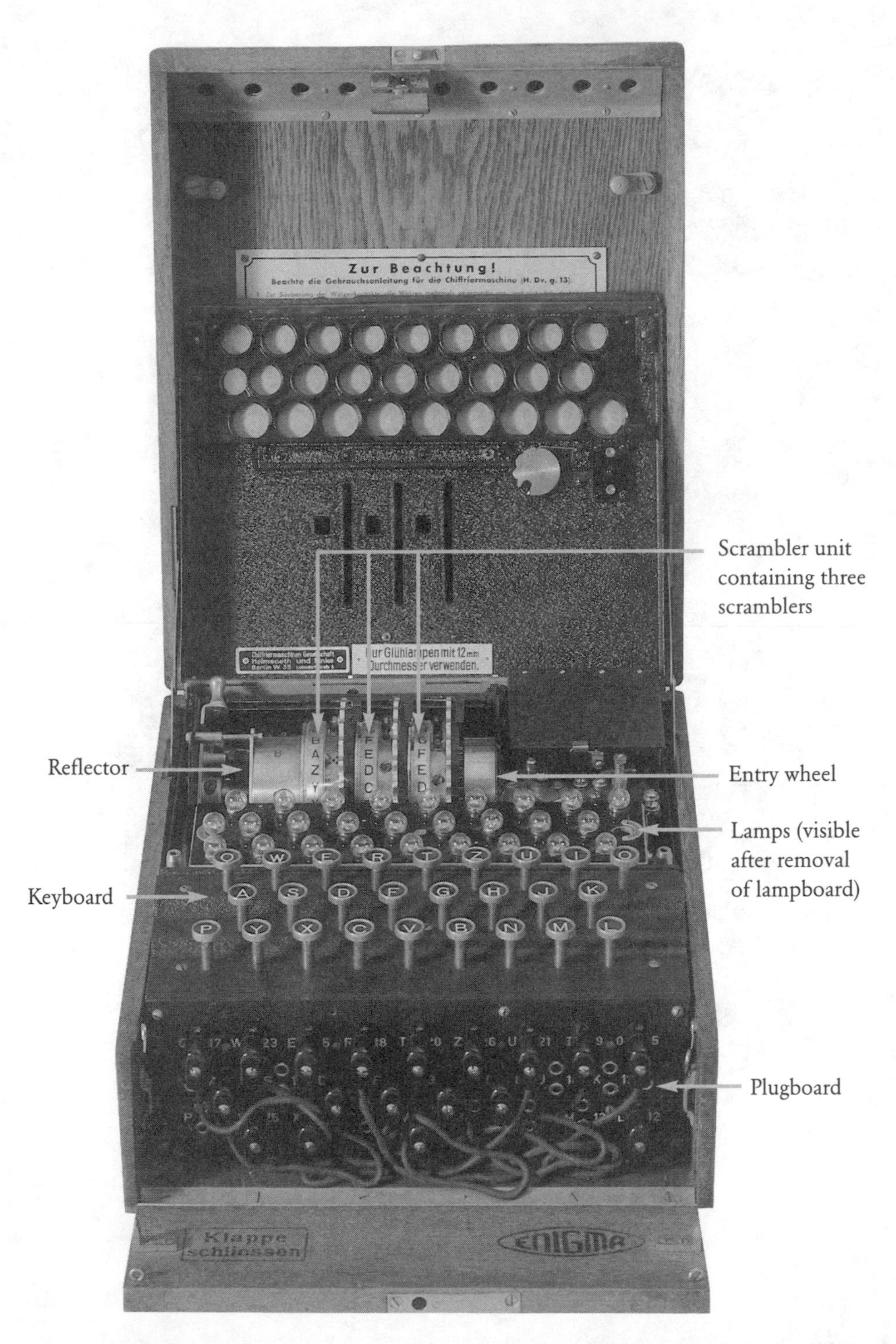

Figure 40 An Enigma machine with the inner lid opened, revealing the three scramblers.

American Black Chamber solved 45,000 cryptograms in a decade, but by the time Hebern built his factory, Herbert Hoover had been elected President and was attempting to usher in a new era of trust in international affairs. He disbanded the Black Chamber, and his Secretary of State, Henry Stimson, declared that 'Gentlemen should not read each other's mail'. If a nation believes that it is wrong to read the messages of others, then it also begins to believe that others will not read its own messages, and it does not see the necessity for fancy cipher machines. Hebern sold only twelve machines at a total price of roughly $1,200, and in 1926 he was brought to trial by dissatisfied shareholders and found guilty under California's Corporate Securities Act.

Fortunately for Scherbius, however, the German military were eventually shocked into appreciating the value of his Enigma machine, thanks to two British documents. The first was Winston Churchill's *The World Crisis*, published in 1923, which included a dramatic account of how the British had gained access to valuable German cryptographic material:

> At the beginning of September 1914, the German light cruiser *Magdeburg* was wrecked in the Baltic. The body of a drowned German under-officer was picked up by the Russians a few hours later, and clasped in his bosom by arms rigid in death, were the cipher and signal books of the German navy and the minutely squared maps of the North Sea and Heligoland Bight. On September 6 the Russian Naval Attaché came to see me. He had received a message from Petrograd telling him what had happened, and that the Russian Admiralty with the aid of the cipher and signal books had been able to decode portions at least of the German naval messages. The Russians felt that as the leading naval Power, the British Admiralty ought to have these books and charts. If we would send a vessel to Alexandrov, the Russian officers in charge of the books would bring them to England.

This material had helped the cryptanalysts in Room 40 to crack Germany's encrypted messages on a regular basis. Finally, almost a decade later, the Germans were made aware of this failure in their communications security. Also in 1923, the British Royal Navy published their official history of the First World War, which reiterated the fact that the interception and cryptanalysis of German communications had provided the Allies with a clear advantage. These proud achievements of British Intelligence were a stark condemnation of those responsible for German security, who then had to admit in their own report that, 'the German fleet command, whose radio messages were intercepted and deciphered by the English, played so to speak with open cards against the British command'.

The German military held an enquiry into how to avoid repeating the cryptographic fiascos of the First World War, and concluded that the Enigma machine offered the best solution. By 1925 Scherbius began mass-producing Enigmas, which went into military service the following year, and were subsequently used by the government and by state-run organisations such as the railways. These Enigmas were distinct from the few machines that Scherbius had previously sold to the business community, because the scramblers had different internal wirings. Owners of a commercial Enigma machine did not therefore have a complete knowledge of the government and military versions.

Over the next two decades, the German military would buy over 30,000 Enigma machines. Scherbius's invention provided the German military with the most secure system of cryptography in the world, and at the outbreak of the Second World War their communications were protected by an unparalleled level of encryption. At times, it seemed that the Enigma machine would play a vital role in ensuring Nazi victory, but instead it was ultimately part of Hitler's downfall. Scherbius did not live long enough to see the successes and failures of his cipher system. In 1929, while driving a team of horses, he lost control of his carriage and crashed into a wall, dying on 13 May from internal injuries.

4 Cracking the Enigma

In the years that followed the First World War, the British cryptanalysts in Room 40 continued to monitor German communications. In 1926 they began to intercept messages which baffled them completely. Enigma had arrived, and as the number of Enigma machines increased, Room 40's ability to gather intelligence diminished rapidly. The Americans and the French also tried to tackle the Enigma cipher, but their attempts were equally dismal, and they soon gave up hope of breaking it. Germany now had the most secure communications in the world.

The speed with which the Allied cryptanalysts abandoned hope of breaking Enigma was in sharp contrast to their perseverance just a decade earlier in the First World War. Confronted with the prospect of defeat, the Allied cryptanalysts had worked night and day to penetrate German ciphers. It would appear that fear was the main driving force, and that adversity is one of the foundations of successful codebreaking. Similarly, it was fear and adversity that galvanised French cryptanalysis at the end of the nineteenth century, faced with the increasing might of Germany. However, in the wake of the First World War the Allies no longer feared anybody. Germany had been crippled by defeat, the Allies were in a dominant position, and as a result they seemed to lose their cryptanalytic zeal. Allied cryptanalysts dwindled in number and deteriorated in quality.

One nation, however, could not afford to relax. After the First World War, Poland re-established itself as an independent state, but it was concerned about threats to its new-found sovereignty. To the east lay Russia, a nation ambitious to spread its communism, and to the west lay Germany, desperate to regain territory ceded to Poland after the war. Sandwiched between these two enemies, the Poles were desperate for intelligence information, and they formed a new cipher bureau, the Biuro

Szyfrów. If necessity is the mother of invention, then perhaps adversity is the mother of cryptanalysis. The success of the Biuro Szyfrów is exemplified by their success during the Russo-Polish War of 1919–20. In August 1920 alone, when the Soviet armies were at the gates of Warsaw, the Biuro deciphered 400 enemy messages. Their monitoring of German communications had been equally effective, until 1926, when they too encountered the Enigma messages.

In charge of deciphering German messages was Captain Maksymilian Ciezki, a committed patriot who had grown up in the town of Szamotuty, a centre of Polish nationalism. Ciezki had access to a commercial version of the Enigma machine, which revealed all the principles of Scherbius's invention. Unfortunately, the commercial version was distinctly different from the military one in terms of the wirings inside each scrambler. Without knowing the wirings of the military machine, Ciezki had no chance of deciphering messages being sent by the German army. He became so despondent that at one point he even employed a clairvoyant in a frantic attempt to conjure some sense from the enciphered intercepts. Not surprisingly, the clairvoyant failed to make the breakthrough the Biuro Szyfrów needed. Instead, it was left to a disaffected German, Hans-Thilo Schmidt, to make the first step towards breaking the Enigma cipher.

Hans-Thilo Schmidt was born in 1888 in Berlin, the second son of a distinguished professor and his aristocratic wife. Schmidt embarked on a career in the German Army and fought in the First World War, but he was not considered worthy enough to remain in the army after the drastic cuts implemented as part of the Treaty of Versailles. He then tried to make his name as a businessman, but his soap factory was forced to close because of the post-war depression and hyperinflation, leaving him and his family destitute.

The humiliation of Schmidt's failures was compounded by the success of his elder brother, Rudolph, who had also fought in the war, and who was retained in the army afterwards. During the 1920s Rudolph rose through the ranks and was eventually promoted to chief of staff of the Signal Corps. He was responsible for ensuring secure communications, and in fact it was Rudolph who officially sanctioned the army's use of the Enigma cipher.

After his business collapsed, Hans-Thilo was forced to ask his brother for help, and Rudolph arranged a job for him in Berlin at the Chiffrierstelle, the office responsible for administrating Germany's encrypted communications. This was Enigma's command centre, a top-secret establishment dealing with highly sensitive information. When Hans-Thilo moved to his new job, he left his family behind in Bavaria, where the cost of living was affordable. He was living alone in expensive Berlin, impoverished and isolated, envious of his perfect brother and resentful towards a nation which had rejected him. The result was inevitable. By selling secret Enigma information to foreign powers, Hans-Thilo Schmidt could earn money and gain revenge, damaging his country's security and undermining his brother's organisation.

On 8 November 1931, Schmidt arrived at the Grand Hotel in Verviers, Belgium, for a liaison with a French secret agent codenamed Rex. In exchange for 10,000 marks (equivalent to £20,000 in today's money), Schmidt allowed Rex to photograph two documents: 'Gebrauchsanweisung für die Chiffriermaschine Enigma' and 'Schlüsselanleitung für die Chiffriermaschine Enigma'. These documents were essentially instructions

Figure 41 Hans-Thilo Schmidt.

for using the Enigma machine, and although there was no explicit description of the wirings inside each scrambler, they contained the information needed to deduce those wirings.

Thanks to Schmidt's treachery, it was now possible for the Allies to create an accurate replica of the German military Enigma machine. However, this was not enough to enable them to decipher messages encrypted by Enigma. The strength of the cipher depends not on keeping the machine secret, but on keeping the initial setting of the machine (the key) secret. If a cryptanalyst wants to decipher an intercepted message, then, in addition to having a replica of the Enigma machine, he still has to find which of the millions of billions of possible keys was used to encipher it. A German memorandum put it thus: 'It is assumed in judging the security of the cryptosystem that the enemy has at his disposition the machine.'

The French Secret Service was clearly up to scratch, having found an informant in Schmidt, and having obtained the documents that suggested the wirings of the military Enigma machine. In comparison, French cryptanalysts were inadequate, and seemed unwilling and unable to exploit this newly acquired information. In the wake of the First World War they suffered from over-confidence and lack of motivation. The Bureau du Chiffre did not even bother trying to build a replica of the military Enigma machine, because they were convinced that achieving the next stage, finding the key required to decipher a particular Enigma message, was impossible.

As it happened, ten years earlier the French had signed an agreement of military cooperation with the Poles. The Poles had expressed an interest in anything connected with Enigma, so in accordance with their decade-old agreement the French simply handed the photographs of Schmidt's documents to their allies, and left the hopeless task of cracking Enigma to the Biuro Szyfrów. The Biuro realised that the documents were only a starting point, but unlike the French they had the fear of invasion to spur them on. The Poles convinced themselves that there must be a shortcut to finding the key to an Enigma-encrypted message, and that if they applied sufficient effort, ingenuity and wit, they could find that shortcut.

As well as revealing the internal wirings of the scramblers, Schmidt's documents also explained in detail the layout of the codebooks used by the Germans. Each month, Enigma operators received a new codebook which specified which key should be used for each day. For example, on the first day of the month, the codebook might specify the following *day-key*:

(1) *Plugboard settings:* A/L – P/R – T/D – B/W – K/F – O/Y.
(2) *Scrambler arrangement:* 2-3-1.
(3) *Scrambler orientations*: Q-C-W.

Together, the scrambler arrangement and orientations are known as the scrambler settings. To implement this particular day-key, the Enigma operator would set up his Enigma machine as follows:

(1) *Plugboard settings*: Swap the letters A and L by connecting them via a lead on the plugboard, and similarly swap P and R, then T and D, then B and W, then K and F, and then O and Y.

(2) *Scrambler arrangement*: Place the 2nd scrambler in the 1st slot of the machine, the 3rd scrambler in the 2nd slot, and the 1st scrambler in the 3rd slot.

(3) *Scrambler orientations*: Each scrambler has an alphabet engraved on its outer rim, which allows the operator to set it in a particular orientation. In this case, the operator would rotate the scrambler in slot 1 so that Q is facing upwards, rotate the scrambler in slot 2 so that C is facing upwards, and rotate the scrambler in slot 3 so that W is facing upwards.

One way of encrypting messages would be for the sender to encrypt all the day's traffic according to the day-key. This would mean that for a whole day at the start of each message all Enigma operators would set their machines according to the same day-key. Then, each time a message needed to be sent, it would be first typed into the machine; the enciphered output would then be recorded, and handed to the radio operator for transmission. At the other end, the receiving radio operator would record the incoming message, hand it to the Enigma operator, who would type it into his machine, which would already be set to the same day-key. The output would be the original message.

This process is reasonably secure, but it is weakened by the repeated use of a single day-key to encrypt the hundreds of messages that might be sent each day. In general, it is true to say that if a single key is used to encipher an enormous quantity of material, then it is easier for a cryptanalyst to deduce it. A large amount of identically encrypted material provides a cryptanalyst with a correspondingly larger chance of identifying the key. For example, harking back to simpler ciphers, it is much easier to break a monoalphabetic cipher with frequency analysis if there are several pages of encrypted material, as opposed to just a couple of sentences.

As an extra precaution, the Germans therefore took the clever step of using the day-key settings to transmit a new *message-key* for each message. The message-keys would have the same plugboard settings and scrambler arrangement as the day-key, but different scrambler orientations. Because the new scrambler orientation would not be in the codebook, the sender had to transmit it securely to the receiver according to the following process. First, the sender sets his machine according to the agreed day-key, which includes a scrambler orientation, say QCW. Next, he randomly picks a new scrambler orientation for the message-key, say PGH. He then enciphers PGH according to the day-key. The message-key is typed into the Enigma twice, just to provide a double-check for the receiver. For example, the sender might encipher the message-key PGHPGH as KIVBJE. Note that the two PGH's are enciphered differently (the first as KIV, the second as BJE) because the Enigma scramblers are rotating after each letter, and changing the overall mode of encryption. The sender then changes his machine to the PGH setting and encrypts the main message according to this message-key. At the receiver's end, the machine is initially set according to the day-key, QCW. The first six letters of the incoming message, KIVBJE, are typed in and reveal PGHPGH. The receiver then knows to reset his scramblers to PGH, the message-key, and can then decipher the main body of the message.

This is equivalent to the sender and receiver agreeing on a main cipher key. Then, instead of using this single main cipher key to encrypt every message, they use it merely to encrypt a new cipher key for each message, and then encrypt the actual message according to the new cipher key. Had the Germans not employed message-keys, then everything – perhaps thousands of messages containing millions of letters – would have been sent using the same day-key. However, if the day-key is only used to transmit the message-keys, then it encrypts only a limited amount of text. If there are 1,000 message-keys sent in a day, then the day-key encrypts only

6,000 letters. And because each message-key is picked at random and is used to encipher only one message, then it encrypts a limited amount of text, perhaps just a few hundred characters.

At first sight the system seemed to be impregnable, but the Polish cryptanalysts were undaunted. They were prepared to explore every avenue in order to find a weakness in the Enigma machine and its use of day- and message-keys. Foremost in the battle against Enigma was a new breed of cryptanalyst. For centuries, it had been assumed that the best cryptanalysts were experts in the structure of language, but the arrival of Enigma prompted the Poles to alter their recruiting policy. Enigma was a mechanical cipher, and the Biuro Szyfrów reasoned that a more scientific mind might stand a better chance of breaking it. The Biuro organised a course on cryptography and invited twenty mathematicians, each of them sworn to an oath of secrecy. The mathematicians were all from the university at Poznán. Although not the most respected academic institution in Poland, it had the advantage of being located in the west of the country, in territory that had been part of Germany until 1918. These mathematicians were therefore fluent in German.

Three of the twenty demonstrated an aptitude for solving ciphers, and were recruited into the Biuro. The most gifted of them was Marian Rejewski, a timid, spectacled twenty-three-year-old who had previously studied statistics in order to pursue a career in insurance. Although a competent student at university, it was within the Biuro Szyfrów that he was to find his true calling. He served his apprenticeship by breaking a series of traditional ciphers before moving on to the more forbidding challenge of Enigma. Working entirely alone, he concentrated all of his energies on the intricacies of Scherbius's machine. As a mathematician, he would try to analyse every aspect of the machine's operation, probing the effect of the scramblers and the plugboard cablings. However, as with all mathematics, his work required inspiration as well as logic. As another wartime mathematical cryptanalyst put it, the creative codebreaker must 'perforce commune daily with dark spirits to accomplish his feats of mental ju-jitsu'.

Rejewski's strategy for attacking Enigma focused on the fact that repetition is the enemy of security: repetition leads to patterns, and cryptanalysts thrive on patterns. The most obvious repetition in the Enigma encryption was the message-key, which was enciphered twice at the beginning of every message. If the operator chose the message key ULJ, then he would encrypt it twice so that ULJULJ might be enciphered as PEFNWZ, which he would then send at the start before the actual message. The Germans had demanded this repetition in order to avoid mistakes caused by radio interference or operator error. But they did not foresee that this would jeopardise the security of the machine.

Each day, Rejewski would find himself with a new batch of intercepted messages. They all began with the six letters of the repeated three-letter message-key, all encrypted according to the same agreed day-key. For example, he might receive four messages that began with the following encrypted message-keys:

	1st	2nd	3rd	4th	5th	6th
1st message	L	O	K	R	G	M
2nd message	M	V	T	X	Z	E
3rd message	J	K	T	M	P	E
4th message	D	V	Y	P	Z	X

In each case, the 1st and 4th letters are encryptions of the same letter, namely the first letter of the message-key. Also, the 2nd and 5th letters are encryptions of the same letter, namely the second letter of the message-key, and the 3rd and 6th letters are encryptions of the same letter, namely the third letter of the message-key. For example, in the first message L and R are encryptions of the same letter, the first letter of the message-key. The reason why this same letter is encrypted differently, first as L and then as R, is that between the two encryptions the first Enigma scrambler has moved on three steps, changing the overall mode of scrambling.

The fact that L and R are encryptions of the same letter allowed Rejewski to deduce some slight constraint on the initial set-up of the machine. The initial scrambler setting, which is unknown, encrypted the first letter of the day-key, which is also unknown, into L, and then another scrambler setting, three steps on from the initial setting, which is still unknown, encrypted the same letter of the day-key, which is also still unknown, into R.

This constraint might seem vague, as it is full of unknowns, but at least it demonstrates that the letters L and R are intimately related by the initial setting of the Enigma machine, the day-key. As each new message is intercepted, it is possible to identify other relationships between the 1st and

4th letters of the repeated message-key. All these relationships are reflections of the initial setting of the Enigma machine. For example, the second message above tells us that M and X are related, the third tells us that J and M are related, and the fourth that D and P are related. Rejewski began to summarise these relationships by tabulating them. For the four messages we have so far, the table would reflect the relationships between (L,R), (M,X), (J,M) and (D,P):

1st letter	A	B	C	D	E	F	G	H	I	J	K	L	M	N	O	P	Q	R	S	T	U	V	W	X	Y	Z
4th letter				P						M		R	X													

If Rejewski had access to enough messages in a single day, then he would be able to complete the alphabet of relationships. The following table shows such a completed set of relationships:

1st letter	A	B	C	D	E	F	G	H	I	J	K	L	M	N	O	P	Q	R	S	T	U	V	W	X	Y	Z
4th letter	F	Q	H	P	L	W	O	G	B	M	V	R	X	U	Y	C	Z	I	T	N	J	E	A	S	D	K

Figure 42 Marian Rejewski.

Rejewski had no idea of the day-key, and he had no idea which message-keys were being chosen, but he did know that they resulted in this table of relationships. Had the day-key been different, then the table of relationships would have been completely different. The next question was whether there existed any way of determining the day-key by looking at the table of relationships. Rejewski began to look for patterns within the table, structures that might indicate the day-key. Eventually, he began to study one particular type of pattern, which featured chains of letters. For example, in the table, A on the top row is linked to F on the bottom row, so next he would look up F on the top row. It turns out that F is linked to W, and so he would look up W on the top row. And it turns out that W is linked to A, which is where we started. The chain has been completed.

With the remaining letters in the alphabet, Rejewski would generate more chains. He listed all the chains, and noted the number of links in each one:

A→ F →W→ A	3 links
B→Q→ Z → K → V → E → L → R → I → B	9 links
C→ H → G → O → Y → D → P → C	7 links
J →M→ X → S → T → N → U → J	7 links

So far, we have only considered the links between the 1st and 4th letters of the six-letter repeated key. In fact, Rejewski would repeat this whole exercise for the relationships between the 2nd and 5th letters, and the 3rd and 6th letters, identifying the chains in each case and the number of links in each chain.

Rejewski noticed that the chains changed each day. Sometimes there were lots of short chains, sometimes just a few long chains. And, of course, the letters within the chains changed. The characteristics of the chains were clearly a result of the day-key setting – a complex consequence of the plugboard settings, the scrambler arrangement and the scrambler orientations. However, there remained the question of how Rejewski could determine the day-key from these chains. Which of 10,000,000,000,000,000 possible day-keys was related to a particular pattern of chains? The number of possibilities was simply too great.

It was at this point that Rejewski had a profound insight. Although the plugboard and scrambler settings both affect the details of the chains, their contributions can to some extent be disentangled. In particular, there is one aspect of the chains which is wholly dependent on the scrambler settings, and which has nothing to do with the plugboard settings: the numbers of links in the chains is purely a consequence of the scrambler settings. For instance, let us take the example above and pretend that the day-key required the letters S and G to be swapped as part of the plugboard settings. If we change this element of the day-key, by removing the cable that swaps S and G, and use it to swap, say, T and K instead, then the chains would change to the following:

A→ F →W→ A 3 links
B→Q→ Z → T → V → E → L → R → I → B 9 links
C→H → S →O → Y → D → P → C 7 links
J →M → X → G → K → N → U → J 7 links

Some of the letters in the chains have changed, but, crucially, the number of links in each chain remains constant. Rejewski had identified a facet of the chains that was solely a reflection of the scrambler settings.

The total number of scrambler settings is the number of scrambler arrangements (6) multiplied by the number of scrambler orientations (17,576) which comes to 105,456. So, instead of having to worry about which of the 10,000,000,000,000,000 day-keys was associated with a particular set of chains, Rejewski could busy himself with a drastically simpler problem: which of the 105,456 scrambler settings was associated with the numbers of links within a set of chains? This number is still large, but it is roughly one hundred billion times smaller than the total number of possible day-keys. In short, the task has become one hundred billion times easier, certainly within the realm of human endeavour.

Rejewski proceeded as follows. Thanks to Hans-Thilo Schmidt's espionage, he had access to replica Enigma machines. His team began the laborious chore of checking each of 105,456 scrambler settings, and cataloguing the chain lengths that were generated by each one. It took an entire year to complete the catalogue, but once the Biuro had accumulated the data, Rejewski could finally begin to unravel the Enigma cipher.

Each day, he would look at the encrypted message-keys, the first six letters of all the intercepted messages, and use the information to build his table of relationships. This would allow him to trace the chains, and establish the number of links in each chain. For example, analysing the 1st and 4th letters might result in four chains with 3, 9, 7 and 7 links. Analysing the 2nd and 5th letters might also result in four chains, with 2, 3, 9 and 12 links. Analysing the 3rd and 6th letters might result in five chains with 5, 5, 5, 3 and 8 links. As yet, Rejewski still had no idea of the day-key, but he knew that it resulted in 3 sets of chains with the following number of chains and links in each one:

4 chains from the 1st and 4th letters, with 3, 9, 7 and 7 links.
4 chains from the 2nd and 5th letters, with 2, 3, 9 and 12 links.
5 chains from the 3rd and 6th letters, with 5, 5, 5, 3 and 8 links.

Rejewski could now go to his catalogue, which contained every scrambler setting indexed according to the sort of chains it would generate. Having found the catalogue entry that contained the right number of chains with the appropriate number of links in each one, he immediately knew the scrambler settings for that particular day-key. The chains were effectively fingerprints, the evidence that betrayed the initial scrambler arrangement and orientations. Rejewski was working just like a detective who might find a fingerprint at the scene of a crime, and then use a database to match it to a suspect.

Although he had identified the scrambler part of the day-key, Rejewski still had to establish the plugboard settings. Although there are about a hundred billion possibilities for the plugboard settings, this was a relatively straightforward task. Rejewski would begin by setting the scramblers in his Enigma replica according to the newly established scrambler part of the day-key. He would then remove all cables from the plugboard, so that the plugboard had no effect. Finally, he would take a piece of intercepted ciphertext and type it in to the Enigma machine. This would largely result in gibberish, because the plugboard cablings were unknown and missing. However, every so often vaguely recognisable phrases would appear, such as **alliveinbelrin** – presumably, this should be 'arrive in Berlin'. If this assumption is correct, then it would imply that the letters **R** and **L** should be connected and swapped by a plugboard cable, while **A**, **I**, **V**, **E**, **B** and **N** should not. By analysing other phrases it would be possible to identify the other five pairs of letters that had been swapped by the plugboard. Having established the plugboard settings, and having already discovered the scrambler settings, Rejewski had the complete day-key, and could then decipher any message sent that day.

Rejewski had vastly simplified the task of finding the day-key by divorcing the problem of finding the scrambler settings from the problem of finding the plugboard settings. On their own, both of these problems were solvable. Originally, we estimated that it would take more than the lifetime of the universe to check every possible Enigma key. However, Rejewski had spent only a year compiling his catalogue of chain lengths, and thereafter he could find the day-key before the day was out. Once he had the day-key, he possessed the same information as the intended receiver and so could decipher messages just as easily.

Following Rejewski's breakthrough, German communications became transparent. Poland was not at war with Germany, but there was a threat of invasion, and Polish relief at conquering Enigma was nevertheless immense. If they could find out what the German generals had in mind for them, there was a chance that they could defend themselves. The fate of the Polish nation had depended on Rejewski, and he did not disappoint his country. Rejewski's attack on Enigma is one of the truly great accomplishments of cryptanalysis. I have had to sum up his work in just a few pages, and so have omitted many of the technical details, and all of the dead ends. Enigma is a complicated cipher machine, and breaking it required immense intellectual force. My simplifications should not mislead you into underestimating Rejewski's extraordinary achievement.

The Polish success in breaking the Enigma cipher can be attributed to three factors: fear, mathematics and espionage. Without the fear of invasion, the Poles would have been discouraged by the apparent invulnerability of the Enigma cipher. Without mathematics, Rejewski would not have been able to analyse the chains. And without Schmidt, codenamed 'Asche', and his documents, the wirings of the scramblers would not have been known, and cryptanalysis could not even have begun. Rejewski did not hesitate to express the debt he owed Schmidt: 'Asche's documents were welcomed like manna from heaven, and all doors were immediately opened.'

The Poles successfully used Rejewski's technique for several years. When Hermann Göring visited Warsaw in 1934, he was totally unaware of the fact that his communications were being intercepted and deciphered.

As he and other German dignitaries laid a wreath at the Tomb of the Unknown Soldier next to the offices of the Biuro Szyfrów, Rejewski could stare down at them from his window, content in the knowledge that he could read their most secret communications.

Even when the Germans made a minor alteration to the way they transmitted messages, Rejewski fought back. His old catalogue of chain lengths was useless, but rather than rewriting the catalogue he devised a mechanised version of his cataloguing system, which could automatically search for the correct scrambler settings. Rejewski's invention was an adaptation of the Enigma machine, able to rapidly check each of the 17,576 settings until it spotted a match. Because of the six possible scrambler arrangements, it was necessary to have six of Rejewski's machines working in parallel, each one representing one of the possible arrangements. Together, they formed a unit that was about a metre high, capable of finding the day-key in roughly two hours. The units were called *bombes*, a name that might reflect the ticking noise they made while checking scrambler settings. Alternatively, it is said that Rejewski got his inspiration for the machines while at a cafe eating a *bombe*, an ice cream shaped into a hemisphere. The bombes effectively mechanised the process of decipherment. It was a natural response to Enigma, which was a mechanisation of encipherment.

For most of the 1930s, Rejewski and his colleagues worked tirelessly to uncover the Enigma keys. Month after month, the team would have to deal with the stresses and strains of cryptanalysis, continually having to fix mechanical failures in the bombes, continually having to deal with the never-ending supply of encrypted intercepts. Their lives became dominated by the pursuit of the day-key, that vital piece of information that would reveal the meaning of the encrypted messages. However, unknown to the Polish codebreakers, much of their work was unnecessary. The chief of the Biuro, Major Gwido Langer, already had the Enigma day-keys, but he kept them hidden, tucked away in his desk.

Langer, via the French, was still receiving information from Schmidt. The German spy's nefarious activities did not end in 1931 with the delivery of the two documents on the operation of Enigma, but continued for another seven years. He met the French secret agent Rex on twenty occasions, often in secluded alpine chalets where privacy was guaranteed. At every meeting, Schmidt handed over one or more codebooks, each one containing a month's worth of day-keys. These were the codebooks that were distributed to all German Enigma operators, and they contained all the information that was needed to encipher and decipher messages. In total, he provided codebooks that contained 38 months' worth of day-keys. The keys would have saved Rejewski an enormous amount of time and effort, short-cutting the necessity for bombes and sparing manpower that could have been used in other sections of the Biuro. However, the remarkably astute Langer decided not to tell Rejewski that the keys existed. By depriving Rejewski of the keys, Langer believed he was preparing him for the inevitable time when the keys would no longer be available. He knew that if war broke out it would be impossible for Schmidt to continue to attend covert meetings, and Rejewski would then be forced to be self-sufficient. Langer thought that Rejewski should practise self-sufficiency in peacetime, as preparation for what lay ahead.

Rejewski's skills eventually reached their limit in December 1938, when German cryptographers increased Enigma's security. Enigma operators were all given two new scramblers, so that the scrambler arrangement might involve any three of the five available scramblers. Previously there were only three scramblers (labelled 1, 2 and 3) to choose from, and only six ways to arrange them, but now that there were two extra scramblers (labelled 4 and 5) to choose from, the number of arrangements rose to 60, as shown in Table 10. Rejewski's first challenge was to work out the internal wirings of the two new scramblers. More worryingly, he also had to build ten times as many bombes, each representing a different scrambler arrangement. The sheer cost of building such a battery of bombes was fifteen times the Biuro's entire annual equipment budget. The following month the situation worsened when the number of plugboard cables increased from six to ten. Instead of twelve letters being swapped before entering the scramblers, there were now twenty swapped letters. The number of possible keys increased to 159,000,000,000,000,000,000.

In 1938 Polish interceptions and decipherments had been at their peak, but by the beginning of 1939 the new scramblers and extra plugboard cables stemmed the flow of intelligence. Rejewski, who had pushed forward the boundaries of cryptanalysis in previous years, was confounded. He had proved that Enigma was not an unbreakable cipher, but without the resources required to check every scrambler setting he could not find the day-key, and decipherment was impossible. Under such desperate circumstances Langer might have been tempted to hand over the keys that had been obtained by Schmidt, but the keys were no longer being delivered. Just before the introduction of the new scramblers, Schmidt had broken off contact with agent Rex. For seven years he had supplied keys which were superfluous because of Polish innovation. Now, just when the Poles needed the keys, they were no longer available.

The new invulnerability of Enigma was a devastating blow to Poland, because Enigma was not merely a means of communication, it was at the heart of Hitler's blitzkrieg strategy. The concept of blitzkrieg ('lightning war') involved rapid, intense, coordinated attack, which meant that large tank divisions would have to communicate with one another and with infantry and artillery. Furthermore, land forces would be backed up by air support from dive-bombing Stukas, which would rely on effective and secure communication between the front-line troops and the airfields. The ethos of blitzkrieg was 'speed of attack through speed of communications'. If the Poles could not break Enigma, they had no hope of stopping the German onslaught, which was clearly only a matter of months away. Germany already occupied the Sudetenland, and on 27 April 1939 it withdrew from its non-aggression treaty with Poland. Hitler's anti-Polish rhetoric became increasingly vitriolic. Langer was determined that if Poland was invaded, then its cryptanalytic breakthroughs, which had so far been kept secret from the Allies, should not be lost. If Poland could not benefit from Rejewski's work, then at least the Allies should have the chance to

Table 10 Possible arrangements with five scramblers.

Arrangements with three scramblers	Extra arrangements available with two extra scramblers								
123	124	125	134	135	142	143	145	152	153
132	154	214	215	234	235	241	243	245	251
213	253	254	314	315	324	325	341	342	345
231	351	352	354	412	413	415	421	423	425
312	431	432	435	451	452	453	512	513	514
321	521	523	524	531	532	534	541	542	543

Figure 43 General Heinz Guderian's command-post vehicle. An Enigma machine can be seen in use in the bottom left.

try and build on it. Perhaps Britain and France, with their extra resources, could fully exploit the concept of the bombe.

On 30 June, Major Langer telegraphed his French and British counterparts, inviting them to Warsaw to discuss some urgent matters concerning Enigma. On 24 July senior French and British cryptanalysts arrived at the Biuro's headquarters, not knowing quite what to expect. Langer ushered them into a room in which stood an object covered with a black cloth. He pulled away the cloth, dramatically revealing one of Rejewski's bombes. The audience were astonished as they heard how Rejewski had been breaking Enigma for years. The Poles were a decade ahead of anybody else in the world. The French were particularly astonished, because the Polish work had been based on the results of French espionage. The French had handed the information from Schmidt to the Poles because they believed it to be of no value, but the Poles had proved them wrong.

As a final surprise, Langer offered the British and French two spare Enigma replicas and blueprints for the bombes, which were to be shipped in diplomatic bags to Paris. From there, on 16 August, one of the Enigma machines was forwarded to London. It was smuggled across the Channel as part of the baggage of the playwright Sacha Guitry and his wife, the actress Yvonne Printemps, so as not to arouse the suspicion of German spies who would be monitoring the ports. Two weeks later, on 1 September, Hitler invaded Poland and the war began.

The Geese that Never Cackled

For thirteen years the British and the French had assumed that the Enigma cipher was unbreakable, but now there was hope. The Polish revelations had demonstrated that the Enigma cipher was flawed, which boosted the morale of Allied cryptanalysts. Polish progress had ground to a halt on the introduction of the new scramblers and extra plugboard cables, but the fact remained that Enigma was no longer considered a perfect cipher.

The Polish breakthroughs also demonstrated to the Allies the value of employing mathematicians as codebreakers. In Britain, Room 40 had always been dominated by linguists and classicists, but now there was a concerted effort to balance the staff with mathematicians and scientists. They were recruited largely via the old-boy network, with those inside

Room 40 contacting their former Oxford and Cambridge colleges. There was also an old-girl network which recruited women undergraduates from places such as Newnham College and Girton College, Cambridge.

The new recruits were not brought to Room 40 in London, but instead went to Bletchley Park, Buckinghamshire, the home of the Government Code and Cypher School (GC&CS), a newly formed codebreaking organisation that was taking over from Room 40. Bletchley Park could house a much larger staff, which was important because a deluge of encrypted intercepts was expected as soon as the war started. During the First World War, Germany had transmitted two million words a month, but it was anticipated that the greater availability of radios in the Second World War could result in the transmission of two million words a day.

At the centre of Bletchley Park was a large Victorian Tudor–Gothic mansion built by the nineteenth-century financier Sir Herbert Leon. The mansion, with its library, dining hall and ornate ballroom, provided the

Figure 44 In August 1939, Britain's senior codebreakers visited Bletchley Park to assess its suitability as the site for the new Government Code and Cypher School. To avoid arousing suspicion from locals, they claimed to be part of Captain Ridley's shooting party.

central administration for the whole of the Bletchley operation. Commander Alastair Denniston, the director of GC&CS, had a ground-floor office overlooking the gardens, a view that was soon spoiled by the erection of numerous huts. These makeshift wooden buildings housed the various codebreaking activities. For example, Hut 6 specialised in attacking the German Army's Enigma communications. Hut 6 passed its decrypts to Hut 3, where intelligence operatives translated the messages, and attempted to exploit the information. Hut 8 specialised in the naval Enigma, and they passed their decrypts to Hut 4 for translation and intelligence gathering. Initially, Bletchley Park had a staff of only two hundred, but within five years the mansion and the huts would house seven thousand men and women.

During the autumn of 1939, the scientists and mathematicians at Bletchley learnt the intricacies of the Enigma cipher and rapidly mastered the Polish techniques. Bletchley had more staff and resources than the Polish Biuro Szyfrów, and were thus able to cope with the larger selection of scramblers and the fact that Enigma was now ten times harder to break. Every twenty-four hours the British codebreakers went through the same routine. At midnight, German Enigma operators would change to a new day-key, at which point whatever breakthroughs Bletchley had achieved the previous day could no longer be used to decipher messages. The codebreakers now had to begin the task of trying to identify the new day-key. It could take several hours, but as soon as they had discovered the Enigma settings for that day, the Bletchley staff could begin to decipher the German messages that had already accumulated, revealing information that was invaluable to the war effort.

Surprise is an invaluable weapon for a commander to have at his disposal. But if Bletchley could break into Enigma, German plans would become transparent and the British would be able to read the minds of the German High Command. If the British could pick up news of an imminent attack, they could send reinforcements or take evasive action. If they could decipher German discussions of their own weaknesses, the Allies would be able to focus their offensives. The Bletchley decipherments were of the utmost importance. For example, when Germany invaded Denmark and Norway in April 1940, Bletchley provided a detailed picture of German operations. Similarly, during the Battle of

Britain, the cryptanalysts were able to give advance warning of bombing raids, including times and locations. They could also give continual updates on the state of the Luftwaffe, such as the number of planes that had been lost and the speed with which they were being replaced. Bletchley would send all this information to MI6 headquarters, who would forward it to the War Office, the Air Ministry and the Admiralty.

In between influencing the course of the war, the cryptanalysts occasionally found time to relax. According to Malcolm Muggeridge, who served in the secret service and visited Bletchley, rounders was a favourite pastime:

> Every day after luncheon when the weather was propitious the cipher-crackers played rounders on the manor-house lawn, assuming the quasi-serious manner dons affect when engaged in activities likely to be regarded as frivolous or insignificant in comparison with their weightier studies. Thus they would dispute some point about the game with the same fervour as they might the question of free will or determinism, or whether the world began with a big bang or a process of continuing creation.

Figure 45 Bletchley's codebreakers relax with a game of rounders.

Once they had mastered the Polish techniques, the Bletchley cryptanalysts began to invent their own shortcuts for finding the Enigma keys. For example, they cottoned on to the fact that the German Enigma operators would occasionally choose obvious message-keys. For each message, the operator was supposed to select a different message-key, three letters chosen at random. However, in the heat of battle, rather than straining their imaginations to pick a random key, the overworked operators would sometimes pick three consecutive letters from the Enigma keyboard (Figure 46), such as **QWE** or **BNM**. These predictable message-keys became known as *cillies*. Another type of cilly was the repeated use of the same message-key, perhaps the initials of the operator's girlfriend – indeed, one such set of initials, C.I.L., may have been the origin of the term. Before cracking Enigma the hard way, it became routine for the cryptanalysts to try out the cillies, and their hunches would sometimes pay off.

Cillies were not weaknesses of the Enigma machine, rather they were weaknesses in the way the machine was being used. Human error at more senior levels also compromised the security of the Enigma cipher. Those responsible for compiling the codebooks had to decide which scramblers would be used each day, and in which positions. They tried to ensure that the scrambler settings were unpredictable by not allowing any scrambler to remain in the same position for two days in a row. So, if we label the scramblers 1, 2, 3, 4 and 5, then on the first day it would be possible to have the arrangement 134, and on the second day it would be possible to have 215, but not 214, because scrambler number 4 is not allowed to remain in the same position for two days in a row. This might seem a sensible strategy because the scramblers are constantly changing position, but enforcing such a rule actually makes life easier for the cryptanalyst. Excluding certain arrangements to avoid a scrambler remaining in the

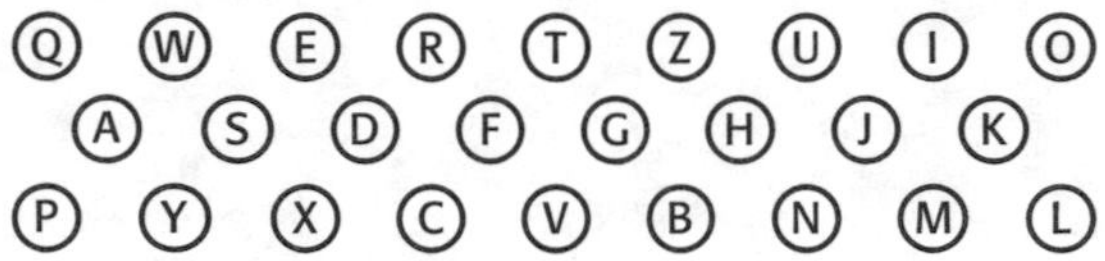

Q W E R T Z U I O
A S D F G H J K
P Y X C V B N M L

Figure 46: Layout of the Enigma keyboard.

same position meant that the codebook compilers reduced by half the number of possible scrambler arrangements. The Bletchley cryptanalysts realised what was happening and made the most of it. Once they identified the scrambler arrangement for one day, they could immediately rule out half the scrambler arrangements for the next day. Hence, their workload was reduced by half.

Similarly, there was a rule that the plugboard settings could not include a swap between any letter and its neighbour, which meant that S could be swapped with any letter except R and T. The theory was that such obvious swappings should be deliberately avoided, but once again the implementation of a rule drastically reduced the number of possible keys.

This search for new cryptanalytic shortcuts was necessary because the Enigma machine continued to evolve during the course of the war. The cryptanalysts were continually forced to innovate, to redesign and refine the bombes, and to devise wholly new strategies. Part of the reason for their success was the bizarre combination of mathematicians, scientists, linguists, classicists, chess grandmasters and crossword addicts within each hut. An intractable problem would be passed around the hut until it reached someone who had the right mental tools to solve it, or reached someone who could at least partially solve it before passing it on again. Gordon Welchman, who was in charge of Hut 6, described his team as 'a pack of hounds trying to pick up the scent'. There were many great cryptanalysts and many significant breakthroughs, and it would take several large volumes to describe the individual contributions in detail. However, if there is one figure who deserves to be singled out, it is Alan Turing, who identified Enigma's greatest weakness and ruthlessly exploited it. Thanks to Turing, it became possible to crack the Enigma cipher under even the most difficult circumstances.

Alan Turing was conceived in the autumn of 1911 in Chatrapur, a town near Madras in southern India, where his father Julius Turing was a member of the Indian civil service. Julius and his wife Ethel were determined that their son should be born in Britain, and returned to London, where Alan was born on 23 June 1912. His father returned to India soon afterwards and his mother followed just fifteen months later, leaving Alan in the care of nannies and friends until he was old enough to attend boarding school.

In 1926, at the age of fourteen, Turing became a pupil at Sherborne School, in Dorset. The start of his first term coincided with the General Strike, but Turing was determined to attend the first day, and he cycled 100 km unaccompanied from Southampton to Sherborne, a feat that was reported in the local newspaper. By the end of his first year at the school he had gained a reputation as a shy, awkward boy whose only skills were in the area of science. The aim of Sherborne was to turn boys into well-rounded men, fit to rule the Empire, but Turing did not share this ambition and had a generally unhappy schooling.

His only real friend at Sherborne was Christopher Morcom, who, like Turing, had an interest in science. Together they discussed the latest scientific news and conducted their own experiments. The relationship fired Turing's intellectual curiosity, but, more importantly, it also had a profound emotional effect on him. Andrew Hodges, Turing's biographer, wrote that 'This was first love . . . It had that sense of surrender, and a heightened awareness, as of brilliant colour bursting upon a black and white world.' Their friendship lasted four years, but Morcom seems to have been unaware of the depth of feeling Turing had for him. Then, during their final year at Sherborne, Turing lost for ever the chance to tell him how he felt. On Thursday 13 February 1930, Christopher Morcom suddenly died of tuberculosis.

Turing was devastated by the loss of the only person he would ever truly love. His way of coming to terms with Morcom's death was to focus on his scientific studies in an attempt to fulfil his friend's potential. Morcom, who appeared to be the more gifted of the two boys, had already won a scholarship to Cambridge University. Turing believed it was his duty also to win a place at Cambridge, and then to make the discoveries his friend would otherwise have made. He asked Christopher's mother for a photograph, and when it arrived he wrote back to thank her: 'He is on my table now, encouraging me to work hard.'

In 1931, Turing gained admission to King's College, Cambridge. He arrived during a period of intense debate about the nature of mathematics and logic, and was surrounded by some of the leading voices, such as Bertrand Russell, Alfred North Whitehead and Ludwig Wittgenstein. At the centre of the argument was the issue of *undecidability*, a controversial notion developed by the logician Kurt Gödel. It had always been assumed

Figure 47 Alan Turing.

that, in theory at least, all mathematical questions could be answered. However, Gödel demonstrated that there could exist a minority of questions which were beyond the reach of logical proof, so-called undecidable questions. Mathematicians were traumatised by the news that mathematics was not the all-powerful discipline they had always believed it to be. They attempted to salvage their subject by trying to find a way of identifying the awkward undecidable questions, so that they could put them safely to one side. It was this objective that eventually inspired Turing to write his most influential mathematical paper, 'On Computable Numbers', published in 1937. In *Breaking the Code*, Hugh Whitemore's play about the life of Turing, a character asks Turing the meaning of his paper. He replies, 'It's about right and wrong. In general terms. It's a technical paper in mathematical logic, but it's also about the difficulty of telling right from wrong. People think – most people think – that in mathematics we always know what is right and what is wrong. Not so. Not any more.'

In his attempt to identify undecidable questions, Turing's paper described an imaginary machine that was designed to perform a particular mathematical operation, or algorithm. In other words, the machine would be capable of running through a fixed, prescribed series of steps which would, for example, multiply two numbers. Turing envisaged that the numbers to be multiplied could be fed into the machine via a paper tape, rather like the punched tape that is used to feed a tune into a Pianola. The answer to the multiplication would be output via another tape. Turing imagined a whole series of these so-called *Turing machines*, each specially designed to tackle a particular task, such as dividing, squaring or factoring. Then Turing took a more radical step.

He imagined a machine whose internal workings could be altered so that it could perform all the functions of all conceivable Turing machines. The alterations would be made by inserting carefully selected tapes, which transformed the single flexible machine into a dividing machine, a multiplying machine, or any other type of machine. Turing called this hypothetical device a *universal Turing machine* because it would be capable of answering any question that could logically be answered. Unfortunately, it turned out that it is not always logically possible to answer a question about the undecidability of another question, and so even the universal Turing machine was unable to identify every undecidable question.

Mathematicians who read Turing's paper were disappointed that Gödel's monster had not been subdued but, as a consolation prize, Turing had given them the blueprint for the modern programmable computer. Turing knew of Babbage's work, and the universal Turing machine can be seen as a reincarnation of Difference Engine No. 2. In fact, Turing had gone much further, and provided computing with a solid theoretical basis, imbuing the computer with a hitherto unimaginable potential. It was still the 1930s though, and the technology did not exist to turn the universal Turing machine into a reality. However, Turing was not at all dismayed that his theories were ahead of what was technically feasible. He merely wanted recognition from within the mathematical community, who indeed applauded his paper as one of the most important breakthroughs of the century. He was still only twenty-six.

This was a particularly happy and successful period for Turing. During the 1930s he rose through the ranks to become a fellow of King's College, home of the world's intellectual elite. He led the life of an archetypal Cambridge don, mixing pure mathematics with more trivial activities. In 1938 he made a point of seeing the film *Snow White and the Seven Dwarfs*, containing the memorable scene in which the Wicked Witch dunks an apple in poison. Afterwards his colleagues heard Turing continually repeating the macabre chant, 'Dip the apple in the brew, Let the sleeping death seep through'.

Turing cherished his years at Cambridge. In addition to his academic success, he found himself in a tolerant and supportive environment. Homosexuality was largely accepted within the university, which meant that he was free to engage in a series of relationships without having to worry about who might find out, and what others might say. Although he had no serious long-term relationships, he seemed to be contented with his life. Then, in 1939, Turing's academic career was brought to an abrupt halt. The Government Code and Cypher School invited him to become a cryptanalyst at Bletchley, and on 4 September 1939, the day after Neville Chamberlain declared war on Germany, Turing moved from the opulence of the Cambridge quadrangle to the Crown Inn at Shenley Brook End.

Each day he cycled 5 km from Shenley Brook End to Bletchley Park, where he spent part of his time in the huts contributing to the routine codebreaking effort, and part of his time in the Bletchley think-tank, formerly Sir Herbert Leon's apple, pear and plum store. The think-tank was where the cryptanalysts brainstormed their way through new problems, or anticipated how to tackle problems that might arise in the future. Turing focused on what would happen if the German military changed their system of exchanging message-keys. Bletchley's early successes relied on Rejewski's work, which exploited the fact that Enigma operators encrypted each message-key twice (for example, if the message-key was YGB, the operator would encipher YGBYGB). This repetition was supposed to ensure that the receiver did not make a mistake, but it created a chink in the security of Enigma. British cryptanalysts guessed it would not be long before the Germans noticed that the repeated key was compromising the Enigma cipher, at which point the Enigma operators would be told to abandon the repetition, thus confounding Bletchley's current codebreaking techniques. It was Turing's job to find an alternative way to attack Enigma, one that did not rely on a repeated message-key.

As the weeks passed, Turing realised that Bletchley was accumulating a vast library of decrypted messages, and he noticed that many of them conformed to a rigid structure. By studying old decrypted messages, he believed he could sometimes predict part of the contents of an undeciphered message, based on when it was sent and its source. For example, experience showed that the Germans sent a regular enciphered weather report shortly after 6 a.m. each day. So, an encrypted message intercepted at 6.05 a.m. would be almost certain to contain wetter, the German word for 'weather'. The rigorous protocol used by any military organisation meant that such messages were highly regimented in style, so Turing could even be confident about the location of wetter within the encrypted message. For example, experience might tell him that the first six letters of a particular ciphertext corresponded to the plaintext letters wetter. When a piece of plaintext can be associated with a piece of ciphertext, this combination is known as a *crib*.

Turing was sure that he could exploit the cribs to crack Enigma. If he had a ciphertext and he knew that a specific section of it, say ETJWPX, represented wetter, then the challenge was to identify the settings of the Enigma machine that would transform wetter into ETJWPX. The straightforward, but impractical, way to do this would be for the cryptanalyst to take an Enigma machine, type in wetter and see if the correct ciphertext emerged. If not, then the cryptanalyst would change the settings of the machine, by swapping plugboard cables, and swapping or reorienting scramblers, and then type in **wetter** again. If the correct ciphertext did not emerge, the cryptanalyst would change the settings again, and again, and again, until he found the right one. The only problem with this trial and error approach was the fact that there were 159,000,000,000,000,000,000 possible settings to check, so finding the one that transformed **wetter** into **ETJWPX** was a seemingly impossible task.

To simplify the problem, Turing attempted to follow Rejewski's strategy of disentangling the settings. He wanted to divorce the problem of finding the scrambler settings (finding which scrambler is in which slot, and what their respective orientations are) from the problem of finding the plugboard cablings. For example, if he could find something in the crib that had nothing to do with the plugboard cablings, then he could feasibly check each of the remaining 1,054,560 possible scrambler combinations (60 arrangements × 17,576 orientations). Having found the correct scrambler settings, he could then deduce the plugboard cablings.

Eventually, his mind settled on a particular type of crib which contained internal loops, similar to the chains exploited by Rejewski. Rejewski's chains linked letters within the repeated message-key. However, Turing's loops had nothing to do with the message-key, as he was working on the assumption that soon the Germans would stop sending repeated message-keys. Instead, Turing's loops connected plaintext and ciphertext letters within a crib. For example, the crib shown in Figure 48 contains a loop.

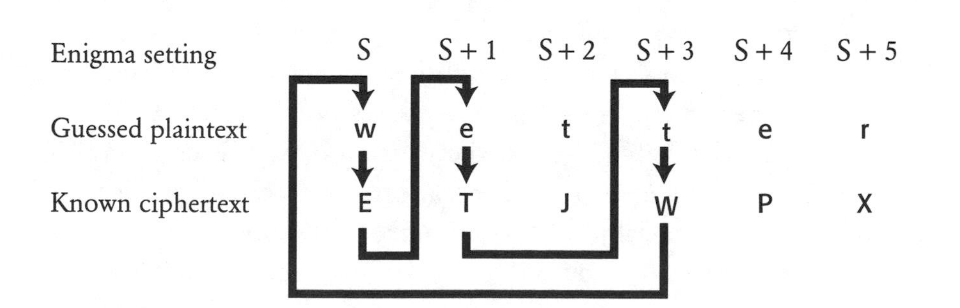

Figure 48 One of Turing's cribs, showing a loop.

Remember, cribs are only guesses, but if we assume that this crib is correct, we can link the letters w→E, e→T, t→W as part of a loop. Although we know none of the Enigma machine settings, we can label the first setting, whatever it is, S. In this first setting we know that w is encrypted as E. After this encryption, the first scrambler clicks around one place to setting S+1, and the letter e is enciphered as T. The scrambler clicks forward another place and encrypts a letter that is not part of the loop, so we ignore this encryption. The scrambler clicks forward one more place and, once again, we reach a letter that is part of the loop. In setting S+3, we know that the letter t is enciphered as W. In summary, we know that

In setting S, Enigma encrypts w as E.
In setting S+1, Enigma encrypts e as T.
In setting S+3, Enigma encrypts t as W.

So far the loop seems like nothing more than a curious pattern, but Turing rigorously followed the implications of the relationships within the loop, and saw that they provided him with the drastic shortcut he needed in order to break Enigma. Instead of working with just one Enigma machine to test every setting, Turing began to imagine three separate machines, each dealing with the encipherment of one element of the loop. The first machine would try to encipher w into E, the second would try to encipher e into T, and the third t into W. The three machines would all have identical settings, except that the second would have its scrambler orientations moved forward one place with respect to the first, a setting labelled S+1, and the third would have its scrambler orientations moved forward three places with respect to the first, a setting labelled S+3. Turing then pictured a frenzied cryptanalyst, continually changing plugboard cables, swapping scrambler arrangements and changing their orientations in order to achieve the correct encryptions. Whatever cables were changed in the first machine would also be changed in the other two. Whatever scrambler arrangements were changed in the first machine would also be changed in the other two. And, crucially, whatever scrambler orientation was set in the first machine, the second would have the same orientation but stepped forward one place, and the third would have the same orientation but stepped forward three places.

Turing does not seem to have achieved much. The cryptanalyst still has to check all 159,000,000,000,000,000,000 possible settings, and, to make matters worse, he now has to do it simultaneously on all three machines instead of just one. However, the next stage of Turing's idea transforms the challenge, and vastly simplifies it. He imagined connecting the three machines by running electrical wires between the inputs and the outputs of each machine, as shown in Figure 49. In effect, the loop in the crib is paralleled by the loop of the electrical circuit. Turing pictured the machines changing their plugboard and scrambler settings, as described above, but only when all the settings are correct for all three machines would the circuit be completed, allowing a current to flow through all three machines. If Turing incorporated a light bulb within the circuit, then the current would illuminate it, signalling that the correct settings had been found. At this point, the three machines still have to check up to 159,000,000,000,000,000,000 possible settings in order to illuminate the bulb. However, everything done so far has merely been preparation for Turing's final logical leap, which would make the task over a hundred million million times easier in one fell swoop.

Turing had constructed his electrical circuit in such a way as to nullify the effect of the plugboard, thereby allowing him to ignore the billions of plugboard settings. Figure 49 shows that the first Enigma has the electric current entering the scramblers and emerging at some unknown letter, which we shall call L_1. The current then flows through the plugboard, which transforms L_1 into E. This letter E is connected via a wire to the letter e in the second Enigma, and as the current flows through the second plugboard it is transformed back to L_1. In other words, the two plugboards cancel each other out. Similarly, the current emerging from the scramblers in the second Enigma enters the plugboard at L_2 before being transformed into T. This letter T is connected via a wire to the letter t in the third Enigma, and as the current flows through the third plugboard it is transformed back to L_2. In short, the plugboards cancel themselves out throughout the whole circuit, so Turing could ignore them completely.

Turing needed only to connect the output of the first set of scramblers, L_1, directly to the input of the second set of scramblers, also L_1, and so on. Unfortunately, he did not know the value of the letter L_1, so he had to connect all 26 outputs of the first set of scramblers to all 26 corresponding inputs in the second set of scramblers, and so on. In effect, there were now

26 electrical loops, and each one would have a light bulb to signal the completion of an electrical circuit. The three sets of scramblers could then simply check each of the 17,576 orientations, with the second set of scramblers always one step ahead of the first set, and the third set of scramblers two steps ahead of the second set. Eventually, when the correct scrambler orientations had been found, one of the circuits would be completed and the bulb would be illuminated. If the scramblers changed orientation every second, it would take just five hours to check all the orientations.

Only two problems remained. First, it could be that the three machines are running with the wrong scrambler arrangement, because the Enigma machine operates with any three of the five available scramblers, placed in any order, giving sixty possible arrangements. Hence, if all 17,576 orientations have been checked, and the lamp has not been illuminated, it is then necessary to try another of the sixty scrambler arrangements, and to keep on trying other arrangements until the circuit is completed. Alternatively, the cryptanalyst could have sixty sets of three Enigmas running in parallel.

The second problem involved finding the plugboard cablings, once the scrambler arrangement and orientations had been established. This is relatively simple. Using an Enigma machine with the correct scrambler arrangement and orientations, the cryptanalyst types in the ciphertext and looks at the emerging plaintext. If the result is **tewwer** rather than **wetter**, then it is clear that plugboard cables should be inserted so as to swap **w** and **t**. Typing in other bits of ciphertext would reveal other plugboard cablings.

The combination of crib, loops and electrically connected machines resulted in a remarkable piece of cryptanalysis, and only Turing, with his unique background in mathematical machines, could ever have come up with it. His musings on the imaginary Turing machines were intended to answer esoteric questions about mathematical undecidability, but this purely academic research had put him in the right frame of mind for designing a practical machine capable of solving very real problems.

Bletchley was able to find £100,000 to turn Turing's idea into working devices, which were dubbed bombes because their mechanical approach bore a passing resemblance to Rejewski's bombe. Each of Turing's bombes was to consist of twelve sets of electrically linked Enigma scramblers, and would thus be able to cope with much longer loops of letters. The complete unit

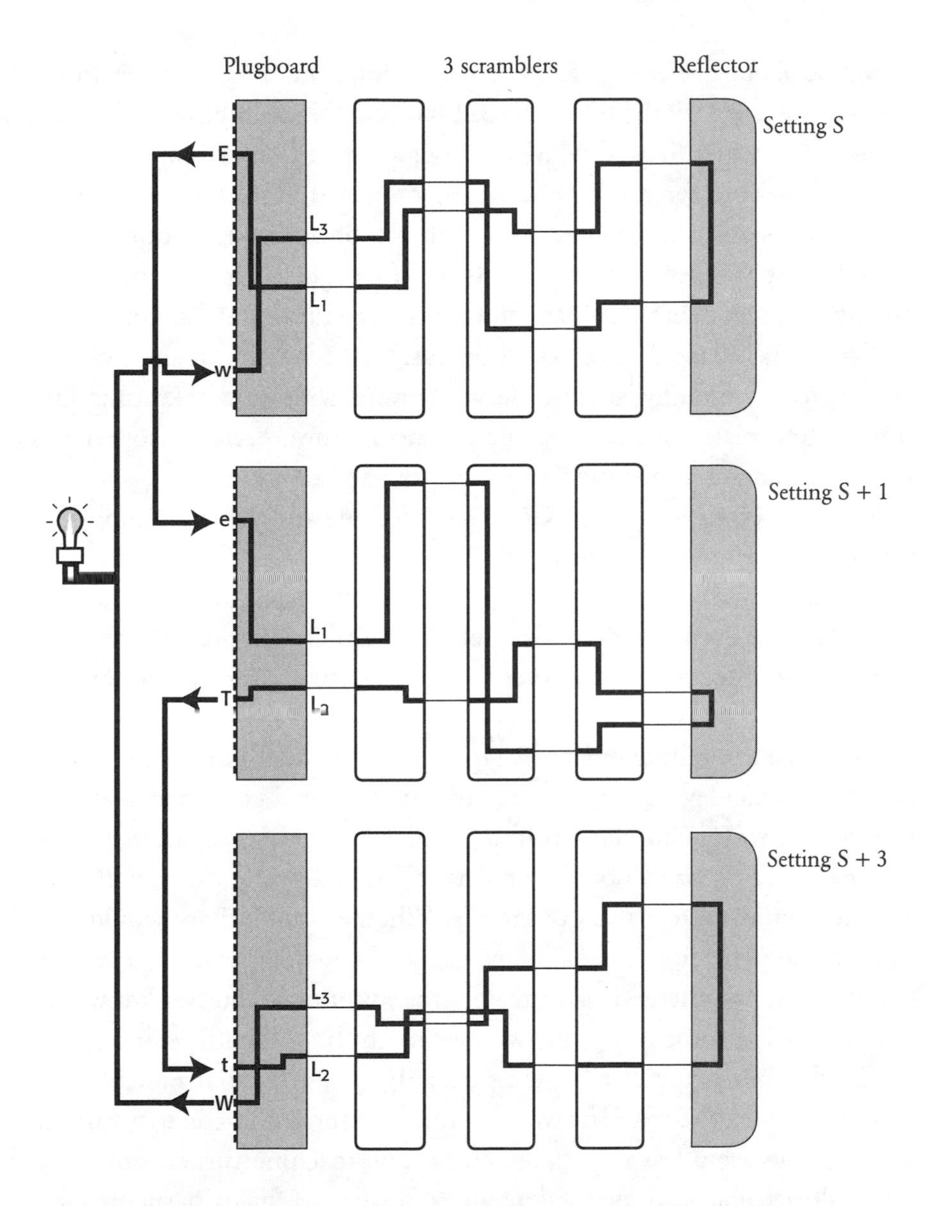

Figure 49 The loop in the crib can be paralleled by an electrical loop. Three Enigma machines are set up in identical ways, except that the second one has its first scrambler moved forward one place (setting S + 1), and the third has its scrambler moved forward two further places (setting S + 3). The output of each Enigma is then connected to the input of the next one. The three sets of scramblers then click around in unison until the circuit is complete and the light illuminates. At this point the correct setting has been found. In the diagram above, the circuit is complete, corresponding to the correct setting.

would be about two metres tall, two metres long and a metre wide. Turing finalised the design at the beginning of 1940, and the job of construction was given to the British Tabulating Machinery factory at Letchworth.

While waiting for the bombes to be delivered, Turing continued his day-to-day work at Bletchley. News of his breakthrough soon spread among the other senior cryptanalysts, who recognised that he was a singularly gifted codebreaker. According to Peter Hilton, a fellow Bletchley codebreaker, 'Alan Turing was obviously a genius, but he was an approachable, friendly genius. He was always willing to take time and trouble to explain his ideas; but he was no narrow specialist, so that his versatile thought ranged over a vast area of the exact sciences.'

However, everything at the Government Code and Cypher School was top secret, so nobody outside of Bletchley Park was aware of Turing's remarkable achievement. For example, his parents had absolutely no idea that Alan was even a codebreaker, let alone Britain's foremost cryptanalyst. He had once told his mother that he was involved in some form of military research, but he did not elaborate. She was merely disappointed that this had not resulted in a more respectable haircut for her scruffy son. Although Bletchley was run by the military, they had conceded that they would have to tolerate the scruffiness and eccentricities of these 'professor-types'. Turing rarely bothered to shave, his nails were stuffed with dirt, and his clothes were a mass of creases. Whether the military would also have tolerated his homosexuality remains unknown. Jack Good, a veteran of Bletchley, commented: 'Fortunately the authorities did not know that Turing was a homosexual. Otherwise we might have lost the war.'

The first prototype bombe, christened *Victory*, arrived at Bletchley on 14 March 1940. The machine was put into operation immediately, but the initial results were less than satisfactory. The machine turned out to be much slower than expected, taking up to a week to find a particular key. There was a concerted effort to increase the bombe's efficiency, and a modified design was submitted a few weeks later. It would take four more months to build the upgraded bombe. In the meantime, the cryptanalysts had to cope with the calamity they had anticipated. On 10 May 1940, the Germans changed their key-exchange protocol. They no longer repeated the message-key, and thereupon the number of successful Enigma decipherments dropped dramatically. The information blackout lasted until

8 August, when the new bombe arrived. Christened *Agnus Dei*, or *Agnes* for short, this machine was to fulfil all Turing's expectations.

Within eighteen months there were fifteen more bombes in operation, exploiting cribs, checking scrambler settings and revealing keys, each one clattering like a million knitting needles. If everything was going well, a bombe might find an Enigma key within an hour. Once the plugboard cablings and the scrambler settings (the message-key) had been established for a particular message, it was easy to deduce the day-key. All the other messages sent that same day could then be deciphered.

Even though the bombes represented a vital breakthrough in cryptanalysis, decipherment had not become a formality. There were many hurdles to overcome before the bombes could even begin to look for a key. For example, to operate a bombe you first needed a crib. The senior codebreakers would give cribs to the bombe operators, but there was no guarantee that the codebreakers had guessed the correct meaning of the ciphertext. And even if they did have the right crib, it might be in the wrong place the cryptanalysts might have guessed that an encrypted message contained a certain phrase, but associated that phrase with the wrong piece of the ciphertext. However, there was a neat trick for checking whether a crib was in the correct position.

In the following crib, the cryptanalyst is confident that the plaintext is right, but he is not sure if he has matched it with the correct letters in the ciphertext.

Guessed plaintext			w	e	t	t	e	r	n	u	l	l	s	e	c	h	s		
Known ciphertext	I	P	R	E	N	L	W	K	M	J	J	S	X	C	P	L	E	J	W	Q

One of the features of the Enigma machine was its inability to encipher a letter as itself, which was a consequence of the reflector. The letter **a** could never be enciphered as **A**, the letter **b** could never be enciphered as **B**, and so on. The particular crib above must therefore be misaligned, because the first **e** in **wetter** is matched with an **E** in the ciphertext. To find the correct alignment, we simply slide the plaintext and the ciphertext relative to each other until no letter is paired with itself. If we shift the plaintext one place to the left, the match still fails because this time the first **s** in **sechs** is matched with **S** in the ciphertext. However, if we shift the plaintext one place to the right there are no illegal encipherments. This crib is therefore likely to be in the right place, and could be used as the basis for a bombe decipherment:

Guessed plaintext				w	e	t	t	e	r	n	u	l	l	s	e	c	h	s		
Known ciphertext	I	P	R	E	N	L	W	K	M	J	J	S	X	C	P	L	E	J	W	Q

The intelligence gathered at Bletchley was passed on to only the most senior military figures and selected members of the war cabinet. Winston Churchill was fully aware of the importance of the Bletchley decipherments, and on 6 September 1941 he visited the codebreakers. On meeting some of the cryptanalysts, he was surprised by the bizarre mixture of people who were providing him with such valuable information; in addition to the mathematicians and linguists, there was an authority on porcelain, a curator from the Prague Museum, the British chess champion and numerous bridge experts. Churchill muttered to Sir Stewart Menzies,

Figure 50 A Bletchley Park bombe in action.

head of the Secret Intelligence Service, 'I told you to leave no stone unturned, but I didn't expect you to take me so literally.' Despite the comment, he had a great fondness for the motley crew, calling them 'the geese who laid golden eggs and never cackled'.

The visit was intended to boost the morale of the codebreakers by showing them that their work was appreciated at the very highest level. It also had the effect of giving Turing and his colleagues the confidence to approach Churchill directly when a crisis loomed. To make the most of the bombes, Turing needed more staff, but his requests had been blocked by Commander Edward Travis, who had taken over as Director of Bletchley, and who felt that he could not justify recruiting more people. On 21 October 1941, the cryptanalysts took the insubordinate step of ignoring Travis and writing directly to Churchill.

> Dear Prime Minister,
>
> Some weeks ago you paid us the honour of a visit, and we believe that you regard our work as important. You will have seen that, thanks largely to the energy and foresight of Commander Travis, we have been well supplied with the 'bombes' for the breaking of the German Enigma codes. We think, however, that you ought to know that this work is being held up, and in some cases is not being done at all, principally because we cannot get sufficient staff to deal with it. Our reason for writing to you direct is that for months we have done everything that we possibly can through the normal channels, and that we despair of any early improvement without your intervention . . .
>
> We are, Sir, Your obedient servants,
>
> A.M. Turing
> W.G. Welchman
> C.H.O'D. Alexander
> P.S. Milner-Barry

Churchill had no hesitation in responding. He immediately issued a memorandum to his principal staff officer:

> **ACTION THIS DAY**
> Make sure they have all they want on extreme priority and report to me that this has been done.

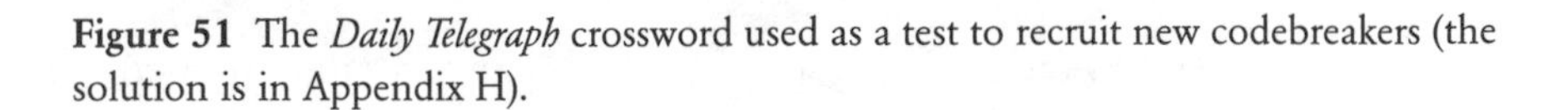

ACROSS

1 A stage company (6)
4 The direct route preferred by the Roundheads (two words–5,3)
9 One of the evergreens (6)
10 Scented (8)
12 Course with an apt finish (5)
13 Much that could be got from a timber merchant (two words–5,4)
15 We have nothing and are in debt (3)
16 Pretend (5)
17 Is this town ready for a flood? (6)
22 The little fellow has some beer: it makes me lose colour, I say (6)
24 Fashion of a famous French family (5)
27 Tree (3)
28 One might of course use this tool to core an apple (9)
31 Once used for unofficial currency (5)
32 Those well brought up help these over stiles (two words–4,4)
33 A sport in a hurry (6)
34 Is the workshop that turns out this part of a motor a hush-hush affair? (8)
35 An illumination functioning (6)

DOWN

1 Official instruction not to forget the servants (8)
2 Said to be a remedy for a burn (two words –5,3)
3 Kind of alias (9)
5 A disagreeable company (5)
6 Debtors may have to this money for their debts unless of course their creditors do it to the debts (5)
7 Boat that should be able to suit anyone (6)
8 Gear (6)
11 Business with the end in sight (6)
14 The right sort of woman to start a dame school (3)
18 "The War" (anag) (6)
19 When hammering take care to hit this (two words)–5,4)
20 Making sound as a bell (8)
21 Half a fortnight of old (8)
23 Bird, dish of coin (3)
25 This sign of the Zodiac has no connection with the Fishes (6)
26 A preservative of teeth (6)
29 Famous sculptor (5)
30 This part of the locomotive engine would sound familiar to the golfer (5)

Figure 51 The *Daily Telegraph* crossword used as a test to recruit new codebreakers (the solution is in Appendix H).

Henceforth there were to be no more barriers to recruitment or materials. By the end of 1942 there were 49 bombes, and a new bombe station was opened at Gayhurst Manor, just north of Bletchley. As part of the recruitment drive, the Government Code and Cypher School placed a letter in the *Daily Telegraph*. They issued an anonymous challenge to its readers, asking if anybody could solve the newspaper's crossword (Figure 51) in under 12 minutes. It was felt that crossword experts might also be good codebreakers, complementing the scientific minds that were already at Bletchley – but of course, none of this was mentioned in the newspaper. The 25 readers who replied were invited to Fleet Street to sit a crossword test. Five of them completed the crossword within the allotted time, and another had only one word missing when the 12 minutes had expired. A few weeks later, all six were interviewed by military intelligence and recruited as codebreakers at Bletchley Park.

Kidnapping Codebooks

So far in this chapter, the Enigma traffic has been treated as one giant communications system, but in fact there were several distinct networks. The German Army in North Africa, for instance, had its own separate network, and their Enigma operators had codebooks that were different from those used in Europe. Hence, if Bletchley succeeded in identifying the North African day-key, it would be able to decipher all the German messages sent from North Africa that day, but the North African day-key would be of no use in cracking the messages being transmitted in Europe. Similarly, the Luftwaffe had its own communications network, and so in order to decipher all Luftwaffe traffic, Bletchley would have to unravel the Luftwaffe day-key.

Some networks were harder to break into than others. The Kriegsmarine network was the hardest of all, because the German Navy operated a more sophisticated version of the Enigma machine. For example, the Naval Enigma operators had a choice of eight scramblers, not just five, which meant that there were almost six times as many scrambler arrangements, and therefore almost six times as many keys for Bletchley to check. The other difference in the Naval Enigma concerned the reflector, which was responsible for sending the electrical signal back through the scramblers. In the standard Enigma the reflector was always fixed in one particular orientation, but in the Naval Enigma the reflector could be fixed in any one of 26 orientations. Hence the number of possible keys was further increased by a factor of 26.

Cryptanalysis of the Naval Enigma was made even harder by the Naval operators, who were careful not to send stereotypical messages, thus depriving Bletchley of cribs. Furthermore, the Kriegsmarine also instituted a more secure system for selecting and transmitting message-keys. Extra scramblers, a variable reflector, non-stereotypical messages and a new system for exchanging message-keys all contributed to making German Naval communications impenetrable.

Bletchley's failure to crack the Naval Enigma meant that the Kriegsmarine were steadily gaining the upper hand in the Battle of the Atlantic. Admiral Karl Dönitz had developed a highly effective two-stage strategy for naval warfare, which began with his U-boats spreading out and scouring the Atlantic in search of Allied convoys. As soon as one of them spotted a target, it would initiate the next stage of the strategy by calling the other U-boats to the scene. The attack would commence only when a large pack of U-boats had been assembled. For this strategy of coordinated attack to succeed, it was essential that the Kriegsmarine had access to secure communication. The Naval Enigma provided such communication, and the U-boat attacks had a devastating impact on the Allied shipping that was supplying Britain with much-needed food and armaments.

As long as U-boat communications remained secure, the Allies had no idea of the locations of the U-boats, and could not plan safe routes for the convoys. It seemed as if the Admiralty's only strategy for pinpointing the location of U-boats was by looking at the sites of sunken British ships. Between June 1940 and June 1941 the Allies lost an average of 50 ships each month, and they were in danger of not being able to build new ships quickly enough to replace them. Besides the intolerable destruction of ships, there was also a terrible human cost – 50,000 Allied seamen died during the war. Unless these losses could be drastically reduced, Britain was in danger of losing the Battle of the Atlantic, which would have meant losing the war. Churchill would later write, 'Amid the torrent of violent events one anxiety reigned supreme. Battles might be won or lost, enterprises might succeed or miscarry, territories might be gained or quitted, but dominating all our power to carry on war, or even keep ourselves alive, lay our mastery of the ocean routes and the free approach and entry to our ports.'

The Polish experience and the case of Hans-Thilo Schmidt had taught Bletchley Park that if intellectual endeavour fails to break a cipher, then it is necessary to rely on espionage, infiltration and theft in order to obtain the enemy keys. Occasionally, Bletchley would make a breakthrough against the Naval Enigma, thanks to a clever ploy by the RAF. British planes would lay mines in a particular location, provoking German vessels to send out warnings to other craft. These Enigma encrypted warnings would inevitably contain a map reference, but crucially this map reference would already be known by the British, so it could be used as a crib. In other words, Bletchley knew that a particular piece of ciphertext represented a particular set of coordinates. Sowing mines to obtain cribs, known as 'gardening', required the RAF to fly special missions, so this could not be done on a regular basis. Bletchley had to find another way of breaking the Naval Enigma.

An alternative strategy for cracking the Naval Enigma depended on stealing keys. One of the most intrepid plans for stealing keys was concocted by Ian Fleming, creator of James Bond and a member of Naval Intelligence during the war. He suggested crashing a captured German bomber in the English Channel, close to a German ship. The German sailors would then approach the plane to rescue their comrades, whereupon the aircrew, British pilots pretending to be German, would board the ship and capture its codebooks. These German codebooks contained the information that was required for establishing the encryption key, and because ships were often away from base for long periods, the codebooks would be valid for at least a month. By capturing such codebooks, Bletchley would be able to decipher the Naval Enigma for an entire month.

After approving Fleming's plan, known as Operation Ruthless, British Intelligence began preparing a Heinkel bomber for the crash-landing, and assembled an aircrew of German-speaking Englishmen. The plan was scheduled for a date early in the month, so as to capture a fresh codebook. Fleming went to Dover to oversee the operation, but unfortunately there was no German shipping in the area so the plan was postponed indefinitely. Four days later, Frank Birch, who headed the Naval section at Bletchley, recorded the reaction of Turing and his colleague Peter Twinn:

'Turing and Twinn came to me like undertakers cheated of a nice corpse two days ago, all in a stew about the cancellation of Operation Ruthless.'

In due course Operation Ruthless was cancelled, but German Naval codebooks were eventually captured during a spate of daring raids on weather ships and U-boats. These so-called 'pinches' gave Bletchley the documents it needed to bring an end to the intelligence blackout. With the Naval Enigma transparent, Bletchley could pinpoint the location of U-boats, and the Battle of the Atlantic began to swing in favour of the Allies. Convoys could be steered clear of U-boats, and British destroyers could even begin to go on the offensive, seeking out and sinking U-boats.

It was vital that the German High Command never suspected that the Allies had pinched Enigma codebooks. If the Germans found that their security had been compromised, they would upgrade their Enigma machines, and Bletchley would be back to square one. As with the Zimmermann telegram episode, the British took various precautions to avoid arousing suspicion, such as sinking a German vessel after pinching its codebooks. This would persuade Admiral Dönitz that the cipher material had found its way to the bottom of the sea, and not fallen into British hands.

Once material had been secretly captured, further precautions had to be taken before exploiting the resulting intelligence. For example, the Enigma decipherments gave the locations of numerous U-boats, but it would have been unwise to have attacked every single one of them, because a sudden unexplained increase in British success would warn Germany that its communications were being deciphered. Consequently, the Allies would allow some U-boats to escape, and would attack others only when a spotter plane had been sent out first, thus justifying the approach of a destroyer some hours later. Alternatively, the Allies might send fake messages describing sightings of U-boats, which likewise provided sufficient explanation for the ensuing attack.

Despite this policy of minimising telltale signs that Enigma had been broken, British actions did sometimes raise concerns among Germany's security experts. On one occasion, Bletchley deciphered an Enigma message giving the exact location of a group of German tankers and supply ships, nine in total. The Admiralty decided not to sink all of the ships in case a clean sweep of targets aroused German suspicions. Instead, they informed destroyers of the exact location of just seven of the ships, which should have allowed the *Gadania* and the *Gonzenheim* to escape unharmed. The seven targeted ships were indeed sunk, but Royal Navy destroyers accidentally encountered the two ships that were supposed to be spared, and also sank them. The destroyers did not know about Enigma or the policy of not arousing suspicion – they merely believed they were doing their duty. Back in Berlin, Admiral Kurt Fricke instigated an investigation into this and other similar attacks, exploring the possibility that the British had broken Enigma. The report concluded that the numerous losses were either the result of natural misfortune, or caused by a British spy who had infiltrated the Kriegsmarine. The breaking of Enigma was considered impossible and inconceivable.

The Anonymous Cryptanalysts

As well as breaking the German Enigma cipher, Bletchley Park also succeeded in deciphering Italian and Japanese messages. The intelligence that emerged from these three sources was given the codename Ultra, and the Ultra Intelligence files were responsible for giving the Allies a clear advantage in all the major arenas of conflict. In North Africa, Ultra helped to destroy German supply lines and informed the Allies of the status of General Rommel's forces, enabling the Eighth Army to fight back against the German advances. Ultra also warned of the German invasion of Greece, which allowed British troops to retreat without heavy losses. In fact, Ultra provided accurate reports on the enemy's situation throughout the entire Mediterranean. This information was particularly valuable when the Allies landed in Italy and Sicily in 1943.

In 1944, Ultra played a major role in the Allied invasion of Europe. For example, in the months prior to D-Day, the Bletchley decipherments provided a detailed picture of the German troop concentrations along the French coast. Sir Harry Hinsley, official historian of British Intelligence during the war, wrote:

> As Ultra accumulated, it administered some unpleasant shocks. In particular, it revealed in the second half of May – following earlier disturbing indications that the Germans were concluding that the area between Le Havre and Cherbourg was a likely, and perhaps even the main, invasion

> area – that they were sending reinforcements to Normandy and the Cherbourg peninsula. But this evidence arrived in time to enable the Allies to modify the plans for the landings on and behind the Utah beach; and it is a singular fact that before the expedition sailed the Allied estimate of the number, identification, and location of the enemy's divisions in the west, fifty-eight in all, was accurate in all but two items that were to be of operational importance.

Throughout the war, the Bletchley codebreakers knew that their decipherments were vital, and Churchill's visit to Bletchley had reinforced this point. But the cryptanalysts were never given any operational details or told how their decipherments were being used. For example, the codebreakers were given no information about the date for D-Day, and they arranged a dance for the evening before the landings. This worried Commander Travis, the Director of Bletchley and the only person on site who was privy to the plans for D-Day. He could not tell the Hut 6 Dance Committee to cancel the event because this would have been a clear hint that a major offensive was in the offing, and as such a breach of security. The dance was allowed to go ahead. As it happened, bad weather postponed the landings for twenty-four hours, so the codebreakers had time to recover from the frivolities. On the day of the landings, the French resistance destroyed landlines, forcing the Germans to communicate solely by radio, which in turn gave Bletchley the opportunity to intercept and decipher even more messages. At the turning point of the war, Bletchley was able to provide an even more detailed picture of German military operations.

Stuart Milner-Barry, one of the Hut 6 cryptanalysts, wrote: 'I do not imagine that any war since classical times, if ever, has been fought in which one side read consistently the main military and naval intelligence of the other.' An American report came to a similar conclusion: 'Ultra created in senior staffs and at the political summit a state of mind which transformed the taking of decisions. To feel that you know your enemy is a vastly comforting feeling. It grows imperceptibly over time if you regularly and intimately observe his thoughts and ways and habits and actions. Knowledge of this kind makes your own planning less tentative and more assured, less harrowing and more buoyant.'

It has been argued, albeit controversially, that Bletchley Park's achievements were the decisive factor in the Allied victory. What is certain is that the Bletchley codebreakers significantly shortened the war. This becomes evident by re-running the Battle of the Atlantic and speculating what might have happened without the benefit of Ultra intelligence. To begin with, more ships and supplies would certainly have been lost to the dominant U-boat fleet, which would have compromised the vital link to America and forced the Allies to divert manpower and resources into the building of new ships. Historians have estimated that this would have delayed Allied plans by several months, which would have meant postponing the D-Day invasion until at least the following year. According to Sir Harry Hinsley, 'the war, instead of finishing in 1945, would have ended in 1948 had the Government Code and Cypher School not been able to read the Enigma cyphers and produce the Ultra intelligence'.

During this period of delay, additional lives would have been lost in Europe, and Hitler would have been able to make greater use of his V-weapons, inflicting damage throughout southern England. The historian David Kahn summarises the impact of breaking Enigma: 'It saved lives. Not only Allied and Russian lives but, by shortening the war, German, Italian, and Japanese lives as well. Some people alive after World War II might not have been but for these solutions. That is the debt that the world owes to the codebreakers; that is the crowning human value of their triumphs.'

After the war, Bletchley's accomplishments remained a closely guarded secret. Having successfully deciphered messages during the war, Britain wanted to continue its intelligence operations, and was reluctant to divulge its capabilities. In fact, Britain had captured thousands of Enigma machines, and distributed them among its former colonies, who believed that the cipher was as secure as it had seemed to the Germans. The British did nothing to disabuse them of this belief, and routinely deciphered their secret communications in the years that followed.

Meanwhile, the Government Code and Cypher School at Bletchley Park was closed and the thousands of men and women who had contributed to the creation of Ultra were disbanded. The bombes were dismantled, and every scrap of paper that related to the wartime decipherments was either locked away or burnt. Britain's codebreaking activities were officially transferred to the newly formed Government

Communications Headquarters (GCHQ) in London, which was moved to Cheltenham in 1952. Although some of the cryptanalysts moved to GCHQ, most of them returned to their civilian lives, sworn to secrecy, unable to reveal their pivotal role in the Allied war effort. While those who had fought conventional battles could talk of their heroic achievements, those who had fought intellectual battles of no less significance had to endure the embarrassment of having to evade questions about their wartime activities. Gordon Welchman recounted how one of the young cryptanalysts working with him in Hut 6 had received a scathing letter from his old headmaster, accusing him of being a disgrace to his school for not being at the front. Derek Taunt, who also worked in Hut 6, summed up the true contribution of his colleagues: 'Our happy band may not have been with King Harry on St Crispin's Day, but we had certainly not been abed and have no reason to think ourselves accurs't for having been where we were.'

After three decades of silence, the secrecy over Bletchley Park eventually came to an end in the early 1970s. Captain F.W. Winterbotham, who had been responsible for distributing the Ultra intelligence, began to badger the British Government, arguing that the Commonwealth countries had stopped using the Enigma cipher and that there was now nothing to be gained by concealing the fact that Britain had broken it. The intelligence services reluctantly agreed, and permitted him to write a book about the work done at Bletchley Park. Published in the summer of 1974, Winterbotham's book *The Ultra Secret* was the signal that Bletchley personnel were at last free to discuss their wartime activities. Gordon Welchman felt enormous relief: 'After the war I still avoided discussions of wartime events for fear that I might reveal information obtained from Ultra rather than from some published account . . . I felt that this turn of events released me from my wartime pledge of secrecy.'

Those who had contributed so much to the war effort could now receive the recognition they deserved. Possibly the most remarkable consequence of Winterbotham's revelations was that Rejewski realised the staggering consequences of his pre-war breakthroughs against Enigma. After the invasion of Poland, Rejewski had escaped to France, and when France was overrun he fled to Britain. It would seem natural that he should have become part of the British Enigma effort, but instead he was relegated to tackling menial ciphers at a minor intelligence unit in Boxmoor, near Hemel Hempstead. It is not clear why such a brilliant mind was excluded from Bletchley Park, but as a result he was completely unaware of the activities of the Government Code and Cypher School. Until the publication of Winterbotham's book, Rejewski had no idea that his ideas had provided the foundation for the routine decipherment of Enigma throughout the war.

For some, the publication of Winterbotham's book came too late. Many years after the death of Alastair Denniston, Bletchley's first director, his daughter received a letter from one of his colleagues: 'Your father was a great man in whose debt all English-speaking people will remain for a very long time, if not for ever. That so few should know exactly what he did is the sad part.'

Alan Turing was another cryptanalyst who did not live long enough to receive any public recognition. Instead of being acclaimed a hero, he was persecuted for his homosexuality. In 1952, while reporting a burglary to the police, he naively revealed that he was having a homosexual relationship. The police felt they had no option but to arrest and charge him with 'Gross Indecency contrary to Section 11 of the Criminal Law Amendment Act 1885'. The newspapers reported the subsequent trial and conviction, and Turing was publicly humiliated.

Turing's secret had been exposed, and his sexuality was now public knowledge. The British Government withdrew his security clearance. He was forbidden to work on research projects relating to the development of the computer. He was forced to consult a psychiatrist and had to undergo hormone treatment, which made him impotent and obese. Over the next two years he became severely depressed, and on 7 June 1954, he went to his bedroom, carrying with him a jar of cyanide solution and an apple. Twenty years earlier he had chanted the rhyme of the Wicked Witch: 'Dip the apple in the brew, Let the sleeping death seep through'. Now he was ready to obey her incantation. He dipped the apple in the cyanide and took several bites. At the age of just forty-two, one of the true geniuses of cryptanalysis committed suicide.

5 The Language Barrier

While British codebreakers were breaking the German Enigma cipher and altering the course of the war in Europe, American codebreakers were having an equally important influence on events in the Pacific arena by cracking the Japanese machine cipher known as Purple. For example, in June 1942 the Americans deciphered a message outlining a Japanese plan to draw U.S. Naval forces to the Aleutian Islands by faking an attack, which would allow the Japanese Navy to take their real objective, Midway Island. Although American ships played along with the plan by leaving Midway, they never strayed far away. When American cryptanalysts intercepted and deciphered the Japanese order to attack Midway, the ships were able to return swiftly and defend the island in one of the most important battles of the entire Pacific war. According to Admiral Chester Nimitz, the American victory at Midway 'was essentially a victory of intelligence. In attempting surprise, the Japanese were themselves surprised.'

Almost a year later, American cryptanalysts identified a message that showed the itinerary for a visit to the northern Solomon Islands by Admiral Isoruko Yamamoto, Commander-in-Chief of the Japanese Fleet. Nimitz decided to send fighter aircraft to intercept Yamamoto's plane and shoot him down. Yamamoto, renowned for being compulsively punctual, approached his destination at exactly 8.00 a.m., just as stated in the intercepted schedule. There to meet him were eighteen American P-38 fighters. They succeeded in killing one of the most influential figures of the Japanese High Command.

Although Purple and Enigma, the Japanese and German ciphers, were eventually broken, they did offer some security when they were initially implemented and provided real challenges for American and British cryptanalysts. In fact, had the cipher machines been used properly – without repeated message-keys, without cillies, without restrictions on plugboard settings and scrambler arrangements, and without stereotypical messages which resulted in cribs – it is quite possible that they might never have been broken at all.

The true strength and potential of machine ciphers was demonstrated by the Typex (or Type X) cipher machine used by the British army and air force, and the SIGABA (or M-143-C) cipher machine used by the American military. Both these machines were more complex than the Enigma machine and both were used properly, and therefore they remained unbroken throughout the war. Allied cryptographers were confident that complicated electromechanical machine ciphers could guarantee secure communication. However, complicated machine ciphers are not the only way of sending secure messages. Indeed, one of the most secure forms of encryption used in the Second World War was also one of the simplest.

During the Pacific campaign, American commanders began to realise that cipher machines, such as SIGABA, had a fundamental drawback. Although electromechanical encryption offered relatively high levels of security, it was painfully slow. Messages had to be typed into the machine letter by letter, the output had to be noted down letter by letter, and then the completed ciphertext had to be transmitted by the radio operator. The radio operator who received the enciphered message then had to pass it on to a cipher expert, who would carefully select the correct key, and type the ciphertext into a cipher machine, to decipher it letter by letter. The time and space required for this delicate operation is available at headquarters or on board a ship, but machine encryption was not ideally suited to more hostile and intense environments, such as the islands of the Pacific. One war correspondent described the difficulties of communication during the heat of jungle battle: 'When the fighting became confined to a small area, everything had to move on a split-second schedule. There was not time for enciphering and deciphering. At such times, the King's English became a last resort – the profaner the better.' Unfortunately for the Americans, many Japanese soldiers had attended American colleges and were fluent in English, including the profanities. Valuable information about American strategy and tactics was falling into the hands of the enemy.

One of the first to react to this problem was Philip Johnston, an engineer based in Los Angeles, who was too old to fight but still wanted to contribute to the war effort. At the beginning of 1942 he began to formulate an encryption system inspired by his childhood experiences. The son of a Protestant missionary, Johnston had grown up on the Navajo reservations of Arizona, and as a result he had become fully immersed in Navajo culture. He was one of the few people outside the tribe who could speak their language fluently, which allowed him to act as an interpreter for discussions between the Navajo and government agents. His work in this capacity culminated in a visit to the White House, when, as a nine-year-old, Johnston translated for two Navajos who were appealing to President Theodore Roosevelt for fairer treatment for their community. Fully aware of how impenetrable the language was for those outside the tribe, Johnston was struck by the notion that Navajo, or any other Native American language, could act as a virtually unbreakable code. If each battalion in the Pacific employed a pair of Native Americans as radio operators, secure communication could be guaranteed.

He took his idea to Lieutenant Colonel James E. Jones, the area signal officer at Camp Elliott, just outside San Diego. Merely by throwing a few Navajo phrases at the bewildered officer, Johnston was able to persuade him that the idea was worthy of serious consideration. A fortnight later he returned with two Navajos, ready to conduct a test demonstration in front of senior marine officers. The Navajos were isolated from each other, and one was given six typical messages in English, which he translated into Navajo and transmitted to his colleague via a radio. The Navajo receiver translated the messages back into English, wrote them down, and handed them over to the officers, who compared them with the originals. The game of Navajo whispers proved to be flawless, and the marine officers authorised a pilot project and ordered recruitment to begin immediately.

Before recruiting anybody, however, Lieutenant Colonel Jones and Philip Johnston had to decide whether to conduct the pilot study with the Navajo, or select another tribe. Johnston had used Navajo men for his original demonstration because he had personal connections with the tribe, but this did not necessarily make them the ideal choice. The most important selection criterion was simply a question of numbers: the marines needed to find a tribe capable of supplying a large number of men who were fluent in English and literate. The lack of government investment meant that the literacy rate was very low on most of the reservations, and attention was therefore focused on the four largest tribes: the Navajo, the Sioux, the Chippewa and the Pima-Papago.

The Navajo was the largest tribe, but also the least literate, while the Pima-Papago was the most literate but much fewer in number. There was little to choose between the four tribes, and ultimately the decision rested on another critical factor. According to the official report on Johnston's idea:

> The Navajo is the only tribe in the United States that has not been infested with German students during the past twenty years. These Germans, studying the various tribal dialects under the guise of art students, anthropologists, etc., have undoubtedly attained a good working knowledge of all tribal dialects except Navajo. For this reason the Navajo is the only tribe available offering complete security for the type of work under consideration. It should also be noted that the Navajo tribal dialect is completely unintelligible to all other tribes and all other people, with the possible exception of as many as 28 Americans who have made a study of the dialect. This dialect is equivalent to a secret code to the enemy, and admirably suited for rapid, secure communication.

At the time of America's entry into the Second World War, the Navajo were living in harsh conditions and being treated as inferior people. Yet their tribal council supported the war effort and declared their loyalty: 'There exists no purer concentration of Americanism than among the First Americans'. The Navajos were so eager to fight that some of them lied about their age, or gorged themselves on bunches of bananas and swallowed great quantities of water in order to reach the minimum weight requirement of 55 kg. Similarly, there was no difficulty in finding suitable candidates to serve as Navajo code talkers, as they were to become known. Within four months of the bombing of Pearl Harbor, 29 Navajos, some as young as fifteen, began an eight-week communications course with the Marine Corps.

Before training could begin, the Marine Corps had to overcome a problem that had plagued the only other code to have been based on a Native American language. In Northern France during the First World War, Captain E.W. Horner of Company D, 141st Infantry, ordered that eight men from the Choctaw tribe be employed as radio operators. Obviously none of the enemy understood their language, so the Choctaw provided secure communications. However, this encryption system was fundamentally flawed because the Choctaw language had no equivalent for modern military jargon. A specific technical term in a message might therefore have to be translated into a vague Choctaw expression, with the risk that this could be misinterpreted by the receiver.

The same problem would have arisen with the Navajo language, but the Marine Corps planned to construct a lexicon of Navajo terms to replace otherwise untranslatable English words, thus removing any ambiguities. The trainees helped to compile the lexicon, tending to choose words describing the natural world to indicate specific military terms. Thus, the names of birds were used for planes, and fish for ships (Table 11). Commanding officers became 'war chiefs', platoons were 'mud-clans', fortifications turned into 'cave dwellings' and mortars were known as 'guns that squat'.

Even though the complete lexicon contained 274 words, there was still the problem of translating less predictable words and the names of people and places. The solution was to devise an encoded phonetic alphabet for spelling out difficult words. For example, the word 'Pacific' would be spelt out as 'pig, ant, cat, ice, fox, ice, cat', which would then be translated into Navajo as **bi-sodih, wol-la-chee, moasi, tkin, ma-e, tkin, moasi**. The complete Navajo alphabet is given in Table 12. Within eight weeks, the trainee code talkers had learnt the entire lexicon and alphabet, thus

Table 11 Navajo codewords for planes and ships.

Fighter plane	Hummingbird	**Da-he-tih-hi**
Observation plane	Owl	**Ne-as-jah**
Torpedo plane	Swallow	**Tas-chizzie**
Bomber	Buzzard	**Jay-sho**
Dive-bomber	Chicken hawk	**Gini**
Bombs	Eggs	**A-ye-shi**
Amphibious vehicle	Frog	**Chal**
Battleship	Whale	**Lo-tso**
Destroyer	Shark	**Ca-lo**
Submarine	Iron fish	**Besh-lo**

obviating the need for codebooks which might fall into enemy hands. For the Navajos, committing everything to memory was trivial because traditionally their language had no written script, so they were used to memorising their folk stories and family histories. As William McCabe, one of the trainees, said, 'In Navajo everything is in the memory – songs, prayers, everything. That's the way we were raised.'

At the end of their training, the Navajos were put to the test. Senders translated a series of messages from English into Navajo, transmitted them, and then receivers translated the messages back into English, using the memorised lexicon and alphabet when necessary. The results were word-perfect. To check the strength of the system, a recording of the transmissions was given to Navy Intelligence, the unit that had cracked Purple, the toughest Japanese cipher. After three weeks of intense cryptanalysis, the Naval codebreakers were still baffled by the messages. They called the Navajo language a 'weird succession of guttural, nasal, tongue-twisting sounds . . . we couldn't even transcribe it, much less crack it'. The Navajo code was judged a success. Two Navajo soldiers, John Benally and Johnny Manuelito, were asked to stay and train the next batch of recruits, while the other 27 Navajo code talkers were assigned to four regiments and sent to the Pacific.

Table 12 The Navajo alphabet code.

A	Ant	Wol-la-chee	N	Nut	Nesh-chee
B	Bear	Shush	O	Owl	Ne-ahs-jsh
C	Cat	Moasi	P	Pig	Bi-sodih
D	Deer	Be	Q	Quiver	Ca-yeilth
E	Elk	Dzeh	R	Rabbit	Gah
F	Fox	Ma-e	S	Sheep	Dibeh
G	Goat	Klizzie	T	Turkey	Than-zie
H	Horse	Lin	U	Ute	No-da-ih
I	Ice	Tkin	V	Victor	A-keh-di-glini
J	Jackass	Tkele-cho-gi	W	Weasel	Gloe-ih
K	Kid	Klizzie-yazzi	X	Cross	Al-an-as-dzoh
L	Lamb	Dibeh-yazzi	Y	Yucca	Tsah-as-zih
M	Mouse	Na-as-tso-si	Z	Zinc	Besh-do-gliz

Japanese forces had attacked Pearl Harbor on 7 December 1941, and not long after they dominated large parts of the western Pacific. Japanese troops overran the American garrison on Guam on 10 December, they took Guadalcanal, one of the islands in the Solomon chain, on 13 December, Hong Kong capitulated on 25 December, and U.S. troops on the Philippines surrendered on 2 January 1942. The Japanese planned to consolidate their control of the Pacific the following summer by building an airfield on Guadalcanal, creating a base for bombers which would enable them to destroy Allied supply lines, thus making any Allied counterattack almost impossible. Admiral Ernest King, Chief of American Naval Operations, urged an attack on the island before the airfield was completed, and on 7 August the 1st Marine Division spearheaded an invasion of Guadalcanal. The initial landing parties included the first group of code talkers to see action.

Although the Navajos were confident that their skills would be a blessing to the marines, their first attempts generated only confusion. Many of

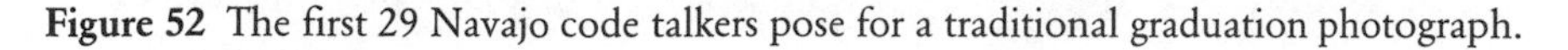

Figure 52 The first 29 Navajo code talkers pose for a traditional graduation photograph.

the regular signal operators were unaware of this new code, and they sent panic messages all over the island, stating that the Japanese were broadcasting on American frequencies. The colonel in charge immediately halted Navajo communications until he could convince himself that the system was worth pursuing. One of the code talkers recalled how the Navajo code was eventually brought back into service:

> The colonel had an idea. He said he would keep us on one condition: that I could out-race his 'white code' – a mechanical ticking cylinder thing. We both sent messages, by white cylinder and by my voice. Both of us received answers and the race was to see who could decode his answer first. I was asked, 'How long will it take you? Two hours?' 'More like two minutes,' I answered. The other guy was still decoding when I got the roger on my return message in about four and a half minutes. I said, 'Colonel, when are you going to give up on that cylinder thing?' He didn't say anything. He just lit up his pipe and walked away.

The code talkers soon proved their worth on the battlefield. During one episode on the island of Saipan, a battalion of marines took over positions previously held by Japanese soldiers, who had retreated. Suddenly a salvo exploded nearby. They were under friendly fire from fellow Americans who were unaware of their advance. The marines radioed back in English explaining their position, but the salvos continued because the attacking American troops suspected that the messages were from Japanese impersonators trying to fool them. It was only when a Navajo message was sent that the attackers saw their mistake and halted the assault. A Navajo message could never be faked, and could always be trusted.

The reputation of the code talkers soon spread, and by the end of 1942 there was a request for 83 more men. The Navajo were to serve in all six Marine Corps divisions, and were sometimes borrowed by other American forces. Their war of words soon turned the Navajos into heroes. Other soldiers would offer to carry their radios and rifles, and they were even given personal bodyguards, partly to protect them from their own comrades. On at least three occasions code talkers were mistaken for Japanese soldiers and captured by fellow Americans. They were released only when colleagues from their own battalion vouched for them.

The impenetrability of the Navajo code was all down to the fact that Navajo belongs to the Na-Dene family of languages, which has no link with any Asian or European language. For example, a Navajo verb is conjugated not solely according to its subject, but also according to its object. The verb ending depends on which category the object belongs to: long (e.g. pipe, pencil), slender and flexible (e.g. snake, thong), granular (e.g. sugar, salt), bundled (e.g. hay), viscous (e.g. mud, faeces) and many others. The verb will also incorporate adverbs, and will reflect whether or not the speaker has experienced what he or she is talking about, or whether it is hearsay. Consequently, a single verb can be equivalent to a whole sentence, making it virtually impossible for foreigners to disentangle its meaning.

Despite its strengths, the Navajo code still suffered from two significant flaws. First, words that were neither in the natural Navajo vocabulary nor in the list of 274 authorised codewords had to be spelt out using the special alphabet. This was time-consuming, so it was decided to add another 234 common terms to the lexicon. For example, nations were given Navajo nicknames: 'Rolled Hat' for Australia, 'Bounded by Water' for Britain, 'Braided Hair' for China, 'Iron Hat' for Germany, 'Floating Land' for the Philippines, and 'Sheep Pain' for Spain.

The second problem concerned those words that would still have to be spelt out. If it became clear to the Japanese that words were being spelt out, they would realise that they could use frequency analysis to identify which Navajo words represented which letters. It would soon become obvious that the most commonly used word was **dzeh**, which means 'elk' and which represents **e**, the most commonly used letter of the English alphabet. Just spelling out the name of the island Guadalcanal and repeating the word **wol-la-chee** (ant) four times would be a big clue as to what word represented the letter **a**. The solution was to add more words to act as extra substitutes (homophones) for the commonly used letters. Two extra words were introduced as alternatives for each of the six commonest letters (**e**, **t**, **a**, **o**, **i**, **n**), and one extra word for the six next commonest letters (**s**, **h**, **r**, **d**, **l**, **u**). The letter **a**, for example, could now also be substituted by the words **be-la-sana** (apple) or **tse-nihl** (axe). Thereafter, Guadalcanal could be spelt with only one repetition: **klizzie**, **shi-da**, **wol-la-chee**, **lha-cha-eh**, **be-la-sana**, **dibeh-yazzie**, **moasi**,

tse-nihl, nesh-chee, tse-nihl, ah-jad (goat, uncle, ant, dog, apple, lamb, cat, axe, nut, axe, leg).

As the war in the Pacific intensified, and as the Americans advanced from the Solomon Islands to Okinawa, the Navajo code talkers played an increasingly vital role. During the first days of the attack on Iwo Jima, more than eight hundred Navajo messages were sent, all without error. According to Major General Howard Conner, 'without the Navajos, the marines would never have taken Iwo Jima'. The contribution of the Navajo code talkers is all the more remarkable when you consider that, in order to fulfil their duties, they often had to confront and defy their own deeply held spiritual fears. The Navajo believe that the spirits of the dead, *chindi*, will seek revenge on the living unless ceremonial rites are performed on the body. The war in the Pacific was particularly bloody, with

Figure 53 Corporal Henry Bake, Jr (left) and Private First Class George H. Kirk using the Navajo code in the dense jungles of Bougainville in 1943.

corpses strewn across the battlefields, and yet the code talkers summoned up the courage to carry on regardless of the *chindi* that haunted them. In Doris Paul's book *The Navajo Code Talkers*, one of the Navajo recounts an incident which typifies their bravery, dedication and composure:

> If you so much as held up your head six inches you were gone, the fire was so intense. And then in the wee hours, with no relief on our side or theirs, there was a dead standstill. It must have gotten so that this one Japanese couldn't take it anymore. He got up and yelled and screamed at the top of his voice and dashed over our trench, swinging a long samurai sword. I imagine he was shot from 25 to 40 times before he fell.
>
> There was a buddy with me in the trench. But that Japanese had cut him across the throat, clear through to the cords on the back of his neck. He was still gasping through his windpipe. And the sound of him trying to breathe was horrible. He died, of course. When the Jap struck, warm blood spattered all over my hand that was holding a microphone. I was calling in code for help. They tell me that in spite of what happened, every syllable of my message came through.

Altogether, there were 420 Navajo code talkers. Although their bravery as fighting men was acknowledged, their special role in securing communications was classified information. The government forbade them to talk about their work, and their unique contribution was not made public. Just like Turing and the cryptanalysts at Bletchley Park, the Navajo were ignored for decades. Eventually, in 1968, the Navajo code was declassified, and the following year the code talkers held their first reunion. Then, in 1982, they were honoured when the U.S. Government named 14 August 'National Navajo Code Talkers Day'. However, the greatest tribute to the work of the Navajo is the simple fact that their code is one of very few throughout history that was never broken. Lieutenant General Seizo Arisue, the Japanese chief of intelligence, admitted that, although they had broken the American Air Force code, they had failed to make any impact on the Navajo code.

Deciphering Lost Languages and Ancient Scripts

The success of the Navajo code was based largely on the simple fact that the mother tongue of one person is utterly meaningless to anybody unacquainted with it. In many ways, the task that confronted Japanese cryptanalysts is similar to that which is faced by archaeologists attempting to decipher a long-forgotten language, perhaps written in an extinct script. If anything, the archaeological challenge is much more severe. For example, while the Japanese had a continuous stream of Navajo words which they could attempt to identify, the information available to the archaeologist can sometimes be just a small collection of clay tablets. Furthermore, the archaeological codebreaker often has no idea of the context or contents of an ancient text, clues which military codebreakers can normally rely on to help them crack a cipher.

Deciphering ancient texts seems an almost hopeless pursuit, yet many men and women have devoted themselves to this arduous enterprise. Their obsession is driven by the desire to understand the writings of our ancestors, allowing us to speak their words and catch a glimpse of their thoughts and lives. Perhaps this appetite for cracking ancient scripts is best summarised by Maurice Pope, the author of *The Story of Decipherment*: 'Decipherments are by far the most glamorous achievements of scholarship. There is a touch of magic about unknown writing, especially when it comes from the remote past, and a corresponding glory is bound to attach itself to the person who first solves its mystery.'

The decipherment of ancient scripts is not part of the ongoing evolutionary battle between codemakers and codebreakers, because, although there are codebreakers in the shape of archaeologists, there are no codemakers. That is to say, in most cases of archaeological decipherment there was no deliberate attempt by the original scribe to hide the meaning of the text. The remainder of this chapter, which is a discussion of archaeological decipherments, is therefore a slight detour from the book's main theme. However, the principles of archaeological decipherment are essentially the same as those of conventional military cryptanalysis. Indeed, many military codebreakers have been attracted by the challenge of unravelling an ancient script. This is probably because archaeological decipherments make a refreshing change from military codebreaking, offering a purely intellectual puzzle rather than a military challenge. In other words, the motivation is curiosity rather than animosity.

The most famous, and arguably the most romantic, of all decipherments was the cracking of Egyptian hieroglyphics. For centuries, hieroglyphics remained a mystery, and archaeologists could do no more than speculate about their meaning. However, thanks to a classic piece of codebreaking, the hieroglyphs were eventually deciphered, and ever since archaeologists have been able to read first-hand accounts of the history, culture and beliefs of the ancient Egyptians. The decipherment of hieroglyphics has bridged the millennia between ourselves and the civilisation of the pharaohs.

The earliest hieroglyphics date back to 3000 BC, and this form of ornate writing endured for the next three and a half thousand years. Although the elaborate symbols of hieroglyphics were ideal for the walls of majestic temples (the Greek word *hieroglyphica* means 'sacred carvings'), they were overly complicated for keeping track of mundane transactions. Hence, evolving in parallel with hieroglyphics was *hieratic*, an everyday script in which each hieroglyphic symbol was replaced by a stylised representation which was quicker and easier to write. In about 600 BC, hieratic was replaced by an even simpler script known as *demotic*, the name being derived from the Greek *demotika* meaning 'popular', which reflects its secular function. Hieroglyphics, hieratic and demotic are essentially the same script – one could almost regard them as merely different fonts.

All three forms of writing are phonetic, which is to say that the characters largely represent distinct sounds, just like the letters in the English alphabet. For over three thousand years the ancient Egyptians used these scripts in every aspect of their lives, just as we use writing today. Then, towards the end of the fourth century AD, within a generation, the Egyptian scripts vanished. The last datable examples of ancient Egyptian writing are to be found on the island of Philae. A hieroglyphic temple inscription was carved in AD 394, and a piece of demotic graffiti has been dated to AD 450. The spread of the Christian Church was responsible for the extinction of the Egyptian scripts, outlawing their use in order to eradicate any link with Egypt's pagan past. The ancient scripts were replaced with Coptic, a script consisting of 24 letters from the Greek alphabet supplemented by six demotic characters used for Egyptian sounds not expressed in Greek. The dominance of Coptic was so complete that the ability to read hieroglyphics, demotic and hieratic vanished. The ancient Egyptian language continued to be spoken, and evolved into what became known as the Coptic language, but in due course both the Coptic language and script were displaced by the spread of Arabic in the eleventh century. The final linguistic link to Egypt's ancient kingdoms had been broken, and the knowledge needed to read the tales of the pharaohs was lost.

Interest in hieroglyphics was reawakened in the seventeenth century, when Pope Sixtus V reorganised the city of Rome according to a new network of avenues, erecting obelisks brought from Egypt at each intersection. Scholars attempted to decipher the meanings of the hieroglyphs on the obelisks, but were hindered by a false assumption: nobody was prepared to accept that the hieroglyphs could possibly represent phonetic characters, or *phonograms*. The idea of phonetic spelling was thought to be too advanced for such an ancient civilisation. Instead, seventeenth-century scholars were convinced that the hieroglyphs were *semagrams* – that these intricate characters represented whole ideas, and were nothing more than primitive picture-writing. The belief that hieroglyphics is merely picture-writing was even commonly held by foreigners who visited Egypt while hieroglyphics was still a living script. Diodorus Siculus, a Greek historian of the first century BC, wrote:

> Now it happens that the forms of the Egyptians' letters take the shape of all kinds of living creatures and of the extremities of the human body and of implements . . . For their writing does not express the intended idea by a combination of syllables, one with another, but by the outward appearance of what has been copied and by the metaphorical meaning impressed upon the memory by practice. . . . So the hawk symbolises for them everything which happens quickly because this creature is just about the fastest of winged animals. And the idea is transferred, through the appropriate metaphorical transfer, to all swift things and to those things to which speed is appropriate.

In the light of such accounts, perhaps it is not so surprising that seventeenth-century scholars attempted to decipher the hieroglyphs by interpreting each one as a whole idea. For example, in 1652 the German Jesuit priest Athanasius Kircher published a dictionary of allegorical interpretations entitled *Œdipus ægyptiacus*, and used it to produce a series of weird and wonderful translations. A handful of hieroglyphs, which we now know merely represent the name of the pharaoh Apries, were translated by Kircher as: 'the benefits of the divine Osiris are to be procured by means of sacred ceremonies and of the chain of the Genii, in order that the benefits of the Nile may be obtained'. Today Kircher's translations seem ludicrous, but their impact on other would-be decipherers was immense. Kircher was more than just an Egyptologist: he wrote a book on cryptography, constructed a musical fountain, invented the magic lantern (a precursor of cinema), and lowered himself into the crater of Vesuvius, earning himself the title of 'father of vulcanology'. The Jesuit priest was widely acknowledged to be the most respected scholar of his age, and consequently his ideas were to influence generations of future Egyptologists.

A century and a half after Kircher, in the summer of 1798, the antiquities of ancient Egypt fell under renewed scrutiny when Napoleon Bonaparte despatched a team of historians, scientists and draughtsmen to follow in the wake of his invading army. These academics, or 'Pekinese dogs' as the soldiers called them, did a remarkable job of mapping, drawing, transcribing, measuring and recording everything they witnessed. In 1799, the French scholars encountered the single most famous slab of stone in the history of archaeology, found by a troop of French soldiers stationed at Fort Julien in the town of Rosetta in the Nile Delta. The soldiers had been given the task of demolishing an ancient wall to clear the way for an extension to the fort. Built into the wall was a stone bearing a remarkable set of inscriptions: the same piece of text had been inscribed on the stone three times, in Greek, demotic and hieroglyphics. The Rosetta Stone, as it became known, appeared to be the equivalent of a cryptanalytic crib, just like the cribs that helped the codebreakers at Bletchley Park to break into Enigma. The Greek, which could easily be read, was in effect a piece of plaintext which could be compared with the demotic and hieroglyphic ciphertexts. The Rosetta Stone was potentially a means of unravelling the meaning of the ancient Egyptian symbols.

The scholars immediately recognised the stone's significance, and sent it to the National Institute in Cairo for detailed study. However, before the institute could embark on any serious research, it became clear that the French army was on the verge of being defeated by the advancing British forces. The French moved the Rosetta Stone from Cairo to the relative safety of Alexandria, but ironically, when the French finally surrendered, Article XVI of the Treaty of Capitulation handed all the antiquities in Alexandria to the British, whereas those in Cairo were allowed to return to France. In 1802, the priceless slab of black basalt (measuring 118 cm in height, 77 cm in width and 30 cm in thickness, and weighing three-

Figure 54 The Rosetta Stone, inscribed in 196 BC and rediscovered in 1799, contains the same text written in three different scripts: hieroglyphics at the top, demotic in the middle and Greek at the bottom.

quarters of a tonne) was sent to Portsmouth on board HMS *L'Egyptienne*, and later that year it took up residence at the British Museum, where it has remained ever since.

The translation of the Greek soon revealed that the Rosetta Stone bore a decree from the general council of Egyptian priests issued in 196 BC. The text records the benefits that the Pharaoh Ptolemy had bestowed upon the people of Egypt, and details the honours that the priests had, in return, piled upon the pharaoh. For example, they declared that 'a festival shall be kept for King Ptolemy, the ever-living, the beloved of Ptah, the god Epiphanes Eucharistos, yearly in the temples throughout the land from the 1st of Troth for five days, in which they shall wear garlands and perform sacrifices and libations and the other usual honours'. If the other two inscriptions contained the identical decree, the decipherment of the hieroglyphic and demotic texts would seem to be straightforward. However, three significant hurdles remained. First, the Rosetta Stone is seriously damaged, as can be seen in Figure 54. The Greek text consists of 54 lines, of which the last 26 are damaged. The demotic consists of 32 lines, of which the beginnings of the first 14 lines are damaged (note that demotic and hieroglyphics are written from right to left). The hieroglyphic text is in the worst condition, with half the lines missing completely, and the remaining 14 lines (corresponding to the last 28 lines of the Greek text) partly missing. The second barrier to decipherment is that the two Egyptian scripts convey the ancient Egyptian language, which nobody had spoken for at least eight centuries. While it was possible to find a set of Egyptian symbols which corresponded to a set of Greek words, which would enable archaeologists to work out the meaning of the Egyptian symbols, it was impossible to establish the sound of the Egyptian words. Unless archaeologists knew how the Egyptian words were spoken, they could not deduce the phonetics of the symbols. Finally, the intellectual legacy of Kircher still encouraged archaeologists to think of Egyptian writing in terms of semagrams, rather than phonograms, and hence few people even considered attempting a phonetic decipherment of hieroglyphics.

One of the first scholars to question the prejudice that hieroglyphics was picture-writing was the English prodigy and polymath Thomas Young. Born in 1773 in Milverton, Somerset, Young was able to read fluently at the age of two. By the age of fourteen he had studied Greek, Latin, French,

Italian, Hebrew, Chaldean, Syriac, Samaritan, Arabic, Persian, Turkish and Ethiopic, and when he became a student at Emmanuel College, Cambridge, his brilliance gained him the sobriquet 'Phenomenon Young'. At Cambridge he studied medicine, but it was said that he was interested only in the diseases, not the patients who had them. Gradually he began to concentrate more on research and less on caring for the sick.

Young performed an extraordinary series of medical experiments, many of them with the object of explaining how the human eye works. He established that colour perception is the result of three separate types of receptors, each one sensitive to one of the three primary colours. Then, by

Figure 55 Thomas Young.

placing metal rings around a living eyeball, he showed that focusing did not require distortion of the whole eye, and postulated that the internal lens did all the work. His interest in optics led him towards physics, and another series of discoveries. He published 'The Undulatory Theory of Light', a classic paper on the nature of light; he created a new and better explanation of tides; he formally defined the concept of energy and he published groundbreaking papers on the subject of elasticity. Young seemed to be able to tackle problems in almost any subject, but this was not entirely to his advantage. His mind was so easily fascinated that he would leap from subject to subject, embarking on a new problem before polishing off the last one.

When Young heard about the Rosetta Stone, it became an irresistible challenge. In the summer of 1814 he set off on his annual holiday to the coastal resort of Worthing, taking with him a copy of the three inscriptions. Young's breakthrough came when he focused on a set of hieroglyphs surrounded by a loop, called a *cartouche*. His hunch was that these hieroglyphs were ringed because they represented something of great significance, possibly the name of the Pharaoh Ptolemy, because his Greek name, Ptolemaios, was mentioned in the Greek text. If this were the case, it would enable Young to discover the phonetics of the corresponding hieroglyphs, because a pharaoh's name would be pronounced roughly the same regardless of the language. The Ptolemy cartouche is repeated six times on the Rosetta Stone, sometimes in a so-called standard version, and sometimes in a longer, more elaborate version. Young assumed that

Table 13 Young's decipherment of (𓊪𓏏𓍯𓃭𓐝𓇌𓋴), the cartouche of Ptolemaios (standard version) from the Rosetta Stone.

Hieroglyph	Young's sound value	Actual sound value
𓊪	p	p
𓏏	t	t
𓍯	optional	o
𓃭	lo or ole	l
𓐝	ma or m	m
𓇌	i	i or y
𓋴	osh or os	s

the longer version was the name of Ptolemy with the addition of titles, so he concentrated on the symbols that appeared in the standard version, guessing sound values for each hieroglyph (Table 13).

Although he did not know it at the time, Young managed to correlate most of the hieroglyphs with their correct sound values. Fortunately, he had placed the first two hieroglyphs (,), which appeared one above the other, in their correct phonetic order. The scribe has positioned the hieroglyphs in this way for aesthetic reasons, at the expense of phonetic clarity. Scribes tended to write in such a way as to avoid gaps and maintain visual harmony; sometimes they would even swap letters around in direct contradiction to any sensible phonetic spelling, merely to increase the beauty of an inscription. After this decipherment, Young discovered a cartouche in an inscription copied from the temple of Karnak at Thebes which he suspected was the name of a Ptolemaic queen, Berenika (or Berenice). He repeated his strategy; the results are shown in Table 14.

Of the thirteen hieroglyphs in both cartouches, Young had identified half of them perfectly, and he got another quarter partly right. He had also correctly identified the feminine termination symbol, placed after the names of queens and goddesses. Although he could not have known the level of his success, the appearance of in both cartouches, representing i on both occasions, should have told Young that he was on the right track, and given him the confidence he needed to press ahead with further decipherments. However, his work suddenly ground to a halt. It

Table 14 Young's decipherment of , the cartouche of Berenika from the temple of Karnak.

Hieroglyph	Young's sound value	Actual sound value
	bir	b
	e	r
	n	n
	i	i
	optional	k
	ke or ken	a
	feminine termination	feminine termination

seems that he had too much reverence for Kircher's argument that hieroglyphs were semagrams, and he was not prepared to shatter that paradigm. He excused his own phonetic discoveries by noting that the Ptolemaic dynasty was descended from Lagus, a general of Alexander the Great. In other words, the Ptolemys were foreigners, and Young hypothesised that their names would have to be spelt out phonetically because there would not be a single natural semagram within the standard list of hieroglyphs. He summarised his thoughts by comparing hieroglyphs with Chinese characters, which Europeans were only just beginning to understand:

> It is extremely interesting to trace some of the steps by which alphabetic writing seems to have arisen out of hieroglyphical; a process which may indeed be in some measure illustrated by the manner in which the modern Chinese express a foreign combination of sounds, the characters being rendered simply 'phonetic' by an appropriate mark, instead of retaining their natural signification; and this mark, in some modern printed books, approaching very near to the ring surrounding the hieroglyphic names.

Young called his achievements 'the amusement of a few leisure hours'. He lost interest in hieroglyphics, and brought his work to a conclusion by summarising it in an article for the 1819 *Supplement to the Encyclopaedia Britannica.*

Meanwhile, in France a promising young linguist, Jean-François Champollion, was prepared to take Young's ideas to their natural conclusion. Although he was still only in his late twenties, Champollion had been fascinated by hieroglyphics for the best part of two decades. The obsession began in 1800, when the French mathematician Jean-Baptiste Fourier, who had been one of Napoleon's original Pekinese dogs, introduced the ten-year-old Champollion to his collection of Egyptian antiquities, many of them decorated with bizarre inscriptions. Fourier explained that nobody could interpret this cryptic writing, whereupon the boy promised that one day he would solve the mystery. Just seven years later, at the age of seventeen, he presented a paper entitled 'Egypt under the Pharaohs'. It was so innovative that he was immediately elected to the Academy in Grenoble. When he heard that he had become a teenage professor, Champollion was so overwhelmed that he immediately fainted.

Figure 56 Jean-François Champollion.

Champollion continued to astonish his peers, mastering Latin, Greek, Hebrew, Ethiopic, Sanskrit, Zend, Pahlevi, Arabic, Syrian, Chaldean, Persian and Chinese, all in order to arm himself for an assault on hieroglyphics. His obsession is illustrated by an incident in 1808, when he bumped into an old friend in the street. The friend casually mentioned that Alexandre Lenoir, a well-known Egyptologist, had published a complete decipherment of hieroglyphics. Champollion was so devastated that he collapsed on the spot. (He appears to have had quite a talent for fainting.) His whole reason for living seemed to depend on being the first to read the script of the ancient Egyptians. Fortunately for Champollion, Lenoir's decipherments were as fantastical as Kircher's seventeenth-century attempts, and the challenge remained.

In 1822, Champollion applied Young's approach to other cartouches. The British naturalist W. J. Bankes had brought an obelisk with Greek and hieroglyphic inscriptions to Dorset, and had recently published a lithograph of these bilingual texts, which included cartouches of Ptolemy and Cleopatra. Champollion obtained a copy, and managed to assign sound values to individual hieroglyphs (Table 15). The letters p, t, o, l and e are common to both names; in four cases they are represented by the same hieroglyph in both Ptolemy and Cleopatra, and only in one case, t, is there a discrepancy. Champollion assumed that the t sound could be represented by two hieroglyphs, just as the hard c sound in English can be

Table 15 Champollion's decipherment of 𓍹𓊪𓏏𓍯𓃭𓐝𓇌𓋴𓍺 and 𓍹𓈎𓃭𓇋𓍯𓊪𓄿𓂧𓂋𓄿𓍺, the cartouches of Ptolemaios and Cleopatra from the Bankes obelisk.

Hieroglyph	Sound Value
𓊪	p
𓏏	t
𓍯	o
𓃭	l
𓐝	m
𓇌	e
𓋴	s

Hieroglyph	Sound Value
𓈎	c
𓃭	l
𓇋	e
𓍯	o
𓊪	p
𓄿	a
𓂧	t
𓂋	r
𓄿	a

represented by c or k, as in 'cat' and 'kid'. Inspired by his success, Champollion began to address cartouches without a bilingual translation, substituting whenever possible the hieroglyph sound values that he had derived from the Ptolemy and Cleopatra cartouches. His first mystery cartouche (Table 16) contained one of the greatest names of ancient times. It was obvious to Champollion that the cartouche, which seemed to read a-l-?-s-e-?-t-r-?, represented the name alksentrs – Alexandros in Greek, or Alexander in English. It also became apparent to Champollion that the scribes were not fond of using vowels, and would often omit them; the scribes assumed that readers would have no problem filling in the missing vowels. With two new hieroglyphs under his belt, the young scholar studied other inscriptions and deciphered a series of cartouches. However, all this progress was merely extending Young's work. All these names, such as Alexander and Cleopatra, were still foreign, supporting the theory that phonetics was invoked only for words outside the traditional Egyptian lexicon.

Then, on 14 September 1822, Champollion received reliefs from the temple of Abu Simbel, containing cartouches that predated the period of Graeco-Roman domination. The significance of these cartouches was that they were old enough to contain traditional Egyptian names, yet they were still spelt out – clear evidence against the theory that spelling was used

Table 16 Champollion's decipherment of [cartouche], the cartouche of Alksentrs (Alexander).

Hieroglyph	Sound Value
	a
	l
	?
	s
	e
	?
	t
	r
	?

only for foreign names. Champollion concentrated on a cartouche containing just four hieroglyphs: 𓍹𓇳𓄟𓋴𓋴𓍺. The first two symbols were unknown, but the repeated pair at the end, 𓋴𓋴, were known from the cartouche of Alexander (**alksentrs**) to both represent the letter **s**. This meant that the cartouche represented (?-?-**s**-**s**). At this point, Champollion brought to bear his vast linguistic knowledge. Although Coptic, the direct descendant of the ancient Egyptian language, had ceased to be a living language in the eleventh century AD, it still existed in a fossilised form in the liturgy of the Christian Coptic Church. Champollion had learnt Coptic as a teenager, and was so fluent that he used it to record entries in his journal. However, until this moment, he had never considered that Coptic might also be the language of hieroglyphics.

Champollion wondered whether the first sign in the cartouche, ☉, might be a semagram representing the sun, i.e. a picture of the sun was the symbol for the word 'sun'. Then, in an act of intuitive genius, he assumed the sound value of the semagram to be that of the Coptic word for sun, **ra**. This gave him the sequence (**ra**-?-**s**-**s**). Only one pharaonic name seemed to fit. Allowing for the irritating omission of vowels, and assuming that the missing letter was **m**, then surely this had to be the name of Rameses, one of the greatest pharaohs, and one of the most ancient. The spell was broken. Even ancient traditional names were phonetically spelt. Champollion dashed into his brother's office and proclaimed 'Je tiens l'affaire!' ('I've got it!'), but once again his intense passion for hieroglyphics got the better of him. He promptly collapsed, and was bedridden for the next five days.

Champollion had demonstrated that the scribes sometimes exploited the rebus principle. In a rebus, still found in children's puzzles, long words are broken into their phonetic components, which are then represented by semagrams. For example, the word 'belief' can be broken down into two syllables, *be-lief*, which can then be rewritten as *bee-leaf*. Instead of writing the word alphabetically, it can be represented by the image of a bee followed by the image of a leaf. In the example discovered by Champollion, only the first syllable (**ra**) is represented by a rebus image, a picture of the sun, while the remainder of the word is spelt more conventionally.

The significance of the sun semagram in the Rameses cartouche is enormous, because it clearly restricts the possibilities for the language spoken by the scribes. For example, the scribes could not have spoken Greek, because this would have meant that the cartouche would be pronounced 'helios-meses'. The cartouche makes sense only if the scribes spoke a form of Coptic, because the cartouche would then be pronounced 'ra-meses'.

Although this was just one more cartouche, its decipherment clearly demonstrated the four fundamental principles of hieroglyphics. First, the language of the script is at least related to Coptic, and, indeed, examination of other hieroglyphics showed that it was Coptic pure and simple. Second, semagrams are used to represent some words, e.g., the word 'sun' is represented by a simple picture of the sun. Third, some long words are built wholly or partly using the rebus principle. Finally, for most of their writing, the ancient scribes relied on using a relatively conventional phonetic alphabet. This final point is the most important one, and Champollion called phonetics the 'soul' of hieroglyphics.

Using his deep knowledge of Coptic, Champollion began an unhindered and prolific decipherment of hieroglyphics beyond the cartouches. Within two years he identified phonetic values for the majority of hieroglyphs, and discovered that some of them represented combinations of two or even three consonants. This sometimes gave scribes the option of spelling a word using several simple hieroglyphs or with just a few multi-consonant hieroglyphs.

Champollion sent his initial results in a letter to Monsieur Dacier, the permanent secretary of the French Académie des Inscriptions. Then, in 1824, at the age of thirty-four, Champollion published all his achievements in a book entitled *Précis du système hiéroglyphique*. For the first time in fourteen centuries it was possible to read the history of the pharaohs, as written by their scribes. For linguists, here was an opportunity to study the evolution of a language and a script across a period of over three thousand years. Hieroglyphics could be understood and traced from the third millennium BC through to the fourth century AD. Furthermore, the evolution of hieroglyphics could be compared to the scripts of hieratic and demotic, which could now also be deciphered.

For several years, politics and envy prevented Champollion's magnificent achievement from being universally accepted. Thomas Young was a particularly bitter critic. On some occasions Young denied that hieroglyphics could be largely phonetic; at other times he accepted the argument, but complained that he himself had reached this conclusion before Champollion, and that the Frenchman had merely filled in the gaps. Much of Young's hostility resulted from Champollion's failure to give him any credit, even though it is likely that Young's initial breakthrough provided the inspiration for the full decipherment.

In July 1828 Champollion embarked on his first expedition to Egypt, which lasted eighteen months. It was a remarkable opportunity for him to see at first hand the inscriptions he had previously seen only in drawings or lithographs. Thirty years earlier, Napoleon's expedition had guessed wildly at the meaning of the hieroglyphs which adorned the temples, but now Champollion could simply read them character by character and reinterpret them correctly. His visit came just in time. Three years later, having written up the notes, drawings and translations from his Egyptian expedition, he suffered a severe stroke. The fainting spells he had suffered throughout his life were perhaps symptomatic of a more serious illness, exacerbated by his obsessive and intense study. He died on 4 March 1832, aged forty-one.

The Mystery of Linear B

In the two centuries since Champollion's breakthrough, Egyptologists have continued to improve their understanding of the intricacies of hieroglyphics. Their level of comprehension is now so high that scholars are able to unravel encrypted hieroglyphics, which are among the world's most ancient ciphertexts. Some of the inscriptions to be found on the tombs of the pharaohs were encrypted using a variety of techniques, including the substitution cipher. Sometimes fabricated symbols would be used in place of the established hieroglyph, and on other occasions a phonetically different but visually similar hieroglyph would be used instead of the correct one. For example, the horned asp hieroglyph, which usually represents f, might be used in place of the serpent, which represents z. Usually these encrypted epitaphs were not intended to be unbreakable, but rather they acted as cryptic puzzles to arouse the curiosity of passersby, who would thus be tempted to linger at a tomb rather than moving on.

Having conquered hieroglyphics, archaeologists went on to decipher many other ancient scripts, including the cuneiform texts of Babylon, the Kök-Turki runes of Turkey and the Brahmi alphabet of India. However, the good news for budding Champollions is that there are several outstanding scripts waiting to be solved, such as the Etruscan and Indus scripts (see Appendix I). The great difficulty in deciphering the remaining scripts is that there are no cribs, nothing which allows the codebreaker to prise open the meanings of these ancient texts. With Egyptian hieroglyphics it was the cartouches that acted as cribs, giving Young and Champollion their first taste of the underlying phonetic foundation. Without cribs, the decipherment of an ancient script might seem to be hopeless, yet there is one notable example of a script that was unravelled without the aid of a crib. Linear B, a Cretan script dating back to the Bronze Age, was deciphered without any helpful clues bequeathed by ancient scribes. It was solved by a combination of logic and inspiration, a potent example of pure cryptanalysis. Indeed, the decipherment of Linear B is generally regarded as the greatest of all archaeological decipherments.

The story of Linear B begins with excavations by Sir Arthur Evans, one of the most eminent archaeologists at the turn of the century. Evans was interested in the period of Greek history described by Homer in his twin epics, the *Iliad* and the *Odyssey*. Homer recounts the history of the Trojan War, the Greek victory at Troy and the ensuing exploits of the conquering hero Odysseus, events which supposedly occurred in the twelfth century BC. Some nineteenth-century scholars had dismissed Homer's epics as nothing more than legends, but in 1872 the German archaeologist Heinrich Schliemann uncovered the site of Troy itself, close to the western coast of Turkey, and suddenly Homer's myths became history. Between 1872 and 1900, archaeologists uncovered further evidence to suggest a rich period of pre-Hellenic history, predating the Greek classical age of Pythagoras, Plato and Aristotle by some six hundred years. The pre-Hellenic period lasted from 2800 to 1100 BC, and it was during the last four centuries of this period that the civilisation reached its peak. On the Greek mainland it was centred around Mycenae, where archaeologists uncovered a vast array of artefacts and treasures. However, Sir Arthur Evans had become perplexed by the failure of archaeologists to uncover any form of writing. He could not accept that such a sophisticated society could have been completely illiterate, and became determined to prove that the Mycenaean civilisation had some form of writing.

After meeting various Athenian dealers in antiquities, Sir Arthur eventually came across some engraved stones, which were apparently seals dating from the pre-Hellenic era. The signs on the seals seemed to be emblematic rather than genuine writing, similar to the symbolism used in heraldry. Yet this discovery gave him the impetus to continue his quest. The seals were said to originate from the island of Crete, and in particular Knossos, where legend told of the palace of King Minos, the centre of an empire that dominated the Aegean. Sir Arthur set out for Crete and began excavating in

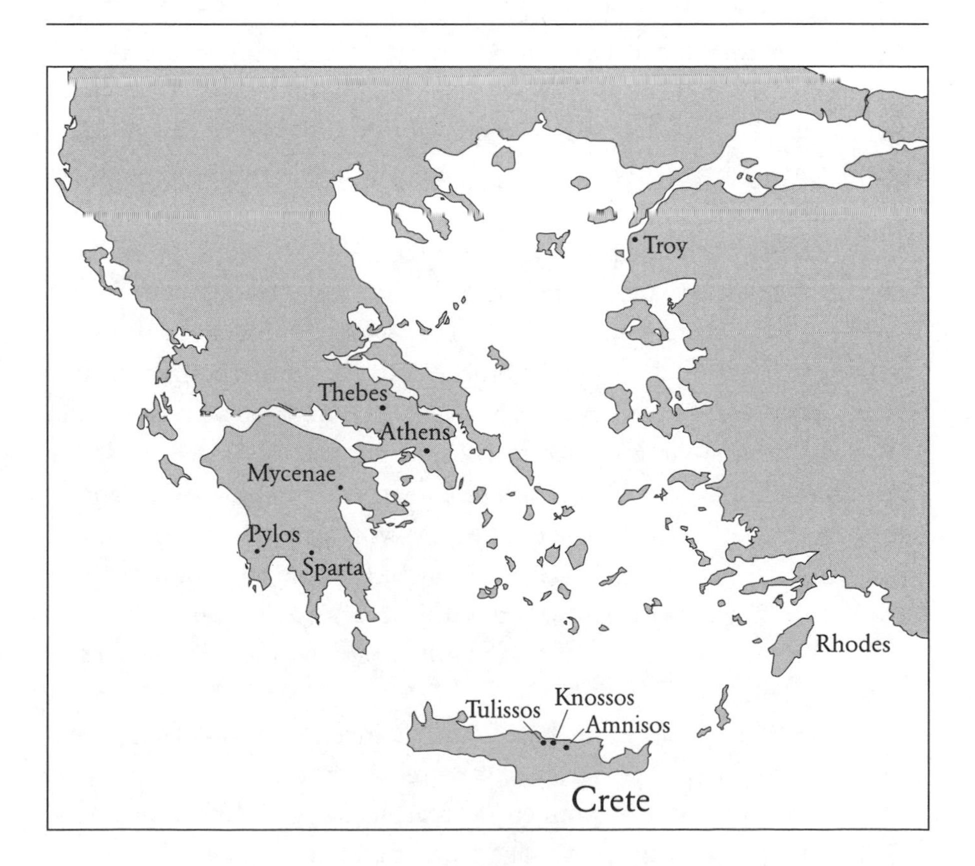

Figure 57 Ancient sites around the Aegean Sea. Having uncovered treasures at Mycenae on mainland Greece, Sir Arthur Evans went in search of inscribed tablets. The first Linear B tablets were discovered on the island of Crete, the centre of the Minoan empire.

March 1900. The results were as spectacular as they were rapid. He uncovered the remains of a luxurious palace, riddled with an intricate network of passageways and adorned with frescoes of young men leaping over ferocious bulls. Evans speculated that the sport of bull-jumping was somehow linked to the legend of the Minotaur, the bull-headed monster that fed on youths, and he suggested that the complexity of the palace passages had inspired the story of the Minotaur's labyrinth.

On 31 March, Sir Arthur began unearthing the treasure that he had desired most of all. Initially he discovered a single clay tablet with an inscription, then a few days later a wooden chest full of them, and then stockpiles of written material beyond all his expectations. All these clay tablets had originally been allowed to dry in the sun, rather than being fired, so that they could be recycled simply by adding water. Over the centuries, rain should have dissolved the tablets, and they should have been lost for ever. However, it appeared that the palace at Knossos had been destroyed by fire, baking the tablets and helping to preserve them for three thousand years. Their condition was so good that it was still possible to discern the fingerprints of the scribes.

The tablets fell into three categories. The first set of tablets, dating from 2000 to 1650 BC, consisted merely of drawings, probably semagrams, apparently related to the symbols on the seals that Sir Arthur Evans had bought from dealers in Athens. The second set of tablets, dating from 1750 to 1450 BC, were inscribed with characters that consisted of simple lines, and hence the script was dubbed Linear A. The third set of tablets, dating from 1450 to 1375 BC, bore a script which seemed to be a refinement of Linear A, and hence called Linear B. Because most of the tablets were Linear B, and because it was the most recent script, Sir Arthur and other archaeologists believed that Linear B gave them their best chance of decipherment.

Many of the tablets seemed to contain inventories. With so many columns of numerical characters it was relatively easy to work out the counting system, but the phonetic characters were far more puzzling. They looked like a meaningless collection of arbitrary doodles. The historian David Kahn described some of the individual characters as 'a Gothic arch enclosing a vertical line, a ladder, a heart with a stem running through it, a bent trident with a barb, a three-legged dinosaur looking behind him, an A with an extra horizontal bar running through it, a backward S, a tall beer glass, half full, with a bow tied on its rim; dozens look like nothing at all'. Only two useful facts could be established about Linear B. First, the direction of the writing was clearly from left to right, as any gap at the end of a line was generally on the right. Second, there were 90 distinct characters, which implied that the writing was almost certainly syllabic. Purely alphabetic scripts tend to have between 20 and 40 characters (Russian, for example, has 36 signs, and Arabic has 28). At the other extreme, scripts that rely on semagrams tend to have hundreds or even thousands of signs (Chinese has over 5,000). Syllabic scripts occupy the middle ground, with between 50 and 100 syllabic characters. Beyond these two facts, Linear B was an unfathomable mystery.

The fundamental problem was that nobody could be sure what language Linear B was written in. Initially, there was speculation that Linear B was a written form of Greek, because seven of the characters bore a close resemblance to characters in the classical Cypriot script, which was known to be a form of Greek script used between 600 and 200 BC. But doubts began to appear. The most common final consonant in Greek is **s**, and consequently the commonest final character in the Cypriot script is 𐠮, which represents the syllable **se** – because the characters are syllabic, a lone consonant has to be represented by a consonant–vowel combination, the vowel remaining silent. This same character also appears in Linear B, but it is rarely found at the end of a word, indicating that Linear B could not be Greek. The general consensus was that Linear B, the older script, represented an unknown and extinct language. When this language died out, the writing remained and evolved over the centuries into the Cypriot script, which was used to write Greek. Therefore, the two scripts looked similar but expressed totally different languages.

Sir Arthur Evans was a great supporter of the theory that Linear B was not a written form of Greek, and instead believed that it represented a native Cretan language. He was convinced that there was strong archaeological evidence to back up his argument. For example, his discoveries on the island of Crete suggested that the empire of King Minos, known as the Minoan empire, was far more advanced than the Mycenaean civilisation on the mainland. The Minoan Empire was not a dominion of the Mycenaean empire, but rather a rival, possibly even the dominant power.

Figure 58 A Linear B tablet, *c.* 1400 BC.

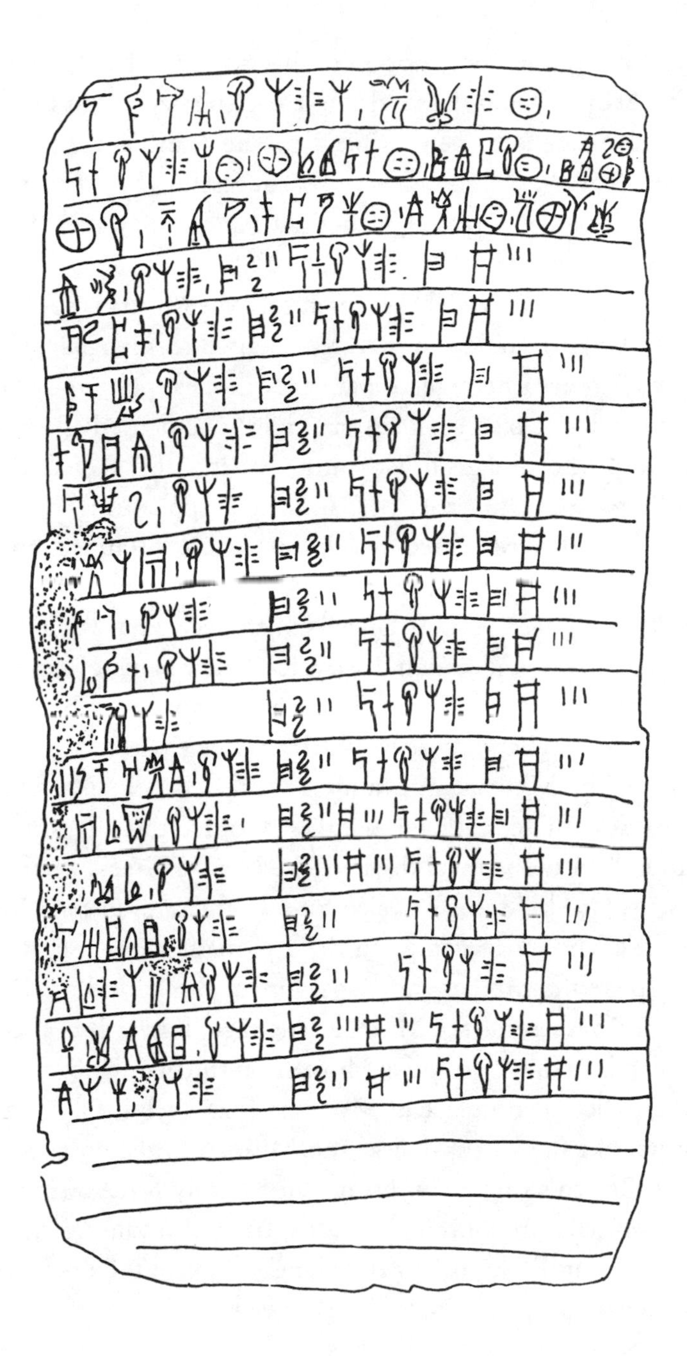

The myth of the Minotaur supported this position. The legend described how King Minos would demand that the Athenians send him groups of youths and maidens to be sacrificed to the Minotaur. In short, Evans concluded that the Minoans were so successful that they would have retained their native language, rather than adopting Greek, the language of their rivals.

Although it became widely accepted that the Minoans spoke their own non-Greek language (and Linear B represented this language), there were one or two heretics who argued that the Minoans spoke and wrote Greek. Sir Arthur did not take such dissent lightly, and used his influence to punish those who disagreed with him. When A.J.B. Wace, Professor of Archaeology at the University of Cambridge, spoke in favour of the theory that Linear B represented Greek, Sir Arthur excluded him from all excavations, and forced him to retire from the British School in Athens.

In 1939, the 'Greek vs. non-Greek' controversy grew when Carl Blegen of the University of Cincinnati discovered a new batch of Linear B tablets at the palace of Nestor at Pylos. This was extraordinary because Pylos is on the Greek mainland, and would have been part of the Mycenaean Empire, not the Minoan. The minority of archaeologists who believed that Linear B was Greek argued that this favoured their hypothesis: Linear B was found on the mainland where they spoke Greek, therefore Linear B represents Greek; Linear B is also found on Crete, so the Minoans also spoke Greek. The Evans camp ran the argument in reverse: the Minoans of Crete spoke the Minoan language; Linear B is found on Crete, therefore Linear B represents the Minoan language; Linear B is also found on the mainland, so they also spoke Minoan on the mainland. Sir Arthur was emphatic: 'There is no place at Mycenae for Greek-speaking dynasts . . . the culture, like the language, was still Minoan to the core.'

In fact, Blegen's discovery did not necessarily force a single language upon the Mycenaeans and the Minoans. In the Middle Ages, many European states, regardless of their native language, kept their records in Latin. Perhaps the language of Linear B was likewise a lingua franca among the accountants of the Aegean, allowing ease of commerce between nations who did not speak a common language.

For four decades, all attempts to decipher Linear B ended in failure. Then, in 1941, at the age of ninety, Sir Arthur died. He did not live to witness the decipherment of Linear B, or to read for himself the meanings of the texts he had discovered. Indeed, at this point, there seemed little prospect of ever deciphering Linear B.

Bridging Syllables

After the death of Sir Arthur Evans the Linear B archive of tablets and his own archaeological notes were available only to a restricted circle of archaeologists, namely those who supported his theory that Linear B represented a distinct Minoan language. However, in the mid-1940s, Alice Kober, a classicist at Brooklyn College, managed to gain access to the material, and began a meticulous and impartial analysis of the script. To

Figure 59 Alice Kober.

those who knew her only in passing, Kober seemed quite ordinary – a dowdy professor, neither charming nor charismatic, with a rather matter-of-fact approach to life. However, her passion for her research was immeasurable. 'She worked with a subdued intensity', recalls Eva Brann, a former student who went on to become an archaeologist at Yale University. 'She once told me that the only way to know when you have done something truly great is when your spine tingles.'

In order to crack Linear B, Kober realised that she would have to abandon all preconceptions. She focused on nothing else but the structure of the overall script and the construction of individual words. In particular, she noticed that certain words formed triplets, inasmuch as they seemed to be the same word reappearing in three slightly varied forms. Within a word triplet, the stems were identical but there were three possible endings. She concluded that Linear B represented a highly inflective language, meaning that word endings are changed in order to reflect gender, tense, case and so on. English is slightly inflective because, for example, we say 'I decipher, you decipher, he deciphers' – in the third person the verb takes an 's'. However, older languages tend to be much more rigid and extreme in their use of such endings. Kober published a paper in which she described the inflective nature of two particular groups of words, as shown in Table 17, each group retaining its respective stems, while taking on different endings according to three different cases.

For ease of discussion, each Linear B symbol was assigned a two-digit number, as shown in Table 18. Using these numbers, the words in Table 17 can be rewritten as in Table 19. Both groups of words could be nouns

Table 17 Two inflective words in Linear B.

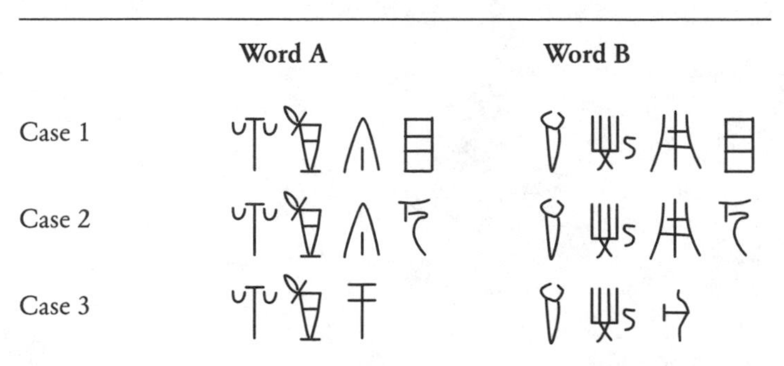

Table 18 Linear B signs and the numbers assigned to them.

01	30	59
02	31	60
03	32	61
04	33	62
05	34	63
06	35	64
07	36	65
08	37	66
09	38	67
10	39	68
11	40	69
12	41	70
13	42	71
14	43	72
15	44	73
16	45	74
17	46	75
18	47	76
19	48	77
20	49	78
21	50	79
22	51	80
23	52	81
24	53	82
25	54	83
26	55	84
27	56	85
28	57	86
29	58	87

changing their ending according to their case – case 1 could be nominative, case 2 accusative, and case 3 dative, for example. It is clear that the first two signs in both groups of words (**25-67-** and **70-52-**) are both stems, as they are repeated regardless of the case. However, the third sign is somewhat more puzzling. If the third sign is part of the stem, then for a given word it should remain constant, regardless of the case, but this does not happen. In word A the third sign is **37** for cases 1 and 2, but **05** for case 3. In word B the third sign is **41** for cases 1 and 2, but **12** for case 3. Alternatively if the third sign is not part of the stem, perhaps it is part of the ending, but this possibility is equally problematic. For a given case the ending should be the same regardless of the word, but for cases 1 and 2 the third sign is **37** in word A, but **41** in word B, and for case 3 the third sign is **05** in word A, but **12** in word B.

The third signs defied expectations because they did not seem to be part of the stem or the ending. Kober resolved the paradox by invoking the theory that every sign represents a syllable, presumably a combination of a consonant followed by a vowel. She proposed that the third syllable could be a bridging syllable, representing part of the stem and part of the ending. The consonant could contribute to the stem, and the vowel to the ending. To illustrate her theory, she gave an example from the Akkadian language, which also has bridging syllables and which is highly inflective. *Sadanu* is a case 1 Akkadian noun, which changes to *sadani* in the second case and *sadu* in the third case (Table 20). It is clear that the three words consist of a stem, **sad-**, and an ending, **-anu** (case 1), **-ani** (case 2), or **-u** (case 3), with **-da-**, **-da-** or **-du** as the bridging syllable. The bridging syllable is the same in cases 1 and 2, but different in case 3. This is exactly the

Table 19 The two inflective Linear B words rewritten in numbers.

	Word A	**Word B**
Case 1	25-67-37-57	70-52-41-57
Case 2	25-67-37-36	70-52-41-36
Case 3	25-67-05	70-52-12

pattern observed in the Linear B words – the third sign in each of Kober's Linear B words must be a bridging syllable.

Merely identifying the inflective nature of Linear B and the existence of bridging syllables meant that Kober had progressed further than anybody else in deciphering the Minoan script, and yet this was just the beginning. She was about to make an even greater deduction. In the Akkadian example, the bridging syllable changes from *-da-* to *-du*, but the consonant is the same in both syllables. Similarly, the Linear B syllables **37** and **05** in word A must share the same consonant, as must syllables **41** and **12** in word B. For the first time since Evans had discovered Linear B, facts were beginning to emerge about the phonetics of the characters. Kober could also establish another set of relationships among the characters. It is clear that Linear B words A and B in case 1 should have the same ending. However, the bridging syllable changes from **37** to **41**. This implies that signs **37** and **41** represent syllables with different consonants but identical vowels. This would explain why the signs are different, while maintaining the same ending for both words. Similarly for the case 3 nouns, the syllables **05** and **12** will have a common vowel but different consonants.

Kober was not able to pinpoint exactly which vowel is common to **05** and **12**, and to **37** and **41**; similarly, she could not identify exactly which consonant is common to **37** and **05**, and which to **41** and **12**. However, regardless of their absolute phonetic values, she had established rigorous relationships between certain characters. She summarised her results in the form of a grid, as in Table 21. What this is saying is that Kober had no idea which syllable was represented by sign **37**, but she knew that its

Table 20 Bridging syllables in the Akkadian noun *sadanu*.

Case 1	sa-da-nu
Case 2	sa-da-ni
Case 3	sa-du

consonant was shared with sign 05 and its vowel with sign 41. Similarly, she had no idea which syllable was represented by sign 12, but knew that its consonant was shared with sign 41 and its vowel with sign 05. She applied her method to other words, eventually constructing a grid of ten signs, two vowels wide and five consonants deep. It is quite possible that Kober would have taken the next crucial step in decipherment, and could even have cracked the entire script. However, she did not live long enough to exploit the repercussions of her work. In 1950, at the age of forty-three, she died of lung cancer.

A Frivolous Digression

Just a few months before she died, Alice Kober received a letter from Michael Ventris, an English architect who had been fascinated by Linear B ever since he was a child. Ventris was born on 12 July 1922, the son of an English Army officer and his half-Polish wife. His mother was largely responsible for encouraging an interest in archaeology, regularly escorting him to the British Museum where he could marvel at the wonders of the ancient world. Michael was a bright child, with an especially prodigious talent for languages. When he began his schooling he went to Gstaad in Switzerland, and became fluent in French and German. Then, at the age of six, he taught himself Polish.

Like Jean-François Champollion, Ventris developed an early love of ancient scripts. At the age of seven he studied a book on Egyptian hieroglyphics, an impressive achievement for one so young, particularly as the book was written in German. This interest in the writings of ancient civilisations continued throughout his childhood. In 1936, at the age of fourteen, it was further ignited when he attended a lecture given by Sir

Table 21 Kober's grid for relationships between Linear B characters.

	Vowel 1	Vowel 2
Consonant I	37	05
Consonant II	41	12

Arthur Evans, the discoverer of Linear B. The young Ventris learned about the Minoan civilisation and the mystery of Linear B, and promised himself that he would decipher the script. That day an obsession was born that remained with Ventris throughout his short but brilliant life.

At the age of just eighteen, he summarised his initial thoughts on Linear B in an article that was subsequently published in the highly respected *American Journal of Archaeology*. When he submitted the article, he had been careful to withhold his age from the journal's editors for fear of not being taken seriously. His article very much supported Sir Arthur's criticism of the Greek hypothesis, stating that 'The theory that Minoan could be Greek is based of course upon a deliberate disregard for historical plausibility.' His own belief was that Linear B was related to Etruscan, a reasonable standpoint because there was evidence that the Etruscans had come from the Aegean before settling in Italy. Although his article made no stab at decipherment, he confidently concluded: 'It can be done.'

Ventris became an architect rather than a professional archaeologist, but remained passionate about Linear B, devoting all of his spare time to studying every aspect of the script. When he heard about Alice Kober's work, he was keen to learn about her breakthrough, and he wrote to her asking for more details. Although she died before she could reply, her ideas lived on in her publications, and Ventris studied them meticulously. He fully appreciated the power of Kober's grid, and attempted to find new words with shared stems and bridging syllables. He extended her grid to include these new signs, encompassing other vowels and consonants. Then, after a year of intense study, he noticed something peculiar – something that seemed to suggest an exception to the rule that all Linear B signs are syllables.

It had been generally agreed that each Linear B sign represented a combination of a consonant with a vowel (CV), and hence spelling would require a word to be broken up into CV components. For example, the English word minute would be spelt as mi-nu-te, a series of three CV syllables. However, many words do not divide conveniently into CV syllables. For example, if we break the word 'visible' into pairs of letters we get vi-si-bl-e, which is problematic because it does not consist of a simple series of CV syllables: there is a double-consonant syllable and a spare -e at the end. Ventris assumed that the Minoans overcame this problem by inserting a silent i to create a cosmetic -bi- syllable, so that the word can now be written as vi-si-bi-le, which is a combination of CV syllables.

However, the word invisible remains problematic. Once again it is necessary to insert silent vowels, this time after the n and the b, turning them into CV syllables. Furthermore, it is also necessary to deal with the single vowel i at the beginning of the word: i-ni-vi-si-bi-le. The initial i cannot easily be turned into a CV syllable because inserting a silent consonant at the start of a word could easily lead to confusion. In short, Ventris concluded that there must be Linear B signs that represent single vowels, to be used in words that begin with a vowel. These signs should be easy to spot because they would appear only at the beginning of words. Ventris worked out how often each sign appears at the beginning, middle

Figure 60 Michael Ventris.

and end of any word. He observed that two particular signs, 08 and 61, were predominantly found at the beginning of words, and concluded that they did not represent syllables, but single vowels.

Ventris published his ideas about vowel signs, and his extensions to the grid, in a series of Work Notes, which he sent out to other Linear B researchers. On 1 June 1952 he published his most significant result, Work Note 20, a turning point in the decipherment of Linear B. He had spent the last two years expanding Kober's grid into the version shown in Table 22. The grid consisted of 5 vowel columns and 15 consonant rows, giving 75 cells in total, with 5 additional cells available for single vowels. Ventris had inserted signs in about half the cells. The grid is a treasure trove of information. For example, from the sixth row it is possible to tell

Table 22 Ventris's expanded grid for relationships between Linear B characters. Although the grid doesn't specify vowels or consonants, it does highlight which characters share common vowels and consonants. For example, all the characters in the first column share the same vowel, labelled 1.

		Vowels				
		1	2	3	4	5
Consonants	I					57
	II	40		75		54
	III	39				03
	IV		36			
	V		14			01
	VI	37	05		69	
	VII	41	12			31
	VIII	30	52	24	55	06
	IX	73	15			80
	X		70	44		
	XI	53				76
	XII		02	27		
	XIII					
	XIV			13		
	XV		32	78		
	Pure vowels		61			08

that the syllabic signs **37**, **05** and **69** share the same consonant, VI, but contain different vowels, 1, 2 and 4. Ventris had no idea of the exact values of consonant VI or vowels 1, 2 and 4, and until this point he had resisted the temptation of assigning sound values to any of the signs. However, he felt that it was now time to follow some hunches, guess a few sound values and examine the consequences.

Ventris had noticed three words that appeared over and over again on several of the Linear B tablets: **08-73-30-12**, **70-52-12** and **69-53-12**. Based on nothing more than intuition, he conjectured that these words might be the names of important towns. Ventris had already speculated that sign **08** was a vowel, and therefore the name of the first town had to begin with a vowel. The only significant name that fitted the bill was Amnisos, an important harbour town. If he was right, then the second and third signs, **73** and **30**, would represent **-mi-** and **-ni-**. These two syllables both contain the same vowel, **i**, so numbers **73** and **30** ought to appear in the same vowel column of the grid. They do. The final sign, **12**, would represent **-so-**, leaving nothing to represent the final **s**. Ventris decided to ignore the problem of the missing final **s** for the time being, and proceeded with the following working translation:

Town 1 = **08-73-30-12** = **a-mi-ni-so** = Amnisos

This was only a guess, but the repercussions on Ventris's grid were enormous. For example, the sign **12**, which seems to represent **-so-**, is in the second vowel column and the seventh consonant row. Hence, if his guess was correct, then all the other syllabic signs in the second vowel column would contain the vowel **o**, and all the other syllabic signs in the seventh consonant row would contain the consonant **s**.

When Ventris examined the second town, he noticed that it also contained sign **12**, **-so-**. The other two signs, **70** and **52**, were in the same vowel column as **-so-**, which implied that these signs also contained the vowel **o**. For the second town he could insert the **-so-**, the **o** where appropriate, and leave gaps for the missing consonants, leading to the following:

Town 2 = **70-52-12** = **?o-?o-so** = ?

Could this be Knossos? The signs could represent **ko-no-so**. Once again, Ventris was happy to ignore the problem of the missing final **s**, at least for the time being. He was pleased to note that sign **52**, which supposedly represented **-no-**, was in the same consonant row as sign **30**, which supposedly represented **-ni-** in Amnisos. This was reassuring, because if they contain the same consonant, **n**, then they should indeed be in the same consonant row. Using the syllabic information from Knossos and Amnisos, he inserted the following letters into the third town:

Town 3 = **69-53-12** = ??-?**i-so**

The only name that seemed to fit was Tulissos (**tu-li-so**), an important town in central Crete. Once again the final **s** was missing, and once again Ventris ignored the problem. He had now tentatively identified three place names and the sound values of eight different signs:

Town 1 = **08-73-30-12**	= **a-mi-ni-so**	=	Amnisos
Town 2 = **70-52-12**	= **ko-no-so**	=	Knossos
Town 3 = **69-53-12**	= **tu-li-so**	=	Tulissos

The repercussions of identifying eight signs were enormous. Ventris could infer consonant or vowel values to many of the other signs in the grid, if they were in the same row or column. The result was that many signs revealed part of their syllabic meaning, and a few could be fully identified. For example, sign **05** is in the same column as **12** (**so**), **52** (**no**) and **70** (**ko**), and so must contain **o** as its vowel. By a similar process of reasoning, sign **05** is in the same row as sign **69** (**tu**), and so must contain **t** as its consonant. In short, the sign **05** represents the syllable **-to-**. Turning to sign **31**, it is in the same column as sign **08**, the **a** column, and it is in the same row as sign **12**, the **s** row. Hence sign **31** represents the syllable **-sa-**.

Deducing the syllabic values of these two signs, **05** and **31**, was particularly important because it allowed Ventris to read two complete words, **05-12** and **05-31**, which often appeared at the bottom of inventories. Ventris already knew that sign **12** represented the syllable **-so-**, because this sign appeared in the word for Tulissos, and hence **05-12** could be read as **to-so**. And the other word, **05-31**, could be read as **to-sa**. This was an astonishing result. Because these words were found at the bottom of inventories, experts had suspected that they meant 'total'. Ventris now read them as **toso** and **tosa**, uncannily similar to the archaic Greek *tossos* and *tossa*, masculine and feminine forms meaning 'so much'. Ever since he was fourteen years old, from the moment he had heard Sir Arthur Evans's talk, he had believed that the language of the Minoans could not be Greek. Now, he was uncovering words which were clear evidence in favour of Greek as the language of Linear B.

It was the ancient Cypriot script that provided some of the earliest evidence against Linear B being Greek, because it suggested that Linear B words rarely end in s, whereas this is a very common ending for Greek words. Ventris had discovered that Linear B words do, indeed, rarely end in s, but perhaps this was simply because the s was omitted as part of some writing convention. Amnisos, Knossos, Tulissos and *tossos* were all spelt without a final s, indicating that the scribes simply did not bother with the final s, allowing the reader to fill in the obvious omission.

Ventris soon deciphered a handful of other words, which also bore a resemblance to Greek, but he was still not absolutely convinced that Linear B was a Greek script. In theory, the few words that he had deciphered could all be dismissed as imports into the Minoan language. A foreigner arriving at a British hotel might overhear such words as 'rendezvous' or 'bon appetit', but would be wrong to assume that the British speak French. Furthermore, Ventris came across words that made no sense to him, providing some evidence in favour of a hitherto unknown language. In Work Note 20 he did not ignore the Greek hypothesis, but he did label it 'a frivolous digression'. He concluded: 'If pursued, I suspect that this line of decipherment would sooner or later come to an impasse, or dissipate itself in absurdities.'

Despite his misgivings, Ventris did pursue the Greek line of attack. While Work Note 20 was still being distributed, he began to discover more Greek words. He could identify *poimen* (shepherd), *kerameus* (potter), *khrusoworgos* (goldsmith) and *khalkeus* (bronzesmith), and he even translated a couple of complete phrases. So far, none of the threatened absurdities blocked his path. For the first time in three thousand years, the silent script of Linear B was whispering once again, and the language it spoke was undoubtedly Greek.

During this period of rapid progress, Ventris was coincidentally asked to appear on BBC radio to discuss the mystery of the Minoan scripts. He decided that this would be an ideal opportunity to go public with his discovery. After a rather prosaic discussion of Minoan history and Linear B, he made his revolutionary announcement: 'During the last few weeks, I have come to the conclusion that the Knossos and Pylos tablets must, after all, be written in Greek – a difficult and archaic Greek, seeing that it is five hundred years older than Homer and written in a rather abbreviated form, but Greek nevertheless.' One of the listeners was John Chadwick, a Cambridge researcher who had been interested in the decipherment of Linear B since the 1930s. During the war he had spent time as a cryptanalyst in Alexandria, where he broke Italian ciphers, before moving to Bletchley Park, where he attacked Japanese ciphers. After the war he tried once again to decipher Linear B, this time employing the

Figure 61 John Chadwick.

techniques he had learnt while working on military codes. Unfortunately, he had little success.

When he heard the radio interview, he was completely taken aback by Ventris's apparently preposterous claim. Chadwick, along with the majority of scholars listening to the broadcast, dismissed the claim as the work of an amateur – which indeed it was. However, as a lecturer in Greek, Chadwick realised that he would be pelted with questions regarding Ventris's claim, and to prepare for the barrage he decided to investigate Ventris's argument in detail. He obtained copies of Ventris's Work Notes, and examined them, fully expecting them to be full of holes. However, within a few days the sceptical scholar became one of the first supporters of Ventris's Greek theory of Linear B. Chadwick soon came to admire the young architect:

> His brain worked with astonishing rapidity, so that he could think out all the implications of a suggestion almost before it was out of your mouth. He had a keen appreciation of the realities of the situation; the Mycenaeans were to him no vague abstractions, but living people whose thoughts he could penetrate. He himself laid stress on the visual approach to the problem; he made himself so familiar with the visual aspect of the texts that large sections were imprinted on his mind simply as visual patterns, long before the decipherment gave them meaning. But a merely photographic memory was not enough, and it was here that his architectural training came to his aid. The architect's eye sees in a building not a mere façade, a jumble of ornamental and structural features: it looks beneath the appearance and distinguishes the significant parts of the pattern, the structural elements and framework of the building. So too Ventris was able to discern among the bewildering variety of the mysterious signs, patterns and regularities which betrayed the underlying structure. It is this quality, the power of seeing order in apparent confusion, that has marked the work of all great men.

However, Ventris lacked one particular expertise, namely a thorough knowledge of archaic Greek. Ventris's only formal education in Greek was as a boy at Stowe School, so he could not fully exploit his breakthrough. For example, he was unable to explain some of the deciphered words because they were not part of his Greek vocabulary. Chadwick's speciality was Greek philology, the study of the historical evolution of the Greek language, and he was therefore well equipped to show that these problematic words fitted in with theories of the most ancient forms of Greek. Together, Chadwick and Ventris formed a perfect partnership.

The Greek of Homer is three thousand years old, but the Greek of Linear B is five hundred years older still. In order to translate it, Chadwick needed to extrapolate back from the established ancient Greek to the words of Linear B, taking into account the three ways in which language develops. First, pronunciation evolves with time. For example, the Greek word for 'bath-pourers' changes from *lewotrokhowoi* in Linear B to *loutrokhooi* by the time of Homer. Second, there are changes in grammar. For example, in Linear B the genitive ending is *-oio*, but this is replaced in classical Greek by *-ou*. Finally, the lexicon can change dramatically. Some words are born, some die, others change their meaning. In Linear B *harmo* means 'wheel', but in later Greek the same word means 'chariot'. Chadwick pointed out that this is similar to the use of 'wheels' to mean a car in modern English.

With Ventris's deciphering skills and Chadwick's expertise in Greek, the duo went on to convince the rest of the world that Linear B is indeed Greek. The rate of translation accelerated as each day passed. In Chadwick's account of their work, *The Decipherment of Linear B*, he writes:

> Cryptography is a science of deduction and controlled experiment; hypotheses are formed, tested and often discarded. But the residue which passes the test grows until finally there comes a point when the experimenter feels solid ground beneath his feet: his hypotheses cohere, and fragments of sense emerge from their camouflage. The code 'breaks'. Perhaps this is best defined as the point when the likely leads appear faster than they can be followed up. It is like the initiation of a chain-reaction in atomic physics; once the critical threshold is passed, the reaction propagates itself.

It was not long before they were able to demonstrate their mastery of the script by writing short notes to each other in Linear B.

An informal test for the accuracy of a decipherment is the number of gods in the text. In the past, those who were on the wrong track would, not surprisingly, generate nonsensical words, which would be explained away as being the names of hitherto unknown deities. However, Chadwick and Ventris claimed only four divine names, all of which were well-established gods.

In 1953, confident of their analysis, they wrote up their work in a paper, modestly entitled 'Evidence for Greek Dialect in the Mycenaean Archives', which was published in *The Journal of Hellenic Studies*. Thereafter, archaeologists around the world began to realise that they were witnessing a revolution. In a letter to Ventris, the German scholar Ernst Sittig summarised the mood of the academic community: 'I repeat: your demonstrations are cryptographically the most interesting I have yet heard of, and are really fascinating. If you are right, the methods of the archaeology, ethnology, history and philology of the last fifty years are reduced *ad absurdum*.'

The Linear B tablets contradicted almost everything that had been claimed by Sir Arthur Evans and his generation. First of all was the simple fact that Linear B was Greek. Second, if the Minoans on Crete wrote Greek and presumably spoke Greek, this would force archaeologists to reconsider their views of Minoan history. It now seemed that the dominant force in the region was Mycenae, and Minoan Crete was a lesser state whose people spoke the language of their more powerful neighbours. However, there is evidence that, before 1450 BC, Minoa was a truly independent state with its own language. It was in around 1450 BC that Linear B replaced Linear A, and although the two scripts look very similar, nobody has yet deciphered Linear A. Linear A therefore probably represents a distinctly different language from Linear B. It seems likely that in roughly 1450 BC the Mycenaeans conquered the Minoans, imposed their own language, and transformed Linear A into Linear B so that it functioned as a script for Greek.

As well as clarifying the broad historical picture, the decipherment of Linear B also fills in some detail. For example, excavations at Pylos have failed to uncover any precious objects in the lavish palace, which was ultimately destroyed by fire. This has led to the suspicion that the palace was deliberately torched by invaders, who first stripped it of valuables. Although the Linear B tablets at Pylos do not specifically describe such an attack, they do hint at preparations for an invasion. One tablet describes the setting up of a special military unit to protect the coast, while another describes the commandeering of bronze ornaments for converting into spearheads. A third tablet, untidier than the other two, describes a particularly elaborate temple ritual, possibly involving human sacrifice. Most Linear B tablets are neatly laid out, implying that scribes would begin with a rough draft which would later be destroyed. The untidy tablet has

Table 23 Linear B signs with their numbers and sound values.

No.	Value	No.	Value	No.	Value
01	*da*	30	*ni*	59	*ta*
02	*ro*	31	*sa*	60	*ra*
03	*pa*	32	*qo*	61	*o*
04	*te*	33	ra_2	62	*pte*
05	*to*	34		63	
06	*na*	35		64	
07	*di*	36	*jo*	65	*ju*
08	*a*	37	*ti*	66	ta_2
09	*se*	38	*e*	67	*ki*
10	*u*	39	*pi*	68	ro_2
11	*po*	40	*wi*	69	*tu*
12	*so*	41	*si*	70	*ko*
13	*me*	42	*wo*	71	*dwe*
14	*do*	43	*ai*	72	*pe*
15	*mo*	44	*ke*	73	*mi*
16	pa_2	45	*de*	74	*ze*
17	*za*	46	*je*	75	*we*
18		47		76	ra_2
19		48	*nwa*	77	*ka*
20	*zo*	49		78	*qe*
21	*qi*	50	*pu*	79	*zu*
22		51	*du*	80	*ma*
23	*mu*	52	*no*	81	*ku*
24	*ne*	53	*ri*	82	
25	a_2	54	*wa*	83	
26	*ru*	55	*nu*	84	
27	*re*	56	pa_3	85	
28	*i*	57	*ja*	86	
29	pu_2	58	*su*	87	

large gaps, half-empty lines and text that spills over to the other side. One possible explanation is that the tablet recorded a bid to invoke divine intervention in the face of an invasion, but before the tablet could be redrafted the palace was overrun.

The bulk of Linear B tablets are inventories, and as such they describe everyday transactions. They indicate the existence of a bureaucracy to rival any in history, with tablets recording details of manufactured goods and agricultural produce. Chadwick likened the archive of tablets to the Domesday Book, and Professor Denys Page described the level of detail thus: 'Sheep may be counted up to a glittering total of twenty-five thousand; but there is still purpose to be served by recording the fact that *one* animal was contributed by Komawens . . . One would suppose that not a seed could be sown, not a gram of bronze worked, not a cloth woven, not a goat reared or a hog fattened, without the filling of a form in the Royal Palace.' These palace records might seem mundane, but they are inherently romantic because they are so intimately associated with the *Odyssey* and *Iliad*. While scribes in Knossos and Pylos recorded their daily transactions, the Trojan War was being fought. The language of Linear B is the language of Odysseus.

On 24 June 1953, Ventris gave a public lecture outlining the decipherment of Linear B. The following day it was reported in *The Times*, next to a comment on the recent conquest of Everest. This led to Ventris and Chadwick's achievement being known as the 'Everest of Greek Archaeology'. The following year, the men decided to write an authoritative three-volume account of their work which would include a description of the decipherment, a detailed analysis of three hundred tablets, a dictionary of 630 Mycenaean words and a list of sound values for nearly all Linear B signs, as given in Table 23. *Documents in Mycenaean Greek* was completed in the summer of 1955, and was ready for publication in the autumn of 1956. However, a few weeks before printing, on 6 September 1956, Michael Ventris was killed. While driving home late at night on the Great North Road near Hatfield, his car collided with a lorry. John Chadwick paid tribute to his colleague, a man who matched the genius of Champollion, and who also died at a tragically young age: 'The work he did lives, and his name will be remembered so long as the ancient Greek language and civilisation are studied.'

6 Alice and Bob Go Public

During the Second World War, British codebreakers had the upper hand over German codemakers, mainly because the men and women at Bletchley Park, following the lead of the Poles, developed some of the earliest codebreaking technology. In addition to Turing's bombes, which were used to crack the Enigma cipher, the British also invented another codebreaking device, Colossus, to combat an even stronger form of encryption, namely the German Lorenz cipher. Of the two types of code breaking machine, it was Colossus that would determine the development of cryptography during the latter half of the twentieth century.

The Lorenz cipher was used to encrypt communications between Hitler and his generals. The encryption was performed by the Lorenz SZ40 machine, which operated in a similar way to the Enigma machine, but the Lorenz was far more complicated, and it provided the Bletchley codebreakers with an even greater challenge. However, two of Bletchley's codebreakers, John Tiltman and Bill Tutte, discovered a weakness in the way that the Lorenz cipher was used, a flaw that Bletchley could exploit and thereby read Hitler's messages.

Breaking the Lorenz cipher required a mixture of searching, matching, statistical analysis and careful judgement, all of which was beyond the technical abilities of the bombes. The bombes were able to carry out a specific task at high speed, but they were not flexible enough to deal with the subtleties of Lorenz. Lorenz-encrypted messages had to be broken by hand, which took weeks of painstaking effort, by which time the messages were largely out of date. Eventually, Max Newman, a Bletchley mathematician, came up with a way to mechanise the cryptanalysis of the Lorenz cipher. Drawing heavily on Alan Turing's concept of the universal machine, Newman designed a machine that was capable of adapting itself to different problems, what we today would call a programmable computer.

Implementing Newman's design was deemed technically impossible, so Bletchley's senior officials shelved the project. Fortunately, Tommy Flowers, an engineer who had taken part in discussions about Newman's design, decided to ignore Bletchley's scepticism, and went ahead with building the machine. At the Post Office's research centre at Dollis Hill, North London, Flowers took Newman's blueprint and spent ten months turning it into the Colossus machine, which he delivered to Bletchley Park on 8 December 1943. It consisted of 1,500 electronic valves, which were considerably faster than the sluggish electromechanical relay switches used in the bombes. But more important than Colossus's speed was the fact that it was programmable. It was this fact that made Colossus the precursor to the modern digital computer.

Colossus, as with everything else at Bletchley Park, was destroyed after the war, and those who worked on it were forbidden to talk about it. When Tommy Flowers was ordered to dispose of the Colossus blueprints, he obediently took them down to the boiler room and burnt them. The plans for the world's first computer were lost for ever. This secrecy meant that other scientists gained the credit for the invention of the computer. In 1945, J. Presper Eckert and John W. Mauchly of the University of Pennsylvania completed ENIAC (Electronic Numerical Integrator And Calculator), consisting of 18,000 electronic valves, capable of performing 5,000 calculations per second. For decades, ENIAC, not Colossus, was considered the mother of all computers.

Having contributed to the birth of the modern computer, cryptanalysts continued after the war to develop and employ computer technology in order to break all sorts of ciphers. They could now exploit the speed and flexibility of programmable computers to search through all possible keys until the correct one was found. In due course, the cryptographers began to fight back, exploiting the power of computers to create increasingly complex ciphers. In short, the computer played a crucial role in the post-war battle between codemakers and codebreakers.

Using a computer to encipher a message is, to a large extent, very similar to traditional forms of encryption. Indeed, there are only three significant differences between computer encryption and the sort of mechanical encryption that was the basis for ciphers like Enigma. The first difference is that a mechanical cipher machine is limited by what can be practically built, whereas a computer can mimic a hypothetical cipher machine of immense complexity. For example, a computer could be programmed to mimic the action of a hundred scramblers, some spinning clockwise, some anticlockwise, some vanishing after every tenth letter, others rotating faster and faster as encryption progresses. Such a mechanical machine would be practically impossible to build, but its 'virtual' computerised equivalent would deliver a highly secure cipher.

The second difference is simply a matter of speed. Electronics can operate far more quickly than mechanical scramblers: a computer programmed to mimic the Enigma cipher could encipher a lengthy message in an instant. Alternatively, a computer programmed to perform a vastly more complex form of encryption could still accomplish the task within a reasonable time.

The third, and perhaps most significant, difference is that a computer scrambles numbers rather than letters of the alphabet. Computers deal only in binary numbers – sequences of ones and zeros known as *binary digits*, or *bits* for short. Before encryption, any message must therefore be converted into binary digits. This conversion can be performed according to various protocols, such as the American Standard Code for Information Interchange, known familiarly by the acronym ASCII, pronounced 'ass-key'. ASCII assigns a 7-digit binary number to each letter of the alphabet. For the time being, it is sufficient to think of a binary number as merely a pattern of ones and zeros that uniquely identifies each letter (Table 24), just as Morse code identifies each letter with a unique series of dots and dashes. There are 128 (2^7) ways to arrange a combination of 7 binary digits, so ASCII can identify up to 128 distinct characters. This allows plenty of room to define all the lower-case letters (e.g. **a** = **1100001**), all necessary punctuation (e.g. **!** = **0100001**), as well as other symbols (e.g. **&** = **0100110**). Once the message has been converted into binary, encryption can begin.

Even though we are dealing with computers and numbers, and not machines and letters, the encryption still proceeds by the age-old principles of substitution and transposition, in which elements of the message are substituted for other elements, or their positions are switched, or both.

Every encipherment, no matter how complex, can be broken down into combinations of these simple operations. The following two examples demonstrate the essential simplicity of computer encipherment by showing how a computer might perform an elementary substitution cipher and an elementary transposition cipher.

First, imagine that we wish to encrypt the message **HELLO**, employing a simple computer version of a transposition cipher. Before encryption can begin, we must translate the message into ASCII according to Table 24:

Plaintext = HELLO = 1001000 1000101 1001100 1001100 1001111

One of the simplest forms of transposition cipher would be to swap the first and second digits, the third and fourth digits, and so on. In this case the final digit would remain unchanged because there are an odd number of digits. In order to see the operation more clearly, I have removed the spaces between the ASCII blocks in the original plaintext to generate a single string, and then lined it up against the resulting ciphertext for comparison:

Plaintext = 10010001000101100110010011001001111
Ciphertext = 01100010001010011001100011000110111

An interesting aspect of transposition at the level of binary digits is that the transposing can happen within the letter. Furthermore, bits of one letter can swap places with bits of the neighbouring letter. For example, by swapping

Table 24 ASCII binary numbers for the capital letters.

A	1000001	N	1001110
B	1000010	O	1001111
C	1000011	P	1010000
D	1000100	Q	1010001
E	1000101	R	1010010
F	1000110	S	1010011
G	1000111	T	1010100
H	1001000	U	1010101
I	1001001	V	1010110
J	1001010	W	1010111
K	1001011	X	1011000
L	1001100	Y	1011001
M	1001101	Z	1011010

the seventh and eighth numbers, the final 0 of H is swapped with the initial 1 of E. The encrypted message is a single string of 35 binary digits, which can be transmitted to the receiver, who then reverses the transposition to recreate the original string of binary digits. Finally, the receiver reinterprets the binary digits via ASCII to regenerate the message HELLO.

Next, imagine that we wish to encrypt the same message, HELLO, this time employing a simple computer version of a substitution cipher. Once again, we begin by converting the message into ASCII before encryption. As usual, substitution relies on a key that has been agreed between sender and receiver. In this case the key is the word DAVID translated into ASCII, and it is used in the following way. Each element of the plaintext is 'added' to the corresponding element of the key. Adding binary digits can be thought of in terms of two simple rules. If the elements in the plaintext and the key are the same, the element in the plaintext is substituted for 0 in the ciphertext. But, if the elements in the message and key are different, the element in the plaintext is substituted for 1 in the ciphertext:

Message	HELLO
Message in ASCII	10010001000101100110010011001001111
Key = DAVID	10001001000001101011010010011000100
Ciphertext	00011000000100001101000001010001011

The resulting encrypted message is a single string of 35 binary digits which can be transmitted to the receiver, who uses the same key to reverse the substitution, thus recreating the original string of binary digits. Finally, the receiver reinterprets the binary digits via ASCII to regenerate the message HELLO.

Computer encryption was restricted to those who had computers, which in the early days meant the government and the military. However, a series of scientific, technological and engineering breakthroughs made computers, and computer encryption, far more widely available. In 1947, AT&T Bell Laboratories invented the transistor, a cheap alternative to the electronic valve. Commercial computing became a reality in 1951 when companies such as Ferranti began to make computers to order. In 1953 IBM launched its first computer, and four years later it introduced Fortran, a programming language that allowed 'ordinary' people to write computer programs. Then, in 1959, the invention of the integrated circuit heralded a new era of computing.

During the 1960s, computers became more powerful, and at the same time they became cheaper. Businesses were increasingly able to afford computers, and could use them to encrypt important communications such as money transfers or delicate trade negotiations. However, as more and more businesses bought computers, and as encryption between businesses spread, cryptographers were confronted with new problems, difficulties that had not existed when cryptography was the preserve of governments and the military. One of the primary concerns was the issue of standardisation. A company might use a particular encryption system to ensure secure internal communication, but it could not send a secret message to an outside organisation unless the receiver used the same system of encryption. Eventually, on 15 May 1973, America's National Bureau of Standards planned to solve the problem, and formally requested proposals for a standard encryption system that would allow business to speak secretly unto business.

One of the more established cipher algorithms, and a candidate for the standard, was an IBM product known as Lucifer. It had been developed by Horst Feistel, a German émigré who had arrived in America in 1934. He was on the verge of becoming a US citizen when America entered the war, which meant that he was placed under house arrest until 1944. For some years after, he suppressed his interest in cryptography to avoid arousing the suspicions of the American authorities. When he did eventually begin research into ciphers, at the Air Force's Cambridge Research Center, he soon found himself in trouble with the National Security Agency (NSA), the organisation with overall responsibility for maintaining the security of military and governmental communications, and which also attempts to intercept and decipher foreign communications. The NSA employs more mathematicians, buys more computer hardware, and intercepts more messages than any other organisation in the world. It is the world leader when it comes to snooping.

The NSA did not object to Feistel's past, they merely wanted to have a monopoly on cryptographic research, and it seems that they arranged for Feistel's research project to be cancelled. In the 1960s Feistel moved to the Mitre Corporation, but the NSA continued to apply pressure and forced him to abandon his work for a second time. Feistel eventually ended up at IBM's Thomas J. Watson Laboratory near New York, where for several years he was able to conduct his research without being harassed. It was there, during the early 1970s, that he developed the Lucifer system.

Lucifer encrypts messages according to the following scrambling operation. First, the message is translated into a long string of binary digits. Second, the string is split into blocks of 64 digits, and encryption is performed separately on each of the blocks. Third, focusing on just one block, the 64 digits are shuffled, and then split into two half-blocks of 32, labelled Left^0 and Right^0. The digits in Right^0 are then put through a 'mangler function', which changes the digits according to a complex substitution. The mangled Right^0 is then added to Left^0 to create a new half-block of 32 digits called Right^1. The original Right^0 is relabelled Left^1. This set of operations is called a 'round'. The whole process is repeated in a second round, but starting with the new half-blocks, Left^1 and Right^1, and ending with Left^2 and Right^2. This process is repeated until there have been 16 rounds in total. The encryption process is a bit like kneading a slab of dough. Imagine a long slab of dough with a message written on it. First, the long slab is divided into blocks that are 64 cm in length. Then, one half of one of the blocks is picked up, mangled, folded over, added to the other half and stretched to make a new block. Then the process is repeated over and over again until the message has been thoroughly mixed up. After 16 rounds of kneading the ciphertext is sent, and is then deciphered at the other end by reversing the process.

The exact details of the mangler function can change, and are determined by a key agreed by sender and receiver. In other words, the same message can be encrypted in a myriad of different ways depending on which key is chosen. The keys used in computer cryptography are simply numbers. Hence, the sender and receiver merely have to agree on a number in order to decide the key. Thereafter, encryption requires the sender to input the key number and the message into Lucifer, which then outputs the ciphertext. Decryption requires the receiver to input the same key number and the ciphertext into Lucifer, which then outputs the original message.

Lucifer was generally held to be one of the strongest commercially available encryption products, and consequently it was used by a variety of organisations. It seemed inevitable that this encryption system would be adopted as the American standard, but once again the NSA interfered with Feistel's work. Lucifer was so strong that it offered the possibility of an encryption standard that was probably beyond the codebreaking capabilities of the NSA; not surprisingly, the NSA did not want to see an encryption standard that they could not break. Hence, it is rumoured that the NSA lobbied to weaken one aspect of Lucifer, the number of possible keys, before allowing it to be adopted as the standard.

The number of possible keys is one of the crucial factors determining the strength of any cipher. A cryptanalyst trying to decipher an encrypted message could attempt to check all possible keys, and the greater the number of possible keys, the longer it will take to find the right one. If there are only 1,000,000 possible keys, a cryptanalyst could use a powerful computer to find the correct one in a matter of minutes, and thereby decipher an intercepted message. However, if the number of possible keys is large enough, finding the correct key becomes impractical. If Lucifer were to become the encryption standard, then the NSA wanted to ensure that it operated with only a restricted number of keys.

The NSA argued in favour of limiting the number of keys to roughly 100,000,000,000,000,000 (technically referred to as 56 bits, because this number consists of 56 digits when written in binary). It seems that the NSA believed that such a key would provide security within the civilian community, because no civilian organisation had a computer powerful enough to check every possible key within a reasonable amount of time. However, the NSA itself, with access to the world's greatest computing resource, would just about be able to break into messages. The 56-bit version of Feistel's Lucifer cipher was officially adopted on 23 November 1976, and was called the Data Encryption Standard (DES). A quarter of a century later, DES remains America's official standard for encryption.

The adoption of DES solved the problem of standardisation, encouraging businesses to use cryptography for security. Furthermore, DES was strong enough to guarantee security against attacks from commercial rivals. It was effectively impossible for a company with a civilian computer to break into a DES-encrypted message because the number of possible keys was sufficiently large. Unfortunately, despite standardisation and despite the strength of DES, businesses still had to deal with one more major issue, a problem known as *key distribution*.

Imagine that a bank wants to send some confidential data to a client via a telephone line, but is worried that there might be somebody tapping the wire. The bank picks a key and uses DES to encrypt the data message. In order to decrypt the message, the client needs not only to have a copy of DES on its computer, but also to know which key was used to encrypt the message. How does the bank inform the client of the key? It cannot send the key via the telephone line, because it suspects that there is an eavesdropper on the line. The only truly secure way to send the key is to hand it over in person, which is clearly a time-consuming task. A less secure but more practical solution is to send the key via a courier. In the 1970s, banks attempted to distribute keys by employing special dispatch riders who had been vetted and who were among the company's most trusted employees. These dispatch riders would race across the world with padlocked briefcases, personally distributing keys to everyone who would receive messages from the bank over the next week. As business networks grew in size, as more messages were sent, and as more keys had to be delivered, the banks found that this distribution process became a horrendous logistical nightmare, and the overhead costs became prohibitive.

The problem of key distribution has plagued cryptographers throughout history. For example, during the Second World War the German High Command had to distribute the monthly book of day-keys to all its Enigma operators, which was an enormous logistical problem. Also, U-boats, which tended to spend extended periods away from base, had to somehow obtain a regular supply of keys. In earlier times, users of the Vigenère cipher had to find a way of getting the keyword from the sender to the receiver. No matter how secure a cipher is in theory, in practice it can be undermined by the problem of key distribution.

To some extent, government and the military have been able to deal with the problem of key distribution by throwing money and resources at it. Their messages are so important that they will go to any lengths to ensure secure key distribution. The U.S. Government keys are managed and distributed by COMSEC, short for Communications Security. In the 1970s, COMSEC was responsible for transporting tonnes of keys every day. When ships carrying COMSEC material came into dock, cryptocustodians would march on board, collect stacks of cards, paper tapes, floppy disks, or whatever other medium the keys might be stored on, and then deliver them to the intended recipients.

Key distribution might seem a mundane issue, but it became the overriding problem for post-war cryptographers. If two parties wanted to communicate securely, they had to rely on a third party to deliver the key, and this became the weakest link in the chain of security. The dilemma for businesses was straightforward – if governments with all their money were struggling to guarantee the secure distribution of keys, then how could civilian companies ever hope to achieve reliable key distribution without bankrupting themselves?

Despite claims that the problem of key distribution was unsolvable, a team of mavericks triumphed against the odds and came up with a brilliant solution in the mid-1970s. They devised an encryption system that appeared to defy all logic. Although computers transformed the implementation of ciphers, the greatest revolution in twentieth-century cryptography has been the development of techniques to overcome the problem of key distribution. Indeed, this breakthrough is considered to be the greatest cryptographic achievement since the invention of the monoalphabetic cipher, over two thousand years ago.

God Rewards Fools

Whitfield Diffie is one of the most ebullient cryptographers of his generation. The mere sight of him creates a striking and somewhat contradictory image. His impeccable suit reflects the fact that for most of the 1990s he has been employed by one of America's giant computer companies – currently his official job title is Distinguished Engineer at Sun Microsystems. However, his shoulder-length hair and long white beard betray the fact that his heart is still stuck in the 1960s. He spends much of his time in front of a computer workstation, but he looks as if he would be equally comfortable in a Bombay ashram. Diffie is aware that his dress and personality can have quite an impact on others, and comments that, 'People always think that I am taller than I really am, and I'm told it's the Tigger effect – "No matter his weight in pounds, shillings and ounces, he always seems bigger because of the bounces."'

Diffie was born in 1944, and spent most of his early years in Queens,

New York. As a child he became fascinated by mathematics, reading books ranging from *The Chemical Rubber Company Handbook of Mathematical Tables* to G.H. Hardy's *Course of Pure Mathematics*. He went on to study mathematics at the Massachusetts Institute of Technology, graduating in 1965. He then took a series of jobs related to computer security, and by the early 1970s he had matured into one of the few truly independent security experts, a freethinking cryptographer, not employed by the government or by any of the big corporations. In hindsight, he was the first cypherpunk.

Diffie was particularly interested in the key-distribution problem, and he realised that whoever could find a solution would go down in history as one of the all-time great cryptographers. Diffie was so captivated by the

Figure 62 Whitfield Diffie.

problem of key distribution that it became the most important entry in his special notebook entitled 'Problems for an Ambitious Theory of Cryptography'. Part of Diffie's motivation came from his vision of a wired world. Back in the 1960s, the U.S. Department of Defense began funding a cutting-edge research organisation called the Advanced Research Projects Agency (ARPA), and one of ARPA's front-line projects was to find a way of connecting military computers across vast distances. This would allow a computer that had been damaged to transfer its responsibilities to another one in the network. The main aim was to make the Pentagon's computer infrastructure more robust in the face of nuclear attack, but the network would also allow scientists to send messages to each other, and perform calculations by exploiting the spare capacity of remote computers. The ARPANet was born in 1969, and by the end of the year there were four connected sites. The ARPANet steadily grew in size, and in 1982 it spawned the Internet. At the end of the 1980s, non-academic and non-governmental users were given access to the Internet, and thereafter the number of users exploded. Today, more than a hundred million people use the Internet to exchange information and send electronic mail messages, or e-mails.

While the ARPANet was still in its infancy, Diffie was far-sighted enough to forecast the advent of the information superhighway and the digital revolution. Ordinary people would one day have their own computers, and these computers would be interconnected via phone lines. Diffie believed that if people then used their computers to exchange e-mails, they deserved the right to encrypt their messages in order to guarantee their privacy. However, encryption required the secure exchange of keys. If governments and large corporations were having trouble coping with key distribution, then the public would find it impossible, and would effectively be deprived of the right to privacy.

Diffie imagined two strangers meeting via the Internet, and wondered how they could send each other an encrypted message. He also considered the scenario of a person wanting to buy a commodity on the Internet. How could that person send an e-mail containing encrypted credit-card details so that only the Internet retailer could decipher them? In both cases, it seemed that the two parties needed to share a key, but how could they securely exchange keys? The number of casual contacts and the amount of spontaneous e-mails among the public would be enormous, and this would mean that key distribution would be impractical. Diffie was fearful that the necessity of key distribution would prevent the public from having access to digital privacy, and he became obsessed with the idea of finding a solution to the problem.

In 1974, Diffie, still an itinerant cryptographer, paid a visit to IBM's Thomas J. Watson Laboratory, where he had been invited to give a talk. He spoke about various strategies for attacking the key-distribution problem, but all his ideas were very tentative, and his audience was sceptical about the prospects for a solution. The only positive response to Diffie's presentation was from Alan Konheim, one of IBM's senior cryptographic experts, who mentioned that someone else had recently visited the laboratory and given a lecture that addressed the issue of key distribution. That speaker was Martin Hellman, a professor from Stanford University in California. That evening Diffie got in his car and began the 5,000 km journey to the West Coast to meet the only person who seemed to share his obsession. The alliance of Diffie and Hellman would become one of the most dynamic partnerships in cryptography.

Martin Hellman was born in 1946 in a Jewish neighbourhood in the Bronx, but at the age of four his family moved to a predominantly Irish Catholic neighbourhood. According to Hellman, this permanently changed his attitude to life: "The other kids went to church and they learned that the Jews killed Christ, so I got called "Christ killer". I also got beat up. To start with, I wanted to be like the other kids, I wanted a Christmas tree and I wanted Christmas presents. But then I realised that I couldn't be like all the other kids, and in self-defence I adopted an attitude of "Who would want to be like everybody else?"" Hellman traces his interest in ciphers to this enduring desire to be different. His colleagues had told him he was crazy to do research in cryptography, because he would be competing with the NSA and their multibillion-dollar budget. How could he hope to discover something that they did not know already? And if he did discover anything, the NSA would classify it.

Just as Hellman was beginning his research, he came across *The Codebreakers* by the historian David Kahn. This book was the first detailed discussion of the development of ciphers, and as such it was the perfect primer for a budding cryptographer. *The Codebreakers* was Hellman's only research companion, until in September 1974 he received an unexpected phone call from Whitfield Diffie, who had just driven across the continent to meet him. Hellman had never heard of Diffie, but grudgingly agreed to a half-hour appointment later that afternoon. By the end of the meeting, Hellman realised that Diffie was the best-informed person he had ever met. The feeling was mutual. Hellman recalls: 'I'd promised my wife I'd be home to watch the kids, so he came home with me and we had dinner together. He left at around midnight. Our personalities are very different – he is much more counter-culture than I am – but eventually the personality clash was very symbiotic. It was just such a breath of fresh air for me. Working in a vacuum had been really hard.'

Since Hellman did not have a great deal of funding, he could not afford to employ his new soulmate as a researcher. Instead, Diffie was enrolled as a graduate student. Together, Hellman and Diffie began to study the key-distribution problem, desperately trying to find an alternative to the tiresome task of physically transporting keys over vast distances. In due course they were joined by Ralph Merkle. Merkle was an intellectual refugee, having emigrated from another research group where the professor had no sympathy for the impossible dream of solving the key-distribution problem. Says Hellman:

> Ralph, like us, was willing to be a fool. And the way to get to the top of the heap in terms of developing original research is to be a fool, because only fools keep trying. You have idea number 1, you get excited, and it flops. Then you have idea number 2, you get excited, and it flops. Then you have idea number 99, you get excited, and it flops. Only a fool would be excited by the 100th idea, but it might take 100 ideas before one really pays off. Unless you're foolish enough to be continually excited, you won't have the motivation, you won't have the energy to carry it through. God rewards fools.

The whole problem of key distribution is a classic catch-22 situation. If two people want to exchange a secret message over the phone, the sender must encrypt it. To encrypt the secret message the sender must use a key, which is itself a secret, so then there is problem of transmitting the secret key to the receiver in order to transmit the secret message. In short, before two people can exchange a secret (an encrypted message) they must already share a secret (the key).

When thinking about the problem of key distribution, it is helpful to consider Alice, Bob and Eve, three fictional characters who have become the industry standard for discussions about cryptography. In a typical situation, Alice wants to send a message to Bob, or vice versa, and Eve is trying to eavesdrop. If Alice is sending private messages to Bob, she will encrypt each one before sending it, using a separate key each time. Alice is continually faced with the problem of key distribution because she has to convey the keys to Bob securely, otherwise he cannot decrypt the messages. One way to solve the problem is for Alice and Bob to meet up once

Figure 63 Martin Hellman.

a week and exchange enough keys to cover the messages that might be sent during the next seven days. Exchanging keys in person is certainly secure, but it is inconvenient and, if either Alice or Bob is taken ill, the system breaks down. Alternatively, Alice and Bob could hire couriers, which would be less secure and more expensive, but at least they have delegated some of the work. Either way, it seems that the distribution of keys is unavoidable. For two thousand years this was considered to be an axiom of cryptography – an indisputable truth. However, Diffie and Hellman were aware of an anecdote that seemed to defy the axiom.

Imagine that Alice and Bob live in a country where the postal system is completely immoral, and postal employees will read any unprotected correspondence. One day, Alice wants to send an intensely personal message to Bob. She puts it inside an iron box, closes it and secures it with a padlock and key. She puts the padlocked box in the post and keeps the key. However, when the box reaches Bob, he is unable to open it because he does not have the key. Alice might consider putting the key inside another box, padlocking it and sending it to Bob, but without the key to the second padlock he is unable to open the second box, so he cannot obtain the key that opens the first box. The only way around the problem seems to be for Alice to make a copy of her key and give it to Bob in advance when they meet for coffee. So far, I have just restated the same old problem in a new scenario. Avoiding key distribution seems logically impossible – surely, if Alice wants to lock something in a box so that only Bob can open it, she must give him a copy of the key. Or, in terms of cryptography, if Alice wants to encipher a message so that only Bob can decipher it, she must give him a copy of the key. Key exchange is an inevitable part of encipherment – or is it?

Now picture the following scenario. As before, Alice wants to send an intensely personal message to Bob. Again, she puts her secret message in an iron box, padlocks it and sends it to Bob. When the box arrives, Bob adds his own padlock and sends the box back to Alice. When Alice receives the box, it is now secured by two padlocks. She removes her own padlock, leaving just Bob's padlock to secure the box. Finally she sends the box back to Bob. And here is the crucial difference: Bob can now open the box because it is secured only with his own padlock, to which he alone has the key.

The implications of this little story are enormous. It demonstrates that a secret message can be securely exchanged between two people without necessarily exchanging a key. For the first time we have a suggestion that key exchange might not be an inevitable part of cryptography. We can reinterpret the story in terms of encryption. Alice uses her own key to encrypt a message to Bob, who encrypts it again with his own key and returns it. When Alice receives the doubly encrypted message, she removes her own encryption and returns it to Bob, who can then remove his own encryption and read the message.

It seems that the problem of key distribution might have been solved, because the doubly encrypted scheme requires no exchange of keys. However, there is a fundamental obstacle to implementing a system in which Alice encrypts, Bob encrypts, Alice decrypts and Bob decrypts. The problem is the order in which the encryptions and decryptions are performed. In general, the order of encryption and decryption is crucial, and should obey the maxim 'last on, first off'. In other words, the last stage of encryption should be the first to be decrypted. In the above scenario, Bob performed the last stage of encryption, so this should have been the first to be decrypted, but it was Alice who removed her encryption first, before Bob removed his. The importance of order is most easily grasped by examining something we do every day. In the morning we put on our socks, and then we put on our shoes, and in the evening we remove our shoes before removing our socks – it is impossible to remove the socks before the shoes. We must obey the maxim 'last on, first off'.

Some very elementary ciphers, such as the Caesar cipher, are so simple that order does not matter. However, in the 1970s it seemed that any form of strong encryption must always obey the 'last on, first off' rule. If a message is encrypted with Alice's key and then with Bob's key, then it must be decrypted with Bob's key before it can be decrypted with Alice's key. Order is crucial even with a monoalphabetic substitution cipher. Imagine that Alice and Bob have their own keys, as shown on the next page, and let us take a look at what happens when the order is incorrect. Alice uses her key to encrypt a message to Bob, then Bob re-encrypts the result using his own key; Alice uses her key to perform a partial decryption, and finally Bob attempts to use his key to perform the full decryption.

Alice's key

a	b	c	d	e	f	g	h	i	j	k	l	m	n	o	p	q	r	s	t	u	v	w	x	y	z
H	F	S	U	G	T	A	K	V	D	E	O	Y	J	B	P	N	X	W	C	Q	R	I	M	Z	L

Bob's key

a	b	c	d	e	f	g	h	i	j	k	l	m	n	o	p	q	r	s	t	u	v	w	x	y	z
C	P	M	G	A	T	N	O	J	E	F	W	I	Q	B	U	R	Y	H	X	S	D	Z	K	L	V

Message	m e e t	m e	a t	n o o n
Encrypted with Alice's key	Y G G C	Y G	H C	J B B J
Encrypted with Bob's key	L N N M	L N	O M	E P P E
Decrypted with Alice's key	Z Q Q X	Z Q	L X	K P P K
Decrypted with Bob's key	w n n t	w n	y t	x b b x

The result is nonsense. However, you can check for yourself that if the decryption order were reversed, and Bob decrypted before Alice, thus obeying the 'last on, first off' rule, then the result would have been the original message. But if order is so important, why did the padlock system seem to work in the anecdote about locked boxes? The answer is that order is not important for padlocks. I can apply twenty padlocks to a box and undo them in any order, and at the end the box will open. Unfortunately, encryption systems are far more sensitive than padlocks when it comes to order.

Although the doubly padlocked box approach would not work for real-world cryptography, it inspired Diffie and Hellman to search for a practical method of circumventing the key-distribution problem. They spent month after month attempting to find a solution. Although every idea ended in failure, they behaved like perfect fools and persevered. Their research concentrated on the examination of various mathematical *functions*. A function is any mathematical operation that turns one number into another number. For example, 'doubling' is a type of function, because it turns the number 3 into 6, or the number 9 into 18. Furthermore, we can think of all forms of computer encryption as functions because they turn one number (the plaintext) into another number (the ciphertext).

Most mathematical functions are classified as two-way functions because they are easy to do, and easy to undo. For example, 'doubling' is a two-way function because it is easy to double a number to generate a new number, and just as easy to undo the function and get from the doubled number back to the original number. For example, if we know that the result of doubling is 26, then it is trivial to reverse the function and deduce that the original number was 13. The easiest way to understand the concept of a two-way function is in terms of an everyday activity. The act of turning on a light switch is a function, because it turns an ordinary light bulb into an illuminated light bulb. This function is two-way because if a switch is turned on, it is easy enough to turn it off and return the light bulb to its original state.

However, Diffie and Hellman were not interested in two-way functions. They focused their attention on one-way functions. As the name suggests, a one-way function is easy to do but very difficult to undo. In other words, two-way functions are reversible, but one-way functions are not reversible. Once again, the best way to illustrate a one-way function is in terms of an everyday activity. Mixing yellow and blue paint to make green paint is a one-way function because it is easy to mix the paint, but impossible to unmix it. Another one-way function is the cracking of an egg, because it is easy to crack an egg but impossible then to return the egg to its original condition. For this reason, one-way functions are sometimes called Humpty Dumpty functions.

Modular arithmetic, sometimes called *clock arithmetic* in schools, is an area of mathematics that is rich in one-way functions. In modular arithmetic, mathematicians consider a finite group of numbers arranged in a loop, rather like the numbers on a clock. For example, Figure 64 shows a clock for modular 7 (or mod 7), which has only the 7 numbers from 0 to 6. To work out 2 + 3, we start at 2 and move around 3 places to reach 5, which is the same answer as in normal arithmetic. To work out 2 + 6 we start at 2 and move around 6 places, but this time we go around the loop and arrive at 1, which is not the result we would get in normal arithmetic. These results can be expressed as:

$$2 + 3 = 5 \pmod 7 \quad \text{and} \quad 2 + 6 = 1 \pmod 7$$

Modular arithmetic is relatively simple, and in fact we do it every day when we talk about time. If it is 9 o'clock now, and we have a meeting 8 hours from now, we would say that the meeting is at 5 o'clock, not

17 o'clock. We have mentally calculated 9 + 8 in (mod 12). Imagine a clock face, look at 9, and then move around 8 spaces, and we end up at 5:

$$9 + 8 = 5 \pmod{12}$$

Rather than visualising clocks, mathematicians often take the shortcut of performing modular calculations according to the following recipe. First, perform the calculation in normal arithmetic. Second, if we want to know the answer in (mod x), we divide the normal answer by x and note the remainder. This remainder is the answer in (mod x). To find the answer to 11×9 (mod 13), we do the following:

$$11 \times 9 = 99$$
$$99 \div 13 = 7, \text{ remainder } 8$$
$$11 \times 9 = 8 \pmod{13}$$

Functions performed in the modular arithmetic environment tend to behave erratically, which in turn sometimes makes them one-way functions. This becomes evident when a simple function in normal arithmetic is compared with the same simple function in modular arithmetic. In the former environment the function will be two-way and easy to reverse; in the latter environment it will be one-way and hard to reverse. As an example, let us take the function 3^x. This means take a number x, then multiply 3 by itself x times in order to get the new number. For example, if $x = 2$, and we perform the function, then:

$$3^x = 3^2 = 3 \times 3 = 9.$$

In other words, the function turns 2 into 9. In normal arithmetic, as the value of x increases so does the result of the function. Hence, if we were given the result of the function it would be relatively easy to work back-

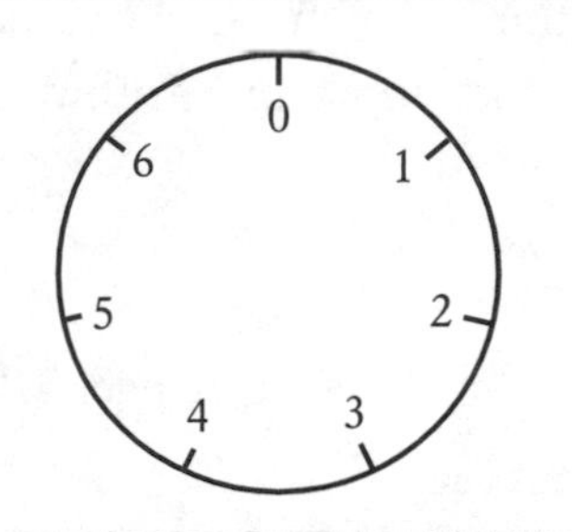

Figure 64 Modular arithmetic is performed on a finite set of numbers, which can be thought of as numbers on a clock face. In this case, we can work out 6 + 5 in modular 7 by starting at 6 and moving around five spaces, which brings us to 4.

wards and deduce the original number. For example, if the result is 81, we can deduce that x is 4, because $3^4 = 81$. If we made a mistake and guessed that x is 5, we could work out that $3^5 = 243$, which tells us that our choice of x is too big. We would then reduce our choice of x to 4, and we would have the right answer. In short, even when we guess wrongly we can home in on the correct value of x, and thereby reverse the function.

However, in modular arithmetic this same function does not behave so sensibly. Imagine that we are told that 3^x in (mod 7) is 1, and we are asked to find the value of x. No value springs to mind, because we are generally unfamiliar with modular arithmetic. We could take a guess that $x = 5$, and we could work out the result of 3^5 (mod 7). The answer turns out to be 5, which is too big, because we are looking for an answer of just 1. We might be tempted to reduce the value of x and try again. But we would be heading in the wrong direction, because the actual answer is $x = 6$.

In normal arithmetic we can test numbers and can sense whether we are getting warmer or colder. The environment of modular arithmetic gives no helpful clues, and reversing functions is much harder. Often, the only way to reverse a function in modular arithmetic is to compile a table by calculating the function for many values of x until the right answer is found. Table 25 shows the result of calculating several values of the function in both normal arithmetic and modular arithmetic. It clearly demonstrates the erratic behaviour of the function when calculated in modular arithmetic. Although drawing up such a table is only a little tedious when we are dealing with relatively small numbers, it would be excruciatingly painful to build a table to deal with a function such as 453^x (mod 21,997). This is a classic example of a one-way function, because I could pick a value for x and calculate the result of the function, but if I gave you a

Table 25 Values of the function 3^x calculated in normal arithmetic (row 2) and modular arithmetic (row 3). The function increases continuously in normal arithmetic, but is highly erratic in modular arithmetic.

x	1	2	3	4	5	6
3^x	3	9	27	81	243	729
3^x(mod 7)	3	2	6	4	5	1

result, say 5,787, you would have enormous difficulty in reversing the function and deducing my choice of x. It took me just seconds to do my calculation and generate 5,787, but it would take you hours to draw up the table and work out my choice of x.

After two years of focusing on modular arithmetic and one-way functions, Hellman's foolishness began to pay off. In the spring of 1976 he hit upon a strategy for solving the key-exchange problem. In half an hour of frantic scribbling, he proved that Alice and Bob could agree a key without meeting, thereby disposing of an axiom that had lasted for centuries. Hellman's idea relied on a one-way function of the form Y^x (mod P). Initially, Alice and Bob agree on values for Y and P. Almost any values are fine, but there are some restrictions, such as Y being smaller than P. These values are not secret, so Alice can telephone Bob and suggest that, say, $Y = 7$ and $P = 11$. Even if the telephone line is insecure and nefarious Eve hears this conversation, it does not matter, as we shall see later. Alice and Bob have now agreed on the one-way function 7^x (mod 11). At this point they can begin the process of trying to establish a secret key without meeting. Because they work in parallel, I explain their actions in the two columns of Table 26.

Having followed the stages in Table 26, you will see that, without meeting, Alice and Bob have agreed on the same key, which they can use to encipher a message. For example, they could use their number, 9, as the key for a DES encryption. (DES actually uses much larger numbers as the key, and the exchange process described in Table 26 would be performed with much larger numbers, resulting in a suitably large DES key.) By using Hellman's scheme, Alice and Bob have been able to agree on a key, yet they did not have to meet up and whisper the key to each other. The extraordinary achievement is that the secret key was agreed via an exchange of information on a normal telephone line. But if Eve tapped this line, then surely she also knows the key?

Let us examine Hellman's scheme from Eve's point of view. If she is tapping the line, she knows only the following facts: that the function is 7^x (mod 11), that Alice sends $\alpha = 2$ and that Bob sends $\beta = 4$. In order to find the key, she must either do what Bob does, which is turn α into the key by knowing B, or do what Alice does, which is turn β into the key by knowing A. However, Eve does not know the value of A or B because

Table 26 The general one-way function is Y^x (mod P). Alice and Bob have chosen values for Y and P, and hence have agreed on the one-way function 7^x (mod 11).

	Alice	**Bob**
Stage 1	Alice chooses a number, say 3, and keeps it secret. We label her number A.	Bob chooses a number, say 6, and keeps it secret. We will label his number B.
Stage 2	Alice puts 3 into the one-way function and works out the result of 7^A (mod 11): 7^3 (mod 11) = 343 (mod 11) = 2	Bob puts 6 into the one-way function and works out the result of 7^B (mod 11): 7^6 (mod 11) = 117,649 (mod 11) = 4
Stage 3	Alice calls the result of this calculation α, and she sends her result, 2, to Bob.	Bob calls the result of this calculation β, and he sends his result, 4, to Alice.
The swap	Ordinarily this would be a crucial moment, because Alice and Bob are exchanging information, and therefore this is an opportunity for Eve to eavesdrop and find out the details of the information. However, it turns out that Eve can listen in without it affecting the ultimate security of the system. Alice and Bob could use the same telephone line that they used to agree the values for Y and P, and Eve could intercept the two numbers that are being exchanged, 2 and 4. However, these numbers are not the key, which is why it does not matter if Eve knows them.	
Stage 4	Alice takes Bob's result, and works out the result of β^A (mod 11): 4^3 (mod 11) = 64 (mod 11) = 9	Alice takes Bob's result, and works out the result of α^B (mod 11): 2^6 (mod 11) = 64 (mod 11) = 9
The key	Miraculously, Alice and Bob have ended up with the same number, 9. This is the key!	

Alice and Bob have not exchanged these numbers, and have kept them secret. Eve is stymied. She has only one hope: in theory, she could work out A from α, because α was a consequence of putting A into a function, and Eve knows the function. Or she could work out B from β, because β was a consequence of putting B into a function, and once again Eve knows the function. Unfortunately for Eve, the function is one-way, so whereas it was easy for Alice to turn A into α and for Bob to turn B into β, it is very difficult for Eve to reverse the process, especially if the numbers are very large.

Bob and Alice exchanged just enough information to allow them to establish a key, but this information was insufficient for Eve to work out the key. As an analogy for Hellman's scheme, imagine a cipher that somehow uses colour as the key. First, let us assume that everybody, including Alice, Bob and Eve, has a three-litre pot containing one litre of yellow paint. If Alice and Bob want to agree a secret key, each of them adds one litre of their own secret colour to their own pot. Alice might add a peculiar shade of purple, while Bob might add crimson. Each sends their own mixed pot to the other. Finally, Alice takes Bob's mixture and adds one litre of her own secret colour, and Bob takes Alice's mixture and adds one litre of his own secret colour. Both pots should now be the same colour, because they both contain one litre of yellow, one litre of purple and one litre of crimson. It is the exact colour of the doubly contaminated pots that is used as the key. Alice has no idea what colour was added by Bob, and Bob has no idea what colour was added by Alice, but they have both achieved the same end. Meanwhile, Eve is furious. Even if she intercepts the intermediate pots she cannot work out the colour of the final pots, which is the agreed key. She might see the colour of the mixed pot containing yellow and Alice's secret colour on its way to Bob, and she might see the colour of the mixed pot containing yellow and Bob's secret colour on its way to Alice, but in order to work out the key she really needs to know Alice and Bob's original secret colours. However, Eve cannot work out Alice and Bob's secret colours by looking at the mixed pots. Even if she takes a sample from one of the mixed paints, she cannot unmix the paint to find out the secret colour, because mixing paint is a one-way function.

Hellman's breakthrough came while he was working at home late one night, so by the time he had finished his calculations it was too late to call

Diffie and Merkle. He had to wait until the following morning to reveal his discovery to the only two other people in the world who had believed that a solution to the key-distribution problem was even possible. 'The muse whispered to me', says Hellman, 'but we all laid the foundations together.' Diffie immediately recognised the power of Hellman's breakthrough: 'Marty explained his system of key exchange in all its unnerving simplicity. Listening to him, I realized that the notion had been at the edge of my mind for some time, but had never really broken through.'

The Diffie–Hellman–Merkle key exchange scheme, as it is known, enables Alice and Bob to establish a secret via public discussion. It is one of the most counterintuitive discoveries in the history of science, and it forced the cryptographic establishment to rewrite the rules of encryption. Diffie, Hellman and Merkle publicly demonstrated their discovery at the National Computer Conference in June 1976, and astonished the audience of cryptoexperts. The following year they filed for a patent. Henceforth, Alice and Bob no longer had to meet in order to exchange a key. Instead, Alice could just call Bob on the phone, exchange a couple of numbers with him, mutually establish a secret key and then proceed to encrypt.

Although Diffie–Hellman–Merkle key exchange was a gigantic leap forward, the system was not perfect because it was inherently inconvenient. Imagine that Alice lives in Hawaii, and that she wants to send an e-mail to Bob in Istanbul. Bob is probably asleep, but the joy of e-mail is that Alice can send a message at any time, and it will be waiting on Bob's computer when he wakes up. However, if Alice wants to encrypt her message, then she needs to agree a key with Bob, and in order to perform the key exchange it is preferable for Alice and Bob to be on-line at the same time – establishing a key requires a mutual exchange of information. In effect, Alice has to wait until Bob wakes up. Alternatively, Alice could transmit her part of the key exchange, and wait 12 hours for Bob's reply, at which point the key is established and Alice can, if she is not asleep herself, encrypt and transmit the message. Either way, Hellman's key exchange system hinders the spontaneity of e-mail.

Hellman had shattered one of the tenets of cryptography and proved that Bob and Alice did not have to meet to agree a secret key. Next, somebody merely had to come up with a more efficient scheme for overcoming the problem of key distribution.

The Birth of Public-key Cryptography

Mary Fisher has never forgotten the first time that Whitfield Diffie asked her out on a date: 'He knew I was a space buff, so he suggested we go and see a launch. Whit explained that he was leaving that evening to see Skylab take off, and so we drove all night, and we got there at about 3 a.m. The bird was on the path, as they used to say in those days. Whit had press credentials, but I didn't. So when they asked for my identification and asked who I was, Whit said "My wife". That was 16 November 1973.' They did eventually marry, and during the early years Mary supported her husband during his cryptographic meditations. Diffie was still being employed as a graduate student, which meant that he received only a meagre salary. Mary, an archaeologist by training, took a job with British Petroleum in order to make ends meet.

While Martin Hellman had been developing his method of key exchange, Whitfield Diffie had been working on a completely different approach to solving the problem of key distribution. He often went through long periods of barren contemplation, and on one occasion in 1975 he became so frustrated that he told Mary that he was just a failed scientist who would never amount to anything. He even told her that she ought to find someone else. Mary told him that she had absolute faith in him, and just two weeks later Diffie came up with his truly brilliant idea.

He can still recall how the idea flashed into his mind, and then almost vanished: 'I walked downstairs to get a Coke, and almost forgot about the idea. I remembered that I'd been thinking about something interesting, but couldn't quite recall what it was. Then it came back in a real adrenaline rush of excitement. I was actually aware for the first time in my work on cryptography of having discovered something really valuable. Everything that I had discovered in the subject up to this point seemed to me to be mere technicalities.' It was mid-afternoon, and he had to wait a couple of hours before Mary returned. 'Whit was waiting at the door', she recalls. 'He said he had something to tell me and he had a funny look on his face. I walked in and he said, "Sit down, please, I want to talk to you. I believe that I have made a great discovery – I know I am the first pe son to have done this." The world stood still for me at that moment. I felt like I was living in a Hollywood film.'

Diffie had concocted a new type of cipher, one that incorporated a so-called *asymmetric key*. So far, all the encryption techniques described in this book have been *symmetric*, which means that the unscrambling process is simply the opposite of scrambling. For example, the Enigma machine uses a certain key setting to encipher a message, and the receiver uses an identical machine in the same key setting to decipher it. Similarly, DES encipherment uses a key to perform 16 rounds of scrambling, and then DES decipherment uses the same key to perform the 16 rounds in reverse. Both sender and receiver effectively have equivalent knowledge, and they both use the same key to encrypt and decrypt – their relationship is symmetric. On the other hand, in an asymmetric key system, as the name suggests, the encryption key and the decryption key are not identical. In an asymmetric cipher, if Alice knows the encryption key she can encrypt a message, but she cannot decrypt a message. In order to decrypt, Alice must have access to the decryption key. This distinction between the encryption and decryption keys is what makes an asymmetric cipher special.

At this point it is worth stressing that although Diffie had conceived of the general concept of an asymmetric cipher, he did not actually have a specific example of one. However, the mere concept of an asymmetric cipher was revolutionary. If cryptographers could find a genuine working asymmetric cipher, a system that fulfilled Diffie's requirements, then the implications for Alice and Bob would be enormous. Alice could create her own pair of keys: an encryption key and a decryption key. If we assume that the asymmetric cipher is a form of computer encryption, then Alice's encryption key is a number, and her decryption key is a different number. Alice keeps the decryption key secret, so it is commonly referred to as Alice's *private-key*. However, she publishes the encryption key so that everybody has access to it, which is why it is commonly referred to as Alice's *public-key*. If Bob wants to send Alice a message, he simply looks up her public-key, which would be listed in something akin to a telephone directory. Bob then uses Alice's public-key to encrypt the message. He sends the encrypted message to Alice, and when it arrives Alice can decrypt it using her private decryption key. Similarly, if Charlie, Dawn or Edward want to send Alice an encrypted message, they too can look up Alice's public encryption key, and in each case only Alice has access to the private decryption key required to decrypt the messages.

The great advantage of this system is that there is no toing and froing, as there is with Diffie–Hellman–Merkle key exchange. Bob does not have to wait to get information from Alice before he can encrypt and send a message to her, he merely has to look up her public encryption key. Furthermore, the asymmetric cipher still overcomes the problem of key distribution. Alice does not have to transport the public encryption key securely to Bob: in complete contrast, she can now publicise her public encryption key as widely as possible. She wants the whole world to know her public encryption key so that anybody can use it to send her encrypted messages. At the same time, even if the whole world knows Alice's public-key, none of them, including Eve, can decrypt any messages encrypted with it, because knowledge of the public-key will not help in decryption. In fact, once Bob has encrypted a message using Alice's public-key, even he cannot decrypt it. Only Alice, who possesses the private-key, can decrypt the message.

This is the exact opposite of a traditional symmetric cipher, in which Alice has to go to great lengths to transport the encryption key securely to Bob. In a symmetric cipher the encryption key is the same as the decryption key, so Alice and Bob must take enormous precautions to ensure that the key does not fall into Eve's hands. This is the root of the key-distribution problem.

Returning to padlock analogies, asymmetric cryptography can be thought of in the following way. Anybody can close a padlock simply by clicking it shut, but only the person who has the key can open it. Locking (encryption) is easy, something everybody can do, but unlocking (decryption) can be done only by the owner of the key. The trivial knowledge of knowing how to click the padlock shut does not tell you how to unlock it. Taking the analogy further, imagine that Alice designs a padlock and key. She guards the key, but she manufactures thousands of replica padlocks and distributes them to post offices all over the world. If Bob wants to send a message, he puts it in a box, goes to the local post office, asks for an 'Alice padlock' and padlocks the box. Now he is unable to unlock the box, but when Alice receives it she can open it with her unique key. The padlock and the process of clicking it shut is equivalent to the public encryption key, because everyone has access to the padlocks, and everyone can use a padlock to seal a message in a box. The padlock's key is equivalent to the private decryption key, because only Alice has it, only she can open the padlock, and only she can gain access to the message in the box.

The system seems simple when it is explained in terms of padlocks, but it is far from trivial to find a mathematical function that does the same job, something that can be incorporated into a workable cryptographic system. To turn asymmetric ciphers from a great idea into a practical invention, somebody had to discover an appropriate mathematical function. Diffie envisaged a special type of one-way function, one that could be reversed under exceptional circumstances. In Diffie's asymmetric system, Bob encrypts the message using the public-key, but he is unable to decrypt it – this is essentially a one-way function. However, Alice is able to decrypt the message because she has the private-key, a special piece of information that allows her to reverse the function. Once again, padlocks are a good analogy – shutting the padlock is a one-way function, because in general it is hard to open the padlock unless you have something special (the key), in which case the function is easily reversed.

Diffie published an outline of his idea in the summer of 1975, whereupon other scientists joined the search for an appropriate one-way function, one that fulfilled the criteria required for an asymmetric cipher. Initially there was great optimism, but by the end of the year nobody had been able to find a suitable candidate. As the months passed, it seemed increasingly likely that special one-way functions did not exist. It seemed that Diffie's idea worked in theory but not in practice. Nevertheless, by the end of 1976 the team of Diffie, Hellman and Merkle had revolutionised the world of cryptography. They had persuaded the rest of the world that there was a solution to the key-distribution problem, and had created Diffie–Hellman–Merkle key exchange – a workable but imperfect system. They had also proposed the concept of an asymmetric cipher – a perfect but as yet unworkable system. They continued their research at Stanford University, attempting to find a special one-way function that would make asymmetric ciphers a reality. However, they failed to make the discovery. The race to find an asymmetric cipher was won by another trio of researchers, based 5,000 km away on the east coast of America.

Prime Suspects

'I walked into Ron Rivest's office', recalls Leonard Adleman, 'and Ron had this paper in his hands. He started saying, "These Stanford guys have this really blah, blah, blah." And I remember thinking, "That's nice, Ron, but I have something else I want to talk about." I was entirely unaware of the history of cryptography and I was distinctly uninterested in what he was saying.' The paper that had made Ron Rivest so excited was by Diffie and Hellman, and it described the concept of asymmetric ciphers. Eventually Rivest persuaded Adleman that there might be some interesting mathematics in the problem, and together they resolved to try to find a one-way function that fitted the requirements of an asymmetric cipher. They were joined in the hunt by Adi Shamir. All three men were researchers on the eighth floor of the MIT Laboratory for Computer Science.

Rivest, Shamir and Adleman formed a perfect team. Rivest is a computer scientist with a tremendous ability to absorb new ideas and apply them in unlikely places. He always kept up with the latest scientific papers, which inspired him to come up with a whole series of weird and wonderful candidates for the one-way function at the heart of an asymmetric cipher. However, each candidate was flawed in some way. Shamir, another computer scientist, has a lightning intellect and an ability to see through the debris and focus on the core of a problem. He too regularly generated ideas for formulating an asymmetric cipher, but his ideas were also inevitably flawed. Adleman, a mathematician with enormous stamina, rigour and patience, was largely responsible for spotting the flaws in the ideas of Rivest and Shamir, ensuring that they did not waste time following false leads. Rivest and Shamir spent a year coming up with new ideas, and Adleman spent a year shooting them down. The threesome began to lose hope, but they were unaware that this process of continual failure was a necessary part of their research, gently steering them away from sterile mathematical territory and towards more fertile ground. In due course, their efforts were rewarded.

In April 1977, Rivest, Shamir and Adleman spent Passover at the house of a student, and had consumed significant amounts of Manischewitz wine before returning to their respective homes some time around midnight. Rivest, unable to sleep, lay on his couch reading a mathematics textbook. He began mulling over the question that had been puzzling him for weeks – is it possible to build an asymmetric cipher? Is it possible to find a one-way function that can be reversed only if the receiver has some special information? Suddenly, the mists began to clear and he had a revelation. He spent the rest of that night formalising his idea, effectively writing a complete scientific paper before daybreak. Rivest had made a breakthrough, but it had grown out of a year-long collaboration with Shamir and Adleman, and it would not have been possible without them. Rivest finished off the paper by listing the authors alphabetically; Adleman, Rivest, Shamir.

The next morning, Rivest handed the paper to Adleman, who went through his usual process of trying to tear it apart, but this time he could find no faults. His only criticism was with the list of authors. 'I told Ron to take my name off the paper', recalls Adleman. 'I told him that it was his invention, not mine. But Ron refused and we got into a discussion about it. We agreed that I would go home and contemplate it for one night, and consider what I wanted to do. I went back the next day and suggested to Ron that I be the third author. I recall thinking that this paper would be the least interesting paper that I will ever be on.' Adleman could not have been more wrong. The system, dubbed RSA (Rivest, Shamir, Adleman) as

Figure 65 Ronald Rivest, Adi Shamir and Leonard Adleman.

opposed to ARS, went on to become the most influential cipher in modern cryptography.

Before exploring Rivest's idea, here is a quick reminder of what scientists were looking for in order to build an asymmetric cipher:

(1) Alice must create a public-key, which she would then publish so that Bob (and everybody else) can use it to encrypt messages to her. Because the public-key is a one-way function, it must be virtually impossible for anybody to reverse it and decrypt Alice's messages.

(2) However, Alice needs to decrypt the messages being sent to her. She must therefore have a private-key, some special piece of information, which allows her to reverse the effect of the public-key. Therefore, Alice (and Alice alone) has the power to decrypt any messages sent to her.

At the heart of Rivest's asymmetric cipher is a one-way function based on the sort of modular functions described earlier in the chapter. Rivest's one-way function can be used to encrypt a message – the message, which is effectively a number, is put into the function, and the result is the ciphertext, another number. I shall not describe Rivest's one-way function in detail (for which see Appendix J), but I shall explain one particular aspect of it, known simply as N, because it is N that makes this one-way function reversible under certain circumstances, and therefore ideal for use as an asymmetric cipher.

N is important because it is a flexible component of the one-way function, which means that each person can choose a different value of N, and personalise the one-way function. In order to choose her personal value of N, Alice picks two prime numbers, p and q, and multiplies them together. A prime number is one that has no divisors except itself and 1. For example, 7 is a prime number because no numbers except 1 and 7 will divide into it without leaving a remainder. Likewise, 13 is a prime number because no numbers except 1 and 13 will divide into it without leaving a remainder. However, 8 is not a prime number, because it can be divided by 2 and 4.

So, Alice could choose her prime numbers to be $p = 17{,}159$ and $q = 10{,}247$. Multiplying these two numbers together gives $N = 17{,}159 \times 10{,}247 = 175{,}828{,}273$. Alice's choice of N effectively becomes her public encryption key, and she could print it on her business card, post it on the Internet, or publish it in a public-key directory along with everybody else's value of *N*. If Bob wants to encrypt a message to Alice, he looks up Alice's value of *N* (175,828,273) and then inserts it into the general form of the one-way function, which would also be public knowledge. Bob now has a one-way function tailored with Alice's public-key, so it could be called Alice's one-way function. To encrypt a message to Alice, he takes Alice's one-way function, inserts the message, notes down the result and sends it to Alice.

At this point the encrypted message is secure because nobody can decipher it. The message has been encrypted with a one-way function, so reversing the one-way function and decrypting the message is, by definition, very difficult. However, the question remains – how can Alice decrypt the message? In order to read messages sent to her, Alice must have a way of reversing the one-way function. She needs to have access to some special piece of information that allows her to decrypt the message. Fortunately for Alice, Rivest designed the one-way function so that it is reversible to someone who knows the values of *p* and *q*, the two prime numbers that are multiplied together to give *N*. Although Alice has told the world that her value for *N* is 175,828,273, she has not revealed her values for *p* and *q*, so only she has the special information required to decrypt her own messages.

We can think of *N* as the public-key, the information that is available to everybody, the information required to encrypt messages to Alice. Whereas, *p* and *q* are the private-key, available only to Alice, the information required to decrypt these messages.

The exact details of how *p* and *q* can be used to reverse the one-way function are outlined in Appendix J. However, there is one question that must be addressed immediately. If everybody knows *N*, the public-key, then surely people can deduce *p* and *q*, the private-key, and read Alice's messages? After all, *N* was created from *p* and *q*. In fact, it turns out that if *N* is large enough, it is virtually impossible to deduce *p* and *q* from *N*, and this is perhaps the most beautiful and elegant aspect of the RSA asymmetric cipher.

Alice created *N* by choosing *p* and *q*, and then multiplying them together. The fundamental point is that this is in itself a one-way function.

To demonstrate the one-way nature of multiplying primes, we can take two prime numbers, such as 9,419 and 1,933, and multiply them together. With a calculator it takes just a few seconds to get the answer, 18,206,927. However, if instead we were given 18,206,927 and asked to find the prime factors (the two numbers that were multiplied to give 18,206,927) it would take us much longer. If you doubt the difficulty of finding prime factors, then consider the following. It took me just ten seconds to generate the number 1,709,023, but it will take you and a calculator the best part of an afternoon to work out the prime factors.

This system of asymmetric cryptography, known as RSA, is said to be a form of *public-key cryptography*. To find out how secure RSA is, we can examine it from Eve's point of view, and try to break a message from Alice to Bob. To encrypt a message to Bob, Alice must look up Bob's public-key. To create his public-key, Bob picked his own prime numbers, p_B and q_B, and multiplied them together to get N_B. He has kept p_B and q_B secret, because these make up his private decryption key, but he has published N_B, which is equal to 408,508,091. So Alice inserts Bob's public-key N_B into the general one-way encryption function, and then encrypts her message to him. When the encrypted message arrives, Bob can reverse the function and decrypt it using his values for p_B and q_B, which make up his private-key. Meanwhile, Eve has intercepted the message en route. Her only hope of decrypting the message is to reverse the one-way function, and this is possible only if she knows p_B and q_B. Bob has kept p_B and q_B secret, but Eve, like everybody else, knows N_B is 408,508,091. Eve then attempts to deduce the values for p_B and q_B by working out which numbers would need to be multiplied together to get 408,508,091, a process known as *factoring*.

Factoring is very time-consuming, but exactly how long would it take Eve to find the factors of 408,508,091? There are various recipes for trying to factor N_B. Although some recipes are faster than others, they all essentially involve checking each prime number to see if it divides into N_B without a remainder. For example, 3 is a prime number, but it is not a factor of 408,508,091 because 3 will not perfectly divide into 408,508,091. So Eve moves on to the next prime number, 5. Similarly, 5 is not a factor, so Eve moves on to the next prime number, and so on. Eventually, Eve arrives at 18,313, the 2,000th prime number, which is indeed a factor of 408,508,091. Having found one factor, it is easy to find the other one, which turns out to be 22,307. If Eve had a calculator and was able to check four primes a minute, then it would have taken her 500 minutes, or more than 8 hours, to find p_B and q_B. In other words, Eve would be able to work out Bob's private-key in less than a day, and could therefore decipher the intercepted message in less than a day.

This is not a very high level of security, but Bob could have chosen much larger prime numbers and increased the security of his private-key. For example, he could have chosen primes that are as big as 10^{65} (this means 1 followed by 65 zeros, or one hundred thousand, million, million, million, million, million, million, million, million, million, million). This would have resulted in a value for N that would have been roughly $10^{65} \times 10^{65}$, which is 10^{130}. A computer could multiply the two primes and generate N in just a second, but if Eve wanted to reverse the process and work out p and q, it would take inordinately longer. Exactly how long depends on the speed of Eve's computer. Security expert Simson Garfinkel estimated that a 100 MHz Intel Pentium computer with 8 MB of RAM would take roughly 50 years to factor a number as big as 10^{130}. Cryptographers tend to have a paranoid streak and consider worst-case scenarios, such as a worldwide conspiracy to crack their ciphers. So, Garfinkel considered what would happen if a hundred million personal computers (the number sold in 1995) ganged up together. The result is that a number as big as 10^{130} could be factored in about 15 seconds. Consequently, it is now generally accepted that for genuine security it is necessary to use even larger primes. For important banking transactions, N tends to be at least 10^{308}, which is ten million billion billion billion billion billion billion billion billion billion billion billion billion billion billion billion billion billion billion billion times bigger than 10^{130}. The combined efforts of a hundred million personal computers would take more than one thousand years to crack such a cipher. With sufficiently large values of p and q, RSA is impregnable.

The only caveat for the security of RSA public-key cryptography is that at some time in the future somebody might find a quick way to factor N. It is conceivable that a decade from now, or even tomorrow, somebody will discover a method for rapid factoring, and thereafter RSA will become useless. However, for over two thousand years mathematicians have tried and failed to find a shortcut, and at the moment factoring remains an enormously time-consuming calculation. Most mathematicians believe that factoring is an inherently difficult task, and that there is some mathematical law that forbids any shortcut. If we assume they are right, then RSA seems secure for the foreseeable future.

The great advantage of RSA public-key cryptography is that it does away with all the problems associated with traditional ciphers and key exchange. Alice no longer has to worry about securely transporting the key to Bob, or that Eve might intercept the key. In fact, Alice does not care who sees the public-key – the more the merrier, because the public-key helps only with encryption, not decryption. The only thing that needs to remain secret is the private-key used for decryption, and Alice can keep this with her at all times.

RSA was first announced in August 1977, when Martin Gardner wrote an article entitled 'A New Kind of Cipher that Would Take Millions of Years to Break' for his 'Mathematical Games' column in *Scientific American*. After explaining how public-key cryptography works, Gardner issued a challenge to his readers. He printed a ciphertext and also provided the public-key that had been used to encrypt it:

N = 114,381,625,757,888,867,669,235,779,976,146,612,010,218,296, 721,242,362,562,561,842,935,706,935,245,733,897,830,597,123,563, 958,705,058,989,075,147,599,290,026,879,543,541.

The challenge was to factor N into p and q, and then use these numbers to decrypt the message. The prize was $100. Gardner did not have space to explain the nitty-gritty of RSA, and instead he asked readers to write to MIT's Laboratory for Computer Science, who in turn would send back a technical memorandum that had just been prepared. Rivest, Shamir and Adleman were astonished by the three thousand requests they received. However, they did not respond immediately, because they were concerned that public distribution of their idea might jeopardise their chances of getting a patent. When the patent issues were eventually resolved, the trio held a celebratory party at which professors and students consumed pizzas and beer while stuffing envelopes with technical memoranda for the readers of *Scientific American*.

As for Gardner's challenge, it would take 17 years before the cipher would be broken. On 26 April 1994, a team of six hundred volunteers announced the factors of N:

q = 3,490,529,510,847,650,949,147,849,619,903,898,133,417,764, 638,493,387,843,990,820,577

p = 32,769,132,993,266,709,549,961,988,190,834,461,413,177, 642,967,992,942,539,798,288,533.

Using these values as the private-key, they were able to decipher the message. The message was a series of numbers, but when converted into letters it read 'the magic words are squeamish ossifrage'. The factoring problem had been split among the volunteers, who came from countries as far apart as Australia, Britain, America and Venezuela. The volunteers used spare time on their workstations, mainframes and supercomputers, each of them tackling a fraction of the problem. In effect, a network of computers around the world were uniting and working simultaneously in order to meet Gardner's challenge. Even bearing in mind the mammoth parallel effort, some readers may still be surprised that RSA was broken in such a short time, but it should be noted that Gardner's challenge used a relatively small value of N – it was only of the order of 10^{129}. Today, users of RSA would pick a much larger value to secure important information. It is now routine to encrypt a message with a sufficiently large value of N so that all the computers on the planet would need longer than the age of the universe to break the cipher.

The Alternative History of Public-Key Cryptography

Over the past twenty years, Diffie, Hellman and Merkle have become world famous as the cryptographers who invented the concept of public-key cryptography, while Rivest, Shamir and Adleman have been credited with developing RSA, the most beautiful implementation of public-key cryptography. However, a recent announcement means that the history books are having to be rewritten. According to the British Government, public-key cryptography was originally invented at the Government

Communications Headquarters (GCHQ) in Cheltenham, the top-secret establishment that was formed from the remnants of Bletchley Park after the Second World War. This is a story of remarkable ingenuity, anonymous heroes and a government cover-up that endured for decades.

The story starts in the late 1960s, when the British military began to worry about the problem of key distribution. Looking ahead to the 1970s, senior military officials imagined a scenario in which miniaturisation of radios and a reduction in cost meant that every soldier could be in continual radio contact with his officer. The advantages of widespread communication would be enormous, but communications would have to be encrypted, and the problem of distributing keys would be insurmountable. This was an era when the only form of cryptography was symmetric, so an individual key would have to be securely transported to every member of the communications network. Any expansion in communications would eventually be choked by the burden of key distribution. At the beginning of 1969, the military asked James Ellis, one of Britain's foremost government cryptographers, to look into ways of coping with the key-distribution problem.

Ellis was a curious and slightly eccentric character. He proudly boasted of travelling halfway round the world before he was even born – he was conceived in Britain, but was born in Australia. Then, while still a baby, he returned to London and grew up in the East End of the 1920s. At school his primary interest was science, and he went on to study physics at Imperial College before joining the Post Office Research Station at Dollis Hill, where Tommy Flowers had built Colossus, the first codebreaking computer. The cryptographic division at Dollis Hill was eventually absorbed into GCHQ, and so on 1 April 1965 Ellis moved to Cheltenham to join the newly formed Communications–Electronics Security Group (CESG), a special section of GCHQ devoted to ensuring the security of British communications. Because he was involved in issues of national security, Ellis was sworn to secrecy throughout his career. Although his wife and family knew that he worked at GCHQ, they were unaware of his discoveries and had no idea that he was one the nation's most distinguished codemakers.

Despite his skills as a codemaker, Ellis was never put in charge of any of the important GCHQ research groups. He was brilliant, but he was also unpredictable, introverted and not a natural teamworker. His colleague Richard Walton recalled:

> He was a rather quirky worker, and he didn't really fit into the day-to-day business of GCHQ. But in terms of coming up with new ideas he was quite exceptional. You had to sort through some rubbish sometimes, but he was very innovative and always willing to challenge the orthodoxy. We would be in real trouble if everybody in GCHQ was like him, but we can tolerate a higher proportion of such people than most organisations. We put up with a number of people like him.

Figure 66 James Ellis.

One of Ellis's greatest qualities was his breadth of knowledge. He read any scientific journal he could get his hands on, and never threw anything away. For security reasons, GCHQ employees must clear their desks each evening and place everything in locked cabinets, which meant that Ellis's cabinets were stuffed full with the most obscure publications imaginable. He gained a reputation as a cryptoguru, and if other researchers found themselves with impossible problems, they would knock on his door in the hope that his vast knowledge and originality would provide a solution. It was probably because of this reputation that he was asked to examine the key-distribution problem.

The cost of key distribution was already enormous, and would become the limiting factor to any expansion in encryption. Even a reduction of 10 per cent in the cost of key distribution would significantly cut the military's security budget. However, instead of merely nibbling away at the problem, Ellis immediately looked for a radical and complete solution. 'He would always approach a problem by asking, "Is this really what we want to do?"' says Walton. 'James being James, one of the first things he did was to challenge the requirement that it was necessary to share secret data, by which I mean the key. There was no theorem that said you had to have a shared secret. This was something that was challengeable.'

Ellis began his attack on the problem by searching through his treasure trove of scientific papers. Many years later, he recorded the moment when he discovered that key distribution was not an inevitable part of cryptography:

> The event which changed this view was the discovery of a wartime Bell Telephone report by an unknown author describing an ingenious idea for secure telephone speech. It proposed that the recipient should mask the sender's speech by adding noise to the line. He could subtract the noise afterwards since he had added it and therefore knew what it was. The obvious practical disadvantages of this system prevented it being actually used, but it has some interesting characteristics. The difference between this and conventional encryption is that in this case the recipient takes part in the encryption process . . . So the idea was born.

Noise is the technical term for any signal that impinges on a communication. Normally it is generated by natural phenomena, and its most irritating feature is that it is entirely random, which means that removing noise from a message is very difficult. If a radio system is well designed, then the level of noise is low and the message is clearly audible, but if the noise level is high and it swamps the message, there is no way to recover the message. Ellis was suggesting that the receiver, Alice, deliberately create noise, which she could measure before adding it to the communication channel that connects her with Bob. Bob could then send a message to Alice, and if Eve tapped the communications channel she would be unable to read the message because it would be swamped in noise. Eve would be unable to disentangle the noise from the message. The only person who can remove the noise and read the message is Alice, because she is in the unique position of knowing the exact nature of the noise, having put it there in the first place. Ellis realised that security had been achieved without exchanging any key. The key was the noise, and only Alice needed to know the details of the noise.

In a memorandum, Ellis detailed his thought processes: 'The next question was the obvious one. Can this be done with ordinary encipherment? Can we produce a secure encrypted message, readable by the authorised recipient without any prior secret exchange of the key? This question actually occurred to me in bed one night, and the proof of the theoretical possibility took only a few minutes. We had an existence theorem. The unthinkable was actually possible.' (An existence theorem shows that a particular concept is possible, but is not concerned with the details of the concept.) In other words, until this moment, searching for a solution to the key-distribution problem was like looking for a needle in a haystack, with the possibility that the needle might not even be there. However, thanks to the existence theorem, Ellis now knew that the needle was in there somewhere.

Ellis's ideas were very similar to those of Diffie, Hellman and Merkle, except that he was several years ahead of them. However, nobody knew of Ellis's work because he was an employee of the British Government and therefore sworn to secrecy. By the end of 1969, Ellis appears to have reached the same impasse that the Stanford trio would reach in 1975. He had proved to himself that public-key cryptography (or non-secret encryption, as he called it) was possible, and he had developed the concept of separate public-keys and private-keys. He also knew that he needed to find a special one-way function, one that could be reversed if the receiver had access to a piece of special information. Unfortunately, Ellis was not a mathematician.

He experimented with a few mathematical functions, but he soon realised that he would be unable to progress any further on his own.

At this point, Ellis revealed his breakthrough to his bosses. Their reactions are still classified material, but in an interview Richard Walton was prepared to paraphrase for me the various memoranda that were exchanged. Sitting with his briefcase on his lap, the lid shielding the papers from my view, he flicked through the documents:

> I can't show you the papers that I have in here because they still have naughty words like TOP SECRET stamped all over them. Essentially, James's idea goes to the top man, who farms it out, in the way that top men do, so that the experts can have a look at it. They state that what James is saying is perfectly true. In other words, they can't write this man off as a crank. At the same time they can't think of a way of implementing his idea in practice. And so they're impressed by James's ingenuity, but uncertain as to how to take advantage of it.

For the next three years, GCHQ's brightest minds struggled to find a one-way function that satisfied Ellis's requirements, but nothing emerged. Then, in September 1973, a new mathematician joined the team. Clifford Cocks had recently graduated from Cambridge University, where he had specialised in number theory, one of the purest forms of mathematics. When he joined GCHQ he knew very little about encryption and the shadowy world of military and diplomatic communication, so he was assigned a mentor, Nick Patterson, who guided him through his first few weeks at GCHQ.

After about six weeks, Patterson told Cocks about 'a really whacky idea'. He outlined Ellis's theory for public-key cryptography, and explained that nobody had yet been able to find a mathematical function that fitted the bill. Patterson was telling Cocks because this was the most titillating cryptographic idea around, not because he expected him to try to solve it. However, as Cocks explains, later that day he set to work: 'There was nothing particular happening, and so I thought I would think about the idea. Because I had been working in number theory, it was natural to think about one-way functions, something you could do but not undo. Prime numbers and factoring was a natural candidate, and that became my starting point.' Cocks was beginning to formulate what would later be known as the RSA asymmetric cipher. Rivest, Shamir and Adleman discovered their formula for public-key cryptography in 1977, but four years earlier the young Cambridge graduate was going through exactly the same thought processes. Cocks recalls: 'From start to finish, it took me no more than half an hour. I was quite pleased with myself. I thought, "Ooh, that's nice. I've been given a problem, and I've solved it."'

Cocks did not fully appreciate the significance of his discovery. He was unaware of the fact that GCHQ's brightest minds had been struggling with the problem for three years, and had no idea that he had made one of the most important cryptographic breakthroughs of the century. Cocks's naivety may have been part of the reason for his success, allowing him to attack the problem with confidence, rather than timidly prodding at it. Cocks told his mentor about his discovery, and it was Patterson who then reported it to the management. Cocks was quite diffident and very much still a rookie, whereas Patterson fully appreciated the context of the

Figure 67 Clifford Cocks.

problem and was more capable of addressing the technical questions that would inevitably arise. Soon complete strangers started approaching Cocks, the wonderkid, and began to congratulate him. One of the strangers was James Ellis, keen to meet the man who had turned his dream into a reality. Because Cocks still did not understand the enormity of his achievement, the details of this meeting did not make a great impact on him, and so now, over two decades later, he has no memory of Ellis's reaction.

When Cocks did eventually realise what he had done, it struck him that his discovery might have disappointed G.H. Hardy, one of the great English mathematicians of the early part of the century. In his *The Mathematician's Apology*, written in 1940, Hardy had proudly stated: 'Real mathematics has no effects on war. No one has yet discovered any war-like purpose to be served by the theory of numbers.' Real mathematics means pure mathematics, such as the number theory that was at the heart of Cocks's work. Cocks proved that Hardy was wrong. The intricacies of number theory could now be used to help generals plan their battles in complete secrecy. Because his work had implications for military communications, Cocks, like Ellis, was forbidden from telling anybody outside GCHQ about what he had done. Working at a top-secret government establishment meant that he could tell neither his parents nor his former colleagues at Cambridge University. The only person he could tell was his wife, Gill, since she was also employed at GCHQ.

Although Cocks's idea was one of GCHQ's most potent secrets, it suffered from the problem of being ahead of its time. Cocks had discovered a mathematical function that permitted public-key cryptography, but there was still the difficulty of implementing the system. Encryption via public-key cryptography requires much more computer power than encryption via a symmetric cipher like DES. In the early 1970s, computers were still relatively primitive and unable to perform the process of public-key encryption within a reasonable amount of time. Hence, GCHQ were not in a position to exploit public-key cryptography. Cocks and Ellis had proved that the apparently impossible was possible, but nobody could find a way of making the possible practical.

At the beginning of the following year, 1974, Cocks explained his work on public-key cryptography to Malcolm Williamson, who had recently joined GCHQ as a cryptographer. The men happened to be old friends.

They had both attended Manchester Grammar School, whose school motto is *Sapere aude*, 'Dare to be wise'. While at school in 1968, the two boys had represented Britain at the Mathematical Olympiad in the Soviet Union. After attending Cambridge University together, they went their separate ways for a couple of years, but now they were reunited at GCHQ. They had been exchanging mathematical ideas since the age of eleven, but Cocks's revelation of public-key cryptography was the most shocking idea that Williamson had ever heard. 'Cliff explained his idea to me,' recalls Williamson, 'and I really didn't believe it. I was very suspicious, because this is a very peculiar thing to be able to do.'

Williamson went away, and began trying to prove that Cocks had made a mistake and that public-key cryptography did not really exist. He probed the mathematics, searching for an underlying flaw. Public-key cryptography seemed too good to be true, and Williamson was so determined to find a mistake that he took the problem home. GCHQ

Figure 68 Malcolm Williamson.

employees are not supposed to take work home, because everything they do is classified, and the home environment is potentially vulnerable to espionage. However, the problem was stuck in Williamson's brain, so he could not avoid thinking about it. Defying orders, he carried his work back to his house. He spent five hours trying to find a flaw. 'Essentially I failed,' says Williamson. 'Instead I came up with another solution to the problem of key distribution.' Williamson was discovering Diffie–Hellman–Merkle key exchange, at roughly the same time that Martin Hellman discovered it. Williamson's initial reaction reflected his cynical disposition: 'This looks great, I thought to myself. I wonder if I can find a flaw in this one. I guess I was in a negative mood that day.'

By 1975, James Ellis, Clifford Cocks and Malcolm Williamson had discovered all the fundamental aspects of public-key cryptography, yet they all had to remain silent. The three Britons had to sit back and watch as their discoveries were rediscovered by Diffie, Hellman, Merkle, Rivest, Shamir and Adleman over the next three years. Curiously, GCHQ discovered RSA before Diffie–Hellman–Merkle key exchange, whereas in the outside world, Diffie–Hellman–Merkle key exchange came first. The scientific press reported the breakthroughs at Stanford and MIT, and the researchers who had been allowed to publish their work in the scientific

Figure 69 Malcolm Williamson (second from left) and Clifford Cocks (extreme right) arriving for the 1968 Mathematical Olympiad.

journals became famous within the community of cryptographers. A quick look on the Internet with a search engine turns up 15 web pages mentioning Clifford Cocks, compared to 1,382 pages that mention Whitfield Diffie. Cocks's attitude is admirably restrained: 'You don't get involved in this business for public recognition.' Williamson is equally dispassionate: 'My reaction was "Okay, that's just the way it is." Basically, I just got on with the rest of my life.'

Williamson's only qualm is that GCHQ failed to patent public-key cryptography. When Cocks and Williamson first made their breakthroughs, there was agreement among GCHQ management that patenting was impossible for two reasons. First, patenting would mean having to reveal the details of their work, which would have been incompatible with GCHQ's aims. Second, in the early 1970s it was far from clear that mathematical algorithms could be patented. When Diffie and Hellman tried to file for a patent in 1976, however, it was evident that they could be patented. At this point, Williamson was keen to go public and block Diffie and Hellman's application, but he was overruled by his senior managers, who were not farsighted enough to see the digital revolution and the potential of public-key cryptography. By the early 1980s Williamson's bosses were beginning to regret their decision, as developments in computers and the embryonic Internet made it clear that RSA and Diffie–Hellman–Merkle key exchange would both be enormously successful commercial products. In 1996, RSA Data Security, Inc., the company responsible for RSA products, was sold for $200 million.

Although the work at GCHQ was still classified, there was one other organisation that was aware of the breakthroughs that had been achieved in Britain. By the early 1980s America's National Security Agency knew about the work of Ellis, Cocks and Williamson, and it is probably via the NSA that Whitfield Diffie heard a rumour about the British discoveries. In September 1982, Diffie decided to see if there was any truth in the rumour, and he travelled with his wife to Cheltenham in order to talk to James Ellis face to face. They met at a local pub, and very quickly Mary was struck by Ellis's remarkable character:

> We sat around talking, and I suddenly became aware that this was the most wonderful person you could possibly imagine. The breadth of his

> mathematical knowledge is not something I could confidently discuss, but he was a true gentleman, immensely modest, a person with great generosity of spirit and gentility. When I say gentility, I don't mean old-fashioned and musty. This man was a *chevalier*. He was a good man, a truly good man. He was a gentle spirit.

Diffie and Ellis discussed various topics, from archaeology to how rats in the barrel improve the taste of cider, but whenever the conversation drifted towards cryptography, Ellis gently changed the subject. At the end of Diffie's visit, as he was ready to drive away, he could no longer resist directly asking Ellis the question that was really on his mind: 'Tell me about how you invented public-key cryptography?' There was a long pause. Ellis eventually whispered: 'Well, I don't know how much I should say. Let me just say that you people did much more with it than we did.'

Although GCHQ were the first to discover public-key cryptography, this should not diminish the achievements of the academics who rediscovered it. It was the academics who were the first to realise the potential of public-key encryption, and it was they who drove its implementation. Furthermore, it is quite possible that GCHQ would never have revealed their work, thus blocking a form of encryption that would enable the digital revolution to reach its full potential. Finally, the discovery by the academics was wholly independent of GCHQ's discovery, and on an intellectual par with it. The academic environment is completely isolated from the top-secret domain of classified research, and academics do not have access to the tools and secret knowledge that may be hidden in the classified world. On the other hand, government researchers always have access to the academic literature. One might think of this flow of information in terms of a one-way function – information flows freely in one direction, but it is forbidden to send information in the opposite direction.

When Diffie told Hellman about Ellis, Cocks and Williamson, his attitude was that the discoveries of the academics should be a footnote in the history of classified research, and that the discoveries at GCHQ should be a footnote in the history of academic research. However, at that stage nobody except GCHQ, NSA, Diffie and Hellman knew about the classified research, and so it could not even be considered as a footnote.

By the mid-1980s, the mood at GCHQ was changing, and the management considered publicly announcing the work of Ellis, Cocks and Williamson. The mathematics of public-key cryptography was already well established in the public domain, and there seemed to be no reason to remain secretive. In fact, there would be distinct benefits if the British revealed their groundbreaking work on public-key cryptography. As Richard Walton recalls:

> We flirted with the idea of coming clean in 1984. We began to see advantages for GCHQ being more publicly acknowledged. It was a time when the government security market was expanding beyond the traditional military and diplomatic customer, and we needed to capture the confidence of those who did not traditionally deal with us. We were in the middle of Thatcherism, and we were trying to counter a sort of 'government is bad, private is good' ethos. So, we had the intention of publishing a paper, but that idea was scuppered by that blighter Peter Wright, who wrote *Spycatcher*. We were just warming up senior management to approve this release, when there was all this hoo-ha about *Spycatcher*. Then the order of the day was 'heads down, hats on'.

Peter Wright was a retired British intelligence officer, and the publication of *Spycatcher*, his memoirs, was a source of great embarrassment to the British government. It would be another 13 years before GCHQ eventually went public – 28 years after Ellis's initial breakthrough. In 1997 Clifford Cocks completed some important unclassified work on RSA, which would have been of interest to the wider community, and which would not be a security risk if it were to be published. As a result, he was asked to present a paper at the Institute of Mathematics and its Applications Conference to be held in Cirencester. The room would be full of cryptography experts. A handful of them would know that Cocks, who would be talking about just one aspect of RSA, was actually its unsung inventor. There was a risk that somebody might ask an embarrassing question, such as 'Did you invent RSA?' If such a question arose, what was Cocks supposed to do? According to GCHQ policy he would have to deny his role in the development of RSA, thus forcing him to lie about an issue that was totally innocuous. The situation was clearly ridiculous, and GCHQ decided that it was time to change its policy. Cocks was given permission to begin his talk by presenting a brief history of GCHQ's contribution to public-key cryptography.

On 18 December 1997, Cocks delivered his talk. After almost three decades of secrecy, Ellis, Cocks and Williamson received the acknowledgement they deserved. Sadly, James Ellis had died just one month earlier on 25 November 1997, at the age of seventy-three. Ellis joined the list of British cipher experts whose contributions would never be recognised during their lifetimes. Charles Babbage's breaking of the Vigenère cipher was never revealed during his lifetime, because his work was invaluable to British forces in the Crimea. Instead, credit for the work went to Friedrich Kasiski. Similarly, Alan Turing's contribution to the war effort was unparalleled, and yet government secrecy demanded that his work on Enigma could not be revealed.

In 1987, Ellis wrote a classified document that recorded his contribution to public-key cryptography, which included his thoughts on the secrecy that so often surrounds cryptographic work:

> Cryptography is a most unusual science. Most professional scientists aim to be the first to publish their work, because it is through dissemination that the work realises its value. In contrast, the fullest value of cryptography is realised by minimising the information available to potential adversaries. Thus professional cryptographers normally work in closed communities to provide sufficient professional interaction to ensure quality while maintaining secrecy from outsiders. Revelation of these secrets is normally only sanctioned in the interests of historical accuracy after it has been demonstrated that no further benefit can be obtained from continued secrecy.

7 Pretty Good Privacy

Just as Whit Diffie predicted in the early 1970s, we are now entering the Information Age, a post-industrial era in which information is the most valuable commodity. The exchange of digital information has become an integral part of our society. Already, tens of millions of e-mails are sent each day, and electronic mail will soon become more popular than conventional mail. The Internet, still in its infancy, has provided the infrastructure for the digital marketplace, and e-commerce is thriving. Money is flowing through cyberspace, and it is estimated that every day half the world's Gross Domestic Product travels through the Society for Worldwide Interbank Financial Telecommunications (SWIFT) network. In the future, democracies that favour referenda will begin to have on-line voting, and governments will use the Internet to help administer their countries, offering facilities such as on-line tax declarations.

However, the success of the Information Age depends on the ability to protect information as it flows around the world, and this relies on the power of cryptography. Encryption can be seen as providing the locks and keys of the Information Age. For two thousand years encryption has been of importance only to governments and the military, but today it also has a role to play in facilitating business, and tomorrow ordinary people will rely on cryptography in order to protect their privacy. Fortunately, just as the Information Age is taking off, we have access to extraordinarily strong encryption. The development of public-key cryptography, particularly the RSA cipher, has given today's cryptographers a clear advantage in their continual power struggle against cryptanalysts. If the value of N is large enough, then finding p and q takes Eve an unreasonable amount of time, and RSA encryption is therefore effectively unbreakable. Most important of all, public-key cryptography is not weakened by any key-distribution

Figure 70 Phil Zimmermann.

problems. In short, RSA guarantees almost unbreakable locks for our most precious pieces of information.

However, as with every technology, there is a dark side to encryption. As well as protecting the communications of law-abiding citizens, encryption also protects the communications of criminals and terrorists. Currently, the police use wire-tapping as a way of gathering evidence in serious cases, such as organised crime and terrorism, but this would be impossible if criminals used unbreakable ciphers. As we enter the twenty-first century, the fundamental dilemma for cryptography is to find a way of allowing the public and business to use encryption in order to exploit the benefits of the Information Age without allowing criminals to abuse encryption and evade arrest. There is currently an active and vigorous debate about the best way forward, and much of the discussion has been inspired by the story of Phil Zimmermann, a man whose attempts to encourage the widespread use of strong encryption have panicked America's security experts, threatened the effectiveness of the billion-dollar National Security Agency, and made him the subject of an FBI inquiry and a grand-jury investigation.

Phil Zimmermann spent the mid-1970s at Florida Atlantic University, where he studied physics and then computer science. On graduation he seemed set for a steady career in the rapidly growing computer industry, but the political events of the early 1980s transformed his life, and he became less interested in the technology of silicon chips and more worried about the threat of nuclear war. He was alarmed by the Soviet invasion of Afghanistan, the election of Ronald Reagan, the instability caused by an ageing Brezhnev and the increasingly tense nature of the Cold War. He even considered taking himself and his family to New Zealand, believing that this would be one of the few places on Earth that would be habitable after a nuclear conflict. But just as he had obtained passports and the necessary immigration papers, he and his wife attended a meeting held by the Nuclear Weapons Freeze Campaign. Rather than flee, the Zimmermanns decided to stay and fight the battle at home, becoming front-line anti-nuclear activists – they educated political candidates on issues of military policy, and were arrested at the Nevada nuclear testing grounds, alongside Carl Sagan and four hundred other protesters.

A few years later, in 1988, Mikhail Gorbachev became head of state of the Soviet Union, heralding perestroika, glasnost and a reduction in tension between East and West. Zimmermann's fears began to subside, but he did not lose his passion for political activism, he merely channelled it in a different direction. He began to focus his attentions on the digital revolution and the necessity for encryption:

> Cryptography used to be an obscure science, of little relevance to everyday life. Historically, it always had a special role in military and diplomatic communications. But in the Information Age, cryptography is about political power, and in particular, about the power relationship between a government and its people. It is about the right to privacy, freedom of speech, freedom of political association, freedom of the press, freedom from unreasonable search and seizure, freedom to be left alone.

These views might seem paranoid, but according to Zimmermann there is a fundamental difference between traditional and digital communication which has important implications for security:

> In the past, if the government wanted to violate the privacy of ordinary citizens, it had to expend a certain amount of effort to intercept and steam open and read paper mail, or listen to and possibly transcribe spoken telephone conversations. This is analogous to catching fish with a hook and a line, one fish at a time. Fortunately for freedom and democracy, this kind of labor-intensive monitoring is not practical on a large scale. Today, electronic mail is gradually replacing conventional paper mail, and is soon to be the norm for everyone, not the novelty it is today. Unlike paper mail, e-mail messages are just too easy to intercept and scan for interesting keywords. This can be done easily, routinely, automatically, and undetectably on a grand scale. This is analogous to driftnet fishing – making a quantitative and qualitative Orwellian difference to the health of democracy.

The difference between ordinary and digital mail can be illustrated by imagining that Alice wants to send out invitations to her birthday party, and that Eve, who has not been invited, wants to know the time and place of the party. If Alice uses the traditional method of posting letters, then it is very difficult for Eve to intercept one of the invitations. To start with, Eve does not know where Alice's invitations entered the postal system, because Alice could use any postbox in the city. Her only hope for intercepting one of the invitations is to somehow identify the address of one of Alice's friends, and infiltrate the local sorting office. She then has to check each and every letter manually. If she does manage to find a letter from Alice, she will have to steam it open in order to get the information she wants, and then return it to its original condition to avoid any suspicion of tampering.

In comparison, Eve's task is made considerably easier if Alice sends her invitations by e-mail. As the messages leave Alice's computer, they will go to a local server, a main entry point for the Internet; if Eve is clever enough, she can hack into that local server without leaving her home. The invitations will carry Alice's e-mail address, and it would be a trivial matter to set up an electronic sieve that looks for e-mails containing Alice's address. Once an invitation has been found, there is no envelope to open, and so no problem in reading it. Furthermore, the invitation can be sent on its way without it showing any sign of having been intercepted. Alice would be oblivious to what was going on. However, there is a way to prevent Eve from reading Alice's e-mails, namely encryption.

More than a hundred million e-mails are sent around the world each day, and they are all vulnerable to interception. Digital technology has aided communication, but it has also given rise to the possibility of those communications being monitored. According to Zimmermann, cryptographers have a duty to encourage the use of encryption and thereby protect the privacy of the individual:

> A future government could inherit a technology infrastructure that's optimized for surveillance, where they can watch the movements of their political opposition, every financial transaction, every communication, every bit of e-mail, every phone call. Everything could be filtered and scanned and automatically recognized by voice recognition technology and transcribed. It's time for cryptography to step out of the shadows of spies and the military, and step into the sunshine and be embraced by the rest of us.

In theory, when RSA was invented in 1977 it offered an antidote to the Big Brother scenario because individuals were able to create their own public- and private-keys, and thereafter send and receive perfectly secure messages. However, in practice there was a major problem because the actual process of RSA encryption required a substantial amount of computing power in comparison with symmetric forms of encryption, such as

DES. Consequently, in the 1980s it was only government, the military and large businesses that owned computers powerful enough to run RSA. Not surprisingly, RSA Data Security, Inc., the company set up to commercialise RSA, developed their encryption products with only these markets in mind.

In contrast, Zimmermann believed that everybody deserved the right to the privacy that was offered by RSA encryption, and he directed his political zeal towards developing an RSA encryption product for the masses. He intended to draw upon his background in computer science to design a product with economy and efficiency in mind, thus not overloading the capacity of an ordinary personal computer. He also wanted his version of RSA to have a particularly friendly interface, so that the user did not have to be an expert in cryptography to operate it. He called his project Pretty Good Privacy, or PGP for short. The name was inspired by Ralph's Pretty Good Groceries, a sponsor of Garrison Keillor's *Prairie Home Companion,* one of Zimmermann's favourite radio shows.

During the late 1980s, working from his home in Boulder, Colorado, Zimmermann gradually pieced together his scrambling software package. His main goal was to speed up RSA encryption. Ordinarily, if Alice wants to use RSA to encrypt a message to Bob, she looks up his public-key and then applies RSA's one-way function to the message. Conversely, Bob decrypts the ciphertext by using his private-key to reverse RSA's one-way function. Both processes require considerable mathematical manipulation, so encryption and decryption can, if the message is long, take several minutes on a personal computer. If Alice is sending a hundred messages a day, she cannot afford to spend several minutes encrypting each one. To speed up encryption and decryption, Zimmermann employed a neat trick that used asymmetric RSA encryption in tandem with old-fashioned symmetric encryption. Traditional symmetric encryption can be just as secure as asymmetric encryption, and it is much quicker to perform, but symmetric encryption suffers from the problem of having to distribute the key, which has to be securely transported from the sender to the receiver. This is where RSA comes to the rescue, because RSA can be used to encrypt the symmetric key.

Zimmermann pictured the following scenario. If Alice wants to send an encrypted message to Bob, she begins by encrypting it with a symmetric cipher. Zimmermann suggested using a cipher known as IDEA, which is similar to DES. To encrypt with IDEA, Alice needs to choose a key, but for Bob to decrypt the message Alice somehow has to get the key to Bob. Alice overcomes this problem by looking up Bob's RSA public-key, and then uses it to encrypt the IDEA key. So, Alice ends up sending two things to Bob: the message encrypted with the symmetric IDEA cipher and the IDEA key encrypted with the asymmetric RSA cipher. At the other end, Bob uses his RSA private-key to decrypt the IDEA key, and then uses the IDEA key to decrypt the message. This might seem convoluted, but the advantage is that the message, which might contain a large amount of information, is being encrypted with a quick symmetric cipher, and only the symmetric IDEA key, which consists of a relatively small amount of information, is being encrypted with a slow asymmetric cipher. Zimmermann planned to have this combination of RSA and IDEA within the PGP product, but the user-friendly interface would mean that the user would not have to get involved in the nuts and bolts of what was going on.

Having largely solved the speed problem, Zimmermann also incorporated a series of handy features into PGP. For example, before using the RSA component of PGP, Alice needs to generate her own private-key and public-key. Key generation is not trivial, because it requires finding a pair of giant primes. However, Alice only has to wiggle her mouse in an erratic manner, and the PGP program will go ahead and create her private-key and public-key – the mouse movements introduce a random factor which PGP utilises to ensure that every user has their own distinct pair of primes, and therefore their own unique private-key and public-key. Thereafter Alice merely has to publicise her public-key.

Another helpful aspect of PGP is its facility for digitally signing an e-mail. Ordinarily e-mail does not carry a signature, which means that it is impossible to verify the true author of an electronic message. For example, if Alice uses e-mail to send a love letter to Bob, she normally encrypts it with his public-key, and when he receives it he decrypts it with his private-key. Bob is initially flattered, but how can he be sure that the love letter is really from Alice? Perhaps the malevolent Eve wrote the e-mail and typed Alice's name at the bottom. Without the reassurance of a handwritten ink signature, there is no obvious way to verify the authorship.

Alternatively, imagine that a bank receives an e-mail from a client, which instructs that all the client's funds should be transferred to a private numbered bank account in the Cayman Islands. Once again, without a handwritten signature, how does the bank know that the e-mail is really from the client? The e-mail could have been written by a criminal attempting to divert the money to his own Cayman Islands bank account. In order to develop trust on the Internet, it is essential that there is some form of reliable digital signature.

The PGP digital signature is based on a principle that was first developed by Whitfield Diffie and Martin Hellman. When they proposed the idea of separate public-keys and private-keys, they realised that, in addition to solving the key-distribution problem, their invention would also provide a natural mechanism for generating e-mail signatures. In Chapter 6 we saw that the public-key is for encrypting and the private-key for decrypting. In fact the process can be swapped around, so that the private-key is used for encrypting and the public-key is used for decrypting. This mode of encryption is usually ignored because it offers no security. If Alice uses her private-key to encrypt a message to Bob, then everybody can decrypt it because everybody has Alice's public-key. However, this mode of operation does verify authorship, because if Bob can decrypt a message using Alice's public-key, then it must have been encrypted using her private-key – only Alice has access to her private-key, so the message must have been sent by Alice.

In effect, if Alice wants to send a love letter to Bob, she has two options. Either she encrypts the message with Bob's public-key to guarantee privacy, or she encrypts it with her own private-key to guarantee authorship. However, if she combines both options she can guarantee privacy and authorship. There are quicker ways to achieve this, but here is one way in which Alice might send her love letter. She starts by encrypting the message using her private-key, then she encrypts the resulting ciphertext using Bob's public-key. We can picture the message surrounded by a fragile inner shell, which represents encryption by Alice's private-key, and a strong outer shell, which represents encryption by Bob's public-key. The resulting ciphertext can only be deciphered by Bob, because only he has access to the private-key necessary to crack the strong outer shell. Having deciphered the outer shell, Bob can then easily decipher the inner shell using Alice's public-key – the inner shell is not meant to protect the message, but it does prove that the message came from Alice, and not an impostor.

By this stage, sending a PGP encrypted message is becoming quite complicated. The IDEA cipher is being used to encrypt the message, RSA is being used to encrypt the IDEA key, and another stage of encryption has to be incorporated if a digital signature is required. However, Zimmermann developed his product in such a way that it would do everything automatically, so that Alice and Bob would not have to worry about the mathematics. To send a message to Bob, Alice would simply write her e-mail and select the PGP option from a menu on her computer screen. Next she would type in Bob's name, then PGP would find Bob's public-key and automatically perform all the encryption. At the same time PGP would do the necessary jiggery-pokery required to digitally sign the message. Upon receiving the encrypted message, Bob would select the PGP option, and PGP would decrypt the message and verify the author. Nothing in PGP was original – Diffie and Hellman had already thought of digital signatures and other cryptographers had used a combination of symmetric and asymmetric ciphers to speed up encryption – but Zimmermann was the first to put everything together in one easy-to-use encryption product, which was efficient enough to run on a moderately sized personal computer.

By the summer of 1991, Zimmermann was well on the way to turning PGP into a polished product. Only two problems remained, neither of them technical. A long-term problem had been the fact that RSA, which is at the heart of PGP, is a patented product, and patent law required Zimmermann to obtain a licence from RSA Data Security, Inc. before he launched PGP. However, Zimmermann decided to put this problem to one side. PGP was intended not as a product for businesses, but rather as something for the individual. He felt that he would not be competing directly with RSA Data Security, Inc., and hoped that the company would give him a free licence in due course.

A more serious and immediate problem was the U.S. Senate's 1991 omnibus anti-crime bill, which contained the following clause: 'It is the sense of Congress that providers of electronic communications services and manufacturers of electronic communications service equipment shall ensure that communications systems permit the government to obtain the plain text contents of voice, data, and other communications when appropriately authorized by law.' The Senate was concerned that developments in digital technology, such as cellular telephones, might prevent law enforcers from performing effective wire-taps. However, as well as forcing companies to guarantee the possibility of wire-tapping, the bill also seemed to threaten all forms of secure encryption.

A concerted effort by RSA Data Security, Inc., the communications industry, and civil liberty groups forced the clause to be dropped, but the consensus was that this was only a temporary reprieve. Zimmermann was fearful that sooner or later the government would again try to bring in legislation that would effectively outlaw encryption such as PGP. He had always intended to sell PGP, but now he reconsidered his options. Rather than waiting and risk PGP being banned by the government, he decided that it was more important for it to be available to everybody before it was too late. In June 1991 he took the drastic step of asking a friend to post PGP on a Usenet bulletin board. PGP is just a piece of software, and so from the bulletin board it could be downloaded by anyone for free. PGP was now loose on the Internet.

Initially, PGP caused a buzz only among aficionados of cryptography. Later it was downloaded by a wider range of Internet enthusiasts. Next, computer magazines ran brief reports and then full-page articles on the PGP phenomenon. Gradually PGP began to permeate the most remote corners of the digital community. For example, human rights groups around the world started to use PGP to encrypt their documents, in order to prevent the information from falling into the hands of the regimes that were being accused of human-rights abuses. Zimmermann began to receive e-mails praising him for his creation. 'There are resistance groups in Burma,' says Zimmermann, 'who are using it in jungle training camps. They've said that it's helped morale there, because before PGP was introduced captured documents would lead to the arrest, torture and execution of entire families.' In October 1993, he received the following e-mail from someone in Latvia on the day that Boris Yeltsin was shelling the Latvian Parliament building: 'Phil, I wish you to know: let it never be, but if dictatorship takes over Russia, your PGP is widespread from Baltic to Far East now and will help democratic people if necessary. Thanks.'

While Zimmermann was gaining fans around the world, back home in America he had been the target of criticism. RSA Data Security, Inc. decided not to give Zimmermann a free licence, and was enraged that its patent was being infringed. Although Zimmermann released PGP as freeware (free software), it contained the RSA system of public-key cryptography, and consequently RSA Data Security, Inc. labelled PGP as 'banditware'. Zimmermann had given something away which belonged to somebody else. The patent wrangle would continue for several years, during which time Zimmermann encountered an even greater problem.

In February 1993, two government investigators paid Zimmermann a visit. After their initial enquiries about patent infringement, they began to ask questions about the more serious accusation of illegally exporting a weapon. Because the U.S. Government included encryption software within its definition of munitions, along with missiles, mortars and machine guns, PGP could not be exported without a licence from the State Department. In other words, Zimmermann was accused of being an arms dealer because he had exported PGP via the Internet. Over the next three years Zimmermann became the subject of a grand-jury investigation and found himself pursued by the FBI.

Encryption for the Masses . . . Or Not?

The investigation into Phil Zimmermann and PGP ignited a debate about the positive and negative effects of encryption in the Information Age. The spread of PGP galvanised cryptographers, politicians, civil libertarians and law enforcers into thinking about the implications of widespread encryption. There were those, like Zimmermann, who believed that the widespread use of secure encryption would be a boon to society, providing individuals with privacy for their digital communications. Ranged against them were those who believed that encryption was a threat to society, because criminals and terrorists would be able to communicate in secret, safe from police wire-taps.

The debate continued throughout the 1990s, and is currently as contentious as ever. The fundamental question is whether or not governments should legislate against cryptography. Cryptographic freedom would allow everyone, including criminals, to be confident that their e-mails are secure. On the other hand, restricting the use of cryptography would allow the police to spy on criminals, but it would also allow the police and everybody else to spy on the average citizen. Ultimately, we, through the governments we elect, will decide the future role of cryptography. This section is devoted to outlining the two sides of the debate. Much of the discussion will refer to policies and policy-makers in America, partly because it is the home of PGP, around which much of the debate has centred, and partly because whatever policy is adopted in America will ultimately have an effect on policies around the globe.

The case against the widespread use of encryption, as argued by law enforcers, centres on the desire to maintain the status quo. For decades, police around the world have conducted legal wire-taps in order to catch criminals. For example, in America in 1918, wire-taps were used to counteract the presence of wartime spies, and in the 1920s they proved especially effective in convicting bootleggers. The view that wire-tapping was a necessary tool of law enforcement became firmly established in the late 1960s, when the FBI realised that organised crime was becoming a growing threat to the nation. Law enforcers were having great difficulty in convicting suspects because the mob made threats against anyone who might consider testifying against them, and there was also the code of *omerta*, or silence. The police felt that their only hope was to gather evidence via wire-taps, and the Supreme Court was sympathetic to this argument. In 1967 it ruled that the police could employ wire-taps as long as they had first obtained a court authorisation.

Twenty years later, the FBI still maintains that 'court ordered wire-tapping is the single most effective investigative technique used by law enforcement to combat illegal drugs, terrorism, violent crime, espionage, and organized crime'. However, police wire-taps would be useless if criminals had access to encryption. A phone call made over a digital line is nothing more than a stream of numbers, and can be encrypted according to the same techniques used to encrypt e-mails. PGPfone, for example, is one of several products capable of encrypting voice communications made over the Internet.

Law enforcers argue that effective wire-tapping is necessary in order to maintain law and order, and that encryption should be restricted so that they can continue with their interceptions. The police have already encountered criminals using strong encryption to protect themselves. A German legal expert said that 'hot businesses such as the arms and drug trades are no longer done by phone, but are being settled in encrypted form on the worldwide data networks'. A White House official indicated a similarly worrying trend in America, claiming that 'organized crime members are some of the most advanced users of computer systems and of strong encryption'. For instance, the Cali cartel arranges its drug deals via encrypted communications. Law enforcers fear that the Internet coupled with cryptography will help criminals to communicate and coordinate their efforts, and they are particularly concerned about the so-called Four Horsemen of the Infocalypse – drug dealers, organised crime, terrorists and paedophiles – the groups who will benefit most from encryption.

In addition to encrypting communications, criminals and terrorists are also encrypting their plans and records, hindering the recovery of evidence. The Aum Shinrikyo sect, responsible for the gas attacks on the Tokyo subway in 1995, were found to have encrypted some of their documents using RSA. Ramsey Yousef, one of the terrorists involved in the World Trade Center bombing, kept plans for future terrorist acts encrypted on his laptop. Besides international terrorist organisations, more run-of-the-mill criminals also benefit from encryption. An illegal gambling syndicate in America, for example, encrypted its accounts for four years. Commissioned in 1997 by the National Strategy Information Center's U.S. Working Group on Organized Crime, a study by Dorothy Denning and William Baugh estimated that there were five hundred criminal cases worldwide involving encryption, and predicted that this number would roughly double each year.

In addition to domestic policing, there are also issues of national security. America's National Security Agency is responsible for gathering intelligence on the nation's enemies by deciphering their communications. The NSA operates a worldwide network of listening stations, in cooperation with Britain, Australia, Canada and New Zealand, who all gather and share information. The network includes sites such as the Menwith Hill Signals Intelligence Base in Yorkshire, the world's largest spy station. Part of Menwith Hill's work involves the Echelon system, which is capable of scanning e-mails, faxes, telexes and telephone calls, searching for particular words. Echelon operates according to a dictionary of suspicious words, such as 'Hezbollah', 'assassin' and 'Clinton', and the system is smart enough to recognise these words in real time. Echelon can earmark questionable messages for further examination, enabling it to monitor messages from particular political groups or terrorist organisations. However, Echelon would effectively be useless if all messages were strongly encrypted. Each of the nations participating in Echelon would lose valuable intelligence on political plotting and terrorist attacks.

On the other side of the debate are the civil libertarians, including groups such as the Center for Democracy and Technology and the Electronic Frontier Foundation. The pro-encryption case is based on the belief that privacy is a fundamental human right, as recognised by Article 12 of the Universal Declaration of Human Rights: 'No one shall be subjected to arbitrary interference with his privacy, family, home or correspondence, nor to attacks upon his honour and reputation. Everyone has the right to the protection of the law against such interference or attacks.'

Civil libertarians argue that the widespread use of encryption is essential for guaranteeing the right to privacy. Otherwise, they fear, the advent of digital technology, which makes monitoring so much easier, will herald a new era of wire-tapping and the abuses that inevitably follow. In the past, governments have frequently used their power in order to conduct wire-taps on innocent citizens. Presidents Lyndon Johnson and Richard Nixon were guilty of unjustified wire-taps, and President John F. Kennedy conducted dubious wire-taps in the first month of his presidency. In the run-up to a bill concerning Dominican sugar imports, Kennedy asked for wire-taps to be placed on several congressmen. His justification was that he believed that they were being bribed, a seemingly valid national-security concern. However, no evidence of bribery was ever found, and the wire-taps merely provided Kennedy with valuable political information, which helped the administration to win the bill.

One of the best-known cases of continuous unjustified wire-tapping concerns Martin Luther King Jr, whose telephone conversations were monitored for several years. For example, in 1963 the FBI obtained information on King via a wire-tap and fed it to Senator James Eastland in order to help him in debates on a civil-rights bill. More generally, the FBI gathered details about King's personal life, which were used to discredit him. Recordings of King telling bawdy stories were sent to his wife and played in front of President Johnson. Then, following King's award of the Nobel prize, embarrassing details about King's life were passed to any organisation that was considering conferring an honour upon him.

Other governments are equally guilty of abusing wire-taps. The Commission Nationale de Contrôle des Interceptions de Securité estimates that there are roughly 100,000 illegal wire-taps conducted in France each year. Possibly the greatest infringement of everybody's privacy is the international Echelon programme. Echelon does not have to justify its interceptions, and it does not focus on particular individuals. Instead, it indiscriminately harvests information, using receivers that detect the telecommunications that bounce off satellites. If Alice sends a harmless transatlantic message to Bob, then it will certainly be intercepted by Echelon, and if the message happens to contain a few words that appear in the Echelon dictionary, then it would be earmarked for further examination, alongside messages from extreme political groups and terrorist gangs. Whereas law enforcers argue that encryption should be banned because it would make Echelon ineffective, the civil libertarians argue that encryption is necessary exactly because it would make Echelon ineffective.

When law enforcers argue that strong encryption will reduce criminal convictions, civil libertarians reply that the issue of privacy is more important. In any case, civil libertarians insist that encryption would not be an enormous barrier to law enforcement because wire-taps are not a crucial element in most cases. For example, in America in 1994 there were roughly a thousand court-sanctioned wire-taps, compared with a quarter of a million federal cases.

Not surprisingly, among the advocates of cryptographic freedom are some of the inventors of public-key cryptography. Whitfield Diffie states that individuals have enjoyed complete privacy for most of history:

> In the 1790s, when the Bill of Rights was ratified, any two people could have a private conversation – with a certainty no one in the world enjoys today – by walking a few meters down the road and looking to see no one was hiding in the bushes. There were no recording devices, parabolic microphones, or laser interferometers bouncing off their eyeglasses. You will note that civilization survived. Many of us regard that period as a golden age in American political culture.

Ron Rivest, one of the inventors of RSA, thinks that restricting cryptography would be foolhardy:

> It is poor policy to clamp down indiscriminately on a technology just because some criminals might be able to use it to their advantage. For example, any U.S. citizen can freely buy a pair of gloves, even though a burglar might use them to ransack a house without leaving fingerprints. Cryptography is a data-protection technology, just as gloves are a hand-protection technology. Cryptography protects data from hackers, corporate spies, and con artists, whereas gloves protect hands from cuts, scrapes, heat, cold, and infection. The former can frustrate FBI wire-tapping, and the latter can thwart FBI fingerprint analysis. Cryptography and gloves are both dirt-cheap and widely available. In fact, you can download good cryptographic software from the Internet for less than the price of a good pair of gloves.

Possibly the greatest allies of the civil-libertarian cause are the big corporations. Internet commerce is still in its infancy, but sales are growing rapidly, with retailers of books, music CDs and computer software leading the way, and with supermarkets, travel companies and other businesses following in their wake. In 1998 a million Britons used the Internet to buy products worth £400 million, a figure that was set to quadruple in 1999. In just a few years from now Internet commerce could dominate the marketplace, but only if businesses can address the issues of security and trust. A business must be able to guarantee the privacy and security of financial transactions, and the only way to do this is to employ strong encryption.

At the moment, a purchase on the Internet can be secured by public-key cryptography. Alice visits a company's website and selects an item. She then fills in an order form which asks her for her name, address and credit-card details. Alice then uses the company's public-key to encrypt the order form. The encrypted order form is transmitted to the company, who are the only people able to decrypt it, because only they have the private-key necessary for decryption. All of this is done automatically by Alice's web browser (e.g. Netscape or Explorer) in conjunction with the company's computer.

As usual, the security of the encryption depends on the size of the key. In America there are no restrictions on key size, but U.S. software companies are still not allowed to export web products that offer strong encryption. Hence, browsers exported to the rest of the world can handle only short keys, and thus offer only moderate security. In fact, if Alice is in London buying a book from a company in Chicago, her Internet transaction is a billion billion billion times less secure than a transaction by Bob in New York buying a book from the same company. Bob's transaction is absolutely secure because his browser supports encryption with a larger key, whereas Alice's transaction could be deciphered by a determined criminal. Fortunately, the cost of the equipment required to decipher Alice's credit-card details is vastly greater than the typical credit-card limit, so such an attack is not cost-effective. However, as the amount of money flowing around the Internet increases, it will eventually become profitable for criminals to decipher credit-card details. In short, if Internet commerce is to thrive, consumers around the world must have proper security, and businesses will not tolerate crippled encryption.

Businesses also desire strong encryption for another reason. Corporations store vast amounts of information on computer databases, including product descriptions, customer details and business accounts. Naturally, corporations want to protect this information from hackers who might infiltrate the computer and steal the information. This protection can be achieved by encrypting stored information, so that it is only accessible to employees who have the decryption key.

To summarise the situation, it is clear that the debate is between two camps: civil libertarians and businesses are in favour of strong encryption, while law enforcers are in favour of severe restrictions. In general, popular opinion appears to be swinging behind the pro-encryption alliance, who have been helped by a sympathetic media and a couple of Hollywood films. In early 1998, *Mercury Rising* told the story of a new, supposedly unbreakable NSA cipher which is inadvertently deciphered by a nine-year-old autistic savant. Alec Baldwin, an NSA agent, sets out to assassinate the boy, who is perceived as a threat to national security. Luckily, the boy has Bruce Willis to protect him. Also in 1998, Hollywood released *Enemy of the State*, which dealt with an NSA plot to murder a politician who supports a bill in favour of strong encryption. The politician is killed, but a lawyer played by Will Smith and an NSA rebel played by Gene Hackman eventually bring the NSA assassins to justice. Both films depict the NSA as more sinister than the CIA, and in many ways the NSA has taken over the role of establishment menace.

While the pro-encryption lobby argues for cryptographic freedom, and the anti-encryption lobby for cryptographic restrictions, there is a third option that might offer a compromise. Over the last decade, cryptographers and policy-makers have been investigating the pros and cons of a scheme known as *key escrow*. The term 'escrow' usually relates to an arrangement in which someone gives a sum of money to a third party, who can then deliver the money to a second party under certain circumstances. For example, a tenant might lodge a deposit with a solicitor, who can then deliver it to a landlord in the event of damage to the property. In terms of cryptography, escrow means that Alice would give a copy of her private-key to an escrow agent, an independent, reliable middleman, who is empowered to deliver the private-key to the police if ever there was sufficient evidence to suggest that Alice was involved in crime.

The most famous trial of cryptographic key escrow was the American Escrowed Encryption Standard, adopted in 1994. The aim was to encourage the adoption of two encryption systems, called *clipper* and *capstone*, to be used for telephone communication and computer communication, respectively. To use clipper encryption, Alice would buy a phone with a pre-installed chip which would hold her secret private-key information. At the very moment she bought the clipper phone, a copy of the private-key in the chip would be split into two halves, and each half would be sent to two separate Federal authorities for storage. The U.S. Government argued that Alice would have access to secure encryption, and her privacy would only be broken if law enforcers could persuade both Federal authorities that there was a case for obtaining her escrowed private-key.

The U.S. Government employed clipper and capstone for its own communications, and made it obligatory for companies involved in government business to adopt the American Escrowed Encryption Standard. Other businesses and individuals were free to use other forms of encryption, but the government hoped that clipper and capstone would gradually become the nation's favourite form of encryption. However, the policy did not work. The idea of key escrow won few supporters outside government. Civil libertarians did not like the idea of Federal authorities having possession of everybody's keys – they made an analogy to real keys, and asked how people would feel if the government had the keys to all our houses. Cryptographic experts pointed out that just one crooked employee could undermine the whole system by selling escrowed keys to the highest bidder. And businesses were worried about confidentiality. For example, a European business in America might fear that its messages were being intercepted by American trade officials in an attempt to obtain secrets that might give American rivals a competitive edge.

Despite the failure of clipper and capstone, many governments remain convinced that key escrow can be made to work, as long as the keys are sufficiently well protected from criminals and as long as there are safeguards to reassure the public that the system is not open to government abuse. Louis J. Freeh, Director of the FBI, said in 1996: 'The law enforcement community fully supports a balanced encryption policy . . . Key escrow is not just the only solution; it is, in fact, a very good solution because it effectively balances fundamental societal concerns involving privacy, information security, electronic commerce, public safety, and national security.' Although the U.S. Government has backtracked on its escrow proposals, many suspect that it will attempt to reintroduce an alternative form of key escrow at some time in the future. Having witnessed the failure of optional escrow, governments might even consider compulsory escrow. Meanwhile, the pro-encryption lobby continues to argue against key escrow. Kenneth Neil Cukier, a technology journalist, has written that: 'The people involved in the crypto debate are all intelligent, honorable and pro-escrow, but they never possess more than two of these qualities at once.'

There are various other options that governments could choose to implement, in order to try to balance the concerns of civil libertarians, business and law enforcement. It is far from clear which will be the preferred option, because at present cryptographic policy is in a state of flux. A steady stream of events around the world is constantly influencing the debate on encryption. In November 1998, the Queen's Speech announced forthcoming British legislation relating to the digital marketplace. In December 1998, 33 nations signed the Wassenaar Arrangement limiting arms exports, which also covers powerful encryption technologies. In January 1999, France repealed its anti-cryptography laws, which had previously been the most restrictive in Western Europe, probably as a result of pressure from the business community. In March 1999, the British Government released a consultation document on a proposed Electronic Commerce Bill.

By the time you read this there will have been several more twists and turns in the debate on cryptographic policy. However, one aspect of future encryption policy seems certain, namely the necessity for *certification authorities*. If Alice wants to send a secure e-mail to a new friend, Zak, she needs Zak's public-key. She might ask Zak to send his public-key to her in the post. Unfortunately, there is then the risk that Eve will intercept Zak's letter to Alice, destroy it and forge a new letter, which actually includes her own public-key instead of Zak's. Alice may then send a sensitive e-mail to Zak, but she will unknowingly have encrypted it with Eve's public-key. If Eve can intercept this e-mail, she can then easily decipher it and read it. In other words, one of the problems with public-key cryptography is being sure that you have the genuine public-key of the person with whom you wish to communicate. Certification authorities are organisations that will verify that a public-key does indeed correspond to a particular person. A certification authority might request a face-to-face meeting with Zak as a way of ensuring that they have correctly catalogued his public-key. If Alice trusts the certification authority, she can obtain from it Zak's public-key, and be confident that the key is valid.

I have explained how Alice could securely buy products from the Internet by using a company's public-key to encrypt the order form. In fact, she would do this only if the public-key had been validated by a certification authority. In 1998, the market leader in certification was Verisign, which has grown into a $30 million company in just four years. As well as ensuring reliable encryption by certifying public-keys, certification authorities can also guarantee the validity of digital signatures. In 1998, Baltimore Technologies in Ireland provided the certification for the digital signatures of President Bill Clinton and Prime Minister Bertie Ahern. This allowed the two leaders to digitally sign a communiqué in Dublin.

Certification authorities pose no risk to security. They would merely have asked Zak to reveal his public-key so that they can validate it for others who wish to send him encrypted messages. However, there are other companies, known as *trusted third parties* (TTPs), that provide a more controversial service known as *key recovery*. Imagine a legal firm that protects all its vital documents by encrypting them with its own public-key, so that only it can decrypt them with its own private-key. Such a system is an effective measure against hackers and anybody else who might attempt to steal information. However, what happens if the employee who stores the private-key forgets it, absconds with it or is knocked over by a bus? Governments are encouraging the formation of TTPs to keep copies of all keys. A company that loses its private-key would then be able to recover it by approaching its TTP.

Trusted third parties are controversial because they would have access to people's private-keys, and hence they would have the power to read their clients' messages. They must be trustworthy, otherwise the system is easily abused. Some argue that TTPs are effectively a reincarnation of key escrow, and that law enforcers would be tempted to bully TTPs into giving up a client's keys during a police investigation. Others maintain that TTPs are a necessary part of a sensible public-key infrastructure.

Nobody can predict what role TTPs will play in the future, and nobody can foresee with certainty the shape of cryptographic policy ten years from now. However, I suspect that in the near future the pro-encryption lobby will initially win the argument, mainly because no country will want to have encryption laws that prohibit e-commerce. However, if this policy does turn out to be a mistake, then it will always be possible to reverse the laws. If there were to be a series of terrorist atrocities, and law enforcers could show that wire-taps would have prevented them, then governments would rapidly gain sympathy for a policy of key escrow. All users of strong encryption would be forced to deposit their keys with a key-escrow agent, and thereafter anybody who sent an encrypted message with a non-escrowed key would be breaking the law. If the penalty for non-escrowed encryption were sufficiently severe, law enforcers could regain control. Later, if governments were to abuse the trust associated with a system of key escrow, the public would call for a return to cryptographic freedom, and the pendulum would swing back. In short, there is no reason why we cannot change our policy to suit the political, economic and social climate. The deciding factor will be whom the public fears the most – criminals or the government.

The Rehabilitation of Zimmermann

In 1993, Phil Zimmermann became the subject of a grand-jury investigation. According to the FBI, he had exported a munition because he was supplying hostile nations and terrorists with the tools they needed to evade the authority of the U.S. Government. As the investigation dragged on, more and more cryptographers and civil libertarians rushed to support Zimmermann, establishing an international fund to finance his legal defence. At the same time, the kudos of being the subject of an FBI inquiry boosted the reputation of PGP, and Zimmermann's creation spread via the Internet even more quickly – after all, this was the encryption software that was so secure that it frightened the Feds.

Pretty Good Privacy had initially been released in haste, and as a result the product was not as polished as it could have been. Soon there was a clamour to develop a revised version of PGP, but clearly Zimmermann was not in a position to continue working on the product. Instead, software engineers in Europe began to rebuild PGP. In general, European attitudes towards encryption were, and still are, more liberal, and there would be no restrictions on exporting a European version of PGP around the world. Furthermore, the RSA patent wrangle was not an issue in Europe, because RSA patents did not apply outside America.

After three years the grand-jury investigation had still not brought Zimmermann to trial. The case was complicated by the nature of PGP and the way it had been distributed. If Zimmermann had loaded PGP onto a computer and then shipped it to a hostile regime, the case against him would have been straightforward because clearly he would have been guilty of exporting a complete working encryption system. Similarly, if he had exported a disk containing the PGP program, then the physical object could have been interpreted as a cryptographic device, and once again the case against Zimmermann would have been fairly solid. On the other hand, if he had printed the computer program and exported it as a book, the case against him would no longer be clear cut, because he would then be considered to have exported knowledge rather than a cryptographic device. However, printed matter can easily be scanned electronically and the information can be fed directly into a computer, which means that a book is as dangerous as a disk. What actually occurred was that Zimmermann gave a copy of PGP to 'a friend', who simply installed it on an American computer, which happened to be connected to the Internet. After that, a hostile regime may or may not have downloaded it. Was Zimmermann really guilty of exporting PGP? Even today, the legal issues surrounding the Internet are subject to debate and interpretation. Back in the early 1990s, the situation was vague in the extreme.

In 1996, after three years of investigation, the U.S. Attorney General's Office dropped its case against Zimmermann. The FBI realised that it was too late – PGP had escaped onto the Internet, and prosecuting Zimmermann would achieve nothing. There was the additional problem that Zimmermann was being supported by major institutions, such as the Massachusetts Institute of Technology Press, which had published PGP in a 600-page book. The book was being distributed around the world, so prosecuting Zimmermann would have meant prosecuting the MIT Press. The FBI was also reluctant to pursue a prosecution because there was a significant chance that Zimmermann would not be convicted. An FBI trial might achieve nothing more than an embarrassing constitutional debate about the right to privacy, thereby stirring up yet more public sympathy in favour of widespread encryption.

Zimmermann's other major problem also disappeared. Eventually he achieved a settlement with RSA and obtained a licence which solved the patent issue. At last, PGP was a legitimate product and Zimmermann was a free man. The investigation had turned him into a cryptographic crusader, and every marketing manager in the world must have envied the notoriety and free publicity that the case gave to PGP. At the end of 1997, Zimmermann sold PGP to Network Associates and he became one of their senior fellows. Although PGP is now sold to businesses, it is still freely available to individuals who do not intend to use it for any commercial purpose. In other words, individuals who merely wish to exercise their right to privacy can still download PGP from the Internet without paying for it.

If you would like to obtain a copy of PGP, there are many sites on the Internet that offer it, and you should find them fairly easily. Probably the most reliable source is at http://www.pgpi.com/, the International PGP Home Page, from where you can download the American and international versions of PGP. At this point, I would like to absolve myself of any responsibility – if you do choose to install PGP, it is up to you check that your computer is capable of running it, that the software is not infected with a virus, and so on. Also, you should check that you are in a country that permits the use of strong encryption. Finally, you should ensure that you are downloading the appropriate version of PGP: individuals living outside America should not download the American version of PGP, because this would violate American export laws. The international version of PGP does not suffer from export restrictions.

I still remember the Sunday afternoon when I first downloaded a copy of PGP from the Internet. Ever since, I have been able to guarantee my e-mails against being intercepted and read, because I can now encrypt sensitive material to Alice, Bob and anybody else who possesses PGP software. My laptop and its PGP software provide me with a level of security that is beyond the combined efforts of all the world's code-breaking establishments.

8 A Quantum Leap into the Future

For two thousand years, codemakers have fought to preserve secrets while codebreakers have tried their best to reveal them. It has always been a neck-and-neck race, with codebreakers battling back when codemakers seemed to be in command, and codemakers inventing new and stronger forms of encryption when previous methods had been compromised. The invention of public-key cryptography and the political debate that surrounds the use of strong cryptography bring us up to the present day, and it is clear that the cryptographers are winning the information war. According to Phil Zimmermann, we live in a golden age of cryptography: 'It is now possible to make ciphers in modern cryptography that are really, really out of reach of all known forms of cryptanalysis. And I think it's going to stay that way.' Zimmermann's view is supported by William Crowell, Deputy Director of the NSA: 'If all the personal computers in the world – approximately 260 million computers – were to be put to work on a single PGP encrypted message, it would take on average an estimated 12 million times the age of the universe to break a single message.'

Previous experience, however, tells us that every so-called unbreakable cipher has, sooner or later, succumbed to cryptanalysis. The Vigenère cipher was called 'le chiffre indéchiffrable', but Babbage broke it; Enigma was considered invulnerable, until the Poles revealed its weaknesses. So, are cryptanalysts on the verge of another breakthrough, or is Zimmermann right? Predicting future developments in any technology is always a precarious task, but with ciphers it is particularly risky. Not only do we have to guess which discoveries lie in the future, but we also have to guess which discoveries lie in the present. The tale of James Ellis and GCHQ warns us that there may already be remarkable breakthroughs hidden behind the veil of government secrecy.

This final chapter examines a few of the futuristic ideas that may enhance or destroy privacy in the twenty-first century. The next section looks at the future of cryptanalysis, and one idea in particular that might enable cryptanalysts to break all today's ciphers. In contrast, the final section of the book looks at the most exciting cryptographic prospect, a system that has the potential to guarantee absolute privacy.

The Future of Cryptanalysis

Despite the enormous strength of RSA and other modern ciphers, cryptanalysts are still able to play a valuable role in intelligence gathering. Their success is demonstrated by the fact that cryptanalysts are in greater demand than ever before – the NSA is still the world's largest employer of mathematicians.

Only a small fraction of the information flowing around the world is securely encrypted, and the remainder is poorly encrypted, or not encrypted at all. This is because the number of Internet users is rapidly increasing, and yet few of these people take adequate precautions in terms of privacy. In turn, this means that national security organisations, law enforcers and anybody else with a curious mind can get their hands on more information than they can cope with.

Even if users employ the RSA cipher properly, there is still plenty that codebreakers can do to glean information from intercepted messages. Codebreakers continue to use old-fashioned techniques like traffic analysis; if codebreakers cannot fathom the contents of a message, at least they might be able to find out who is sending it, and to whom it is being sent, which in itself can be telling. A more recent development is the so-called *tempest attack*, which aims to detect the distinct electromagnetic signals emitted by a computer each time a letter is typed. If Eve parks a van outside Alice's house, she can use sensitive tempest equipment to identify each individual keystroke that Alice makes on her computer. This would allow Eve to intercept the message as it is typed into the computer, before it is encrypted. To defend against tempest attacks, companies are already supplying shielding material that can be used to line the walls of a room to prevent the escape of electromagnetic signals. In America, it is necessary to obtain a government licence before buying such shielding material, which suggests that organisations such as the FBI regularly rely on tempest surveillance.

Other attacks include the use of viruses and Trojan horses. Eve might design a virus that infects PGP software and sits quietly inside Alice's computer. When Alice uses her private-key to decrypt a message, the virus would wake up and make a note of it. The next time that Alice connects to the Internet, the virus would surreptitiously send the private-key to Eve, thereby allowing her to decipher all subsequent messages sent to Alice. The Trojan horse, another software trick, involves Eve designing a program that appears to act like a genuine encryption product, but which actually betrays the user. For example, Alice might believe that she is downloading an authentic copy of PGP, whereas in reality she is downloading a Trojan horse version. This modified version looks just like the genuine PGP program, but contains instructions to send plaintext copies of all Alice's correspondence to Eve. As Phil Zimmermann puts it: 'Anyone could modify the source code and produce a lobotomized zombie imitation of PGP that looks real but does the bidding of its diabolical master. This Trojan horse version of PGP could then be widely circulated, claiming to be from me. How insidious! You should make every effort to get your copy of PGP from a reliable source, whatever that means.'

A variation on the Trojan horse is a brand-new piece of encryption software that seems secure, but which actually contains a *backdoor*, something that allows its designers to decrypt everybody's messages. In 1998, a report by Wayne Madsen revealed that the Swiss cryptographic company Crypto AG had built backdoors into some of its products, and had provided the U.S. Government with details of how to exploit these backdoors. As a result, America was able to read the communications of several countries. In 1991 the assassins who killed Shahpour Bakhtiar, the exiled former Iranian prime minister, were caught thanks to the interception and backdoor decipherment of Iranian messages encrypted using Crypto AG equipment.

Although traffic analysis, tempest attacks, viruses and Trojan horses are all useful techniques for gathering information, cryptanalysts realise that their real goal is to find a way of cracking the RSA cipher, the cornerstone of modern encryption. The RSA cipher is used to protect the most important military, diplomatic, commercial and criminal communications

– exactly the messages that intelligence gathering organisations want to decipher. If they are to challenge strong RSA encryption, cryptanalysts will need to make a major theoretical or technological breakthrough.

A theoretical breakthrough would be a fundamentally new way of finding Alice's private-key. Alice's private-key consists of p and q, and these are found by factoring the public-key, N. The standard approach is to check each prime number one at a time to see if it divides into N, but we know that this takes an unreasonable amount of time. Cryptanalysts have tried to find a shortcut to factoring, a method that drastically reduces the number of steps required to find p and q, but so far all attempts to develop a fast-factoring recipe have ended in failure. Mathematicians have been studying factoring for centuries, and modern factoring techniques are not significantly better than ancient techniques. Indeed, it could be that the laws of mathematics forbid the existence of a significant shortcut for factoring.

Without much hope of a theoretical breakthrough, cryptanalysts have been forced to look for a technological innovation. If there is no obvious way to reduce the number of steps required for factoring, then cryptanalysts need a technology that will perform these steps more quickly. Silicon chips will continue to get faster as the years pass, doubling in speed roughly every eighteen months, but this is not enough to make a real impact on the speed of factoring – cryptanalysts require a technology that is billions of times faster than current computers. Consequently, cryptanalysts are looking towards a radically new form of computer, the *quantum computer*. If scientists could build a quantum computer, it would be able to perform calculations with such enormous speed that it would make a modern supercomputer look like a broken abacus.

The remainder of this section discusses the concept of a quantum computer, and therefore it introduces some of the principles of quantum physics, sometimes called quantum mechanics. Before going any further, please heed a warning originally given by Niels Bohr, one of the fathers of quantum mechanics: 'Anyone who can contemplate quantum mechanics without getting dizzy hasn't understood it.' In other words, prepare to meet some rather bizarre ideas.

In order to explain the principles of quantum computing, it helps to return to the end of the eighteenth century and the work of Thomas Young, the English polymath who made the first breakthrough in deciphering Egyptian hieroglyphics. A fellow of Emmanuel College, Cambridge, Young would often spend his afternoons relaxing near the college duckpond. On one particular day, so the story goes, he noticed two ducks happily swimming alongside each other. He observed that the two ducks left two trails of ripples behind them, which interacted and formed a peculiar pattern of rough and calm patches. The two sets of ripples fanned out behind the two ducks, and when a peak from one duck met a trough from the other duck, the result was a tiny patch of calm water – the peak and the trough cancelled each other out. Alternatively, if two peaks arrived at the same spot simultaneously, then the result was an even higher peak, and if two troughs arrived at the same spot simultaneously, the result was an even deeper trough. He was particularly fascinated, because the ducks reminded him of an experiment concerning the nature of light which he conducted in 1799.

In Young's earlier experiment he had shone light at a partition in which there were two narrow vertical slits, as shown in Figure 71(a). On a screen some distance beyond the slits, Young expected to see two bright stripes, projections of the slits. Instead he observed that the light fanned out from the two slits and formed a pattern of several light and dark stripes on the screen. The striped pattern of light on the screen had puzzled him, but now he believed he could explain it wholly in terms of what he had seen on the duckpond.

Young began by assuming that light was a form of wave. If the light emanating from the two slits behaved like waves, then it was just like the ripples behind the two ducks. Furthermore, the light and dark stripes on the screen were caused by the same interactions that caused the water waves to form high peaks, deep troughs and patches of calm. Young could imagine points on the screen where a trough met a peak, resulting in cancellation and a dark stripe, and points on the screen where two peaks (or two troughs) met, resulting in reinforcement and a bright stripe, as shown in Figure 71(b). The ducks had provided Young with a deeper insight into the true nature of light, and he eventually published 'The Undulatory Theory of Light', an all-time classic among physics papers.

Nowadays, we know that light does indeed behave like a wave, but we know that it can also behave like a particle. Whether we perceive light as a wave or as a particle depends on the circumstances, and this ambiguity

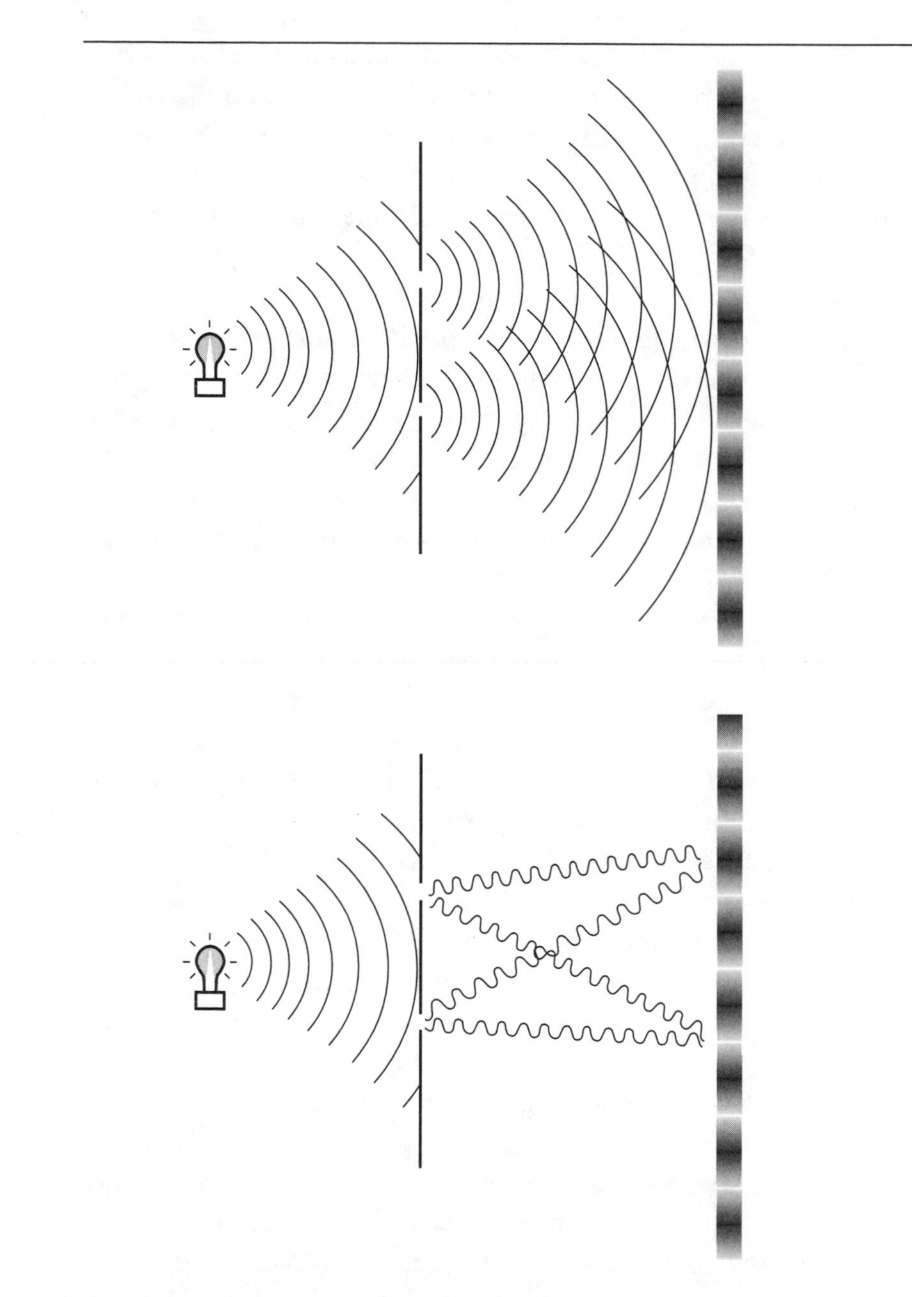

Figure 71 Young's slits experiment viewed from above. Diagram (a) shows light fanning out from the two slits in the partition, interacting and creating a striped pattern on the screen. Diagram (b) shows how individual waves interact. If a trough meets a peak at the screen, the result is a dark stripe. If two troughs (or two peaks) meet at the screen, the result is a bright stripe.

of light is known as wave–particle duality. We do not need to discuss this duality any further, except to say that modern physics thinks of a beam of light as consisting of countless individual particles, known as photons, which exhibit wave-like properties. Looked at this way, we can interpret Young's experiment in terms of photons flooding the slits, and then interacting on the other side of the partition.

So far, there is nothing particularly strange about Young's experiment. However, modern technology allows physicists to repeat Young's experiment using a filament that is so dim that it emits single photons of light. Photons are produced individually at a rate of, say, one per minute, and each photon travels alone towards the partition. Sometimes a photon will pass through one of the two slits, and strike the screen. Although our eyes are not sensitive enough to see the individual photons, they can be observed with the help of a special detector, and over a period of hours we could build up an overall picture of where the photons are striking the screen. With only one photon at a time passing through the slits, we would not expect to see the striped pattern observed by Young, because that phenomenon seems to depend on two photons simultaneously travelling through different slits and interacting with each other on the other side. Instead we might expect to see just two light stripes, simply projections of the slits in the partition. However, for some extraordinary reason, even with single photons the result on the screen is still a pattern of light and dark stripes, just as if photons had been interacting.

This weird result defies common sense. There is no way to explain the phenomenon in terms of the classical laws of physics, by which we mean the traditional laws that were developed to explain how everyday objects behave. Classical physics can explain the orbits of planets or the trajectory of a cannonball, but cannot fully describe the world of the truly tiny, such as the trajectory of a photon. In order to explain such photon phenomena, physicists resort to quantum theory, an explanation of how objects behave at the microscopic level. However, even quantum theorists cannot agree on how to interpret this experiment. They tend to split into two opposing camps, each with their own interpretation.

The first camp posits an idea known as *superposition*. The superpositionists begin by stating that we know only two things for certain about the photon – it leaves the filament and it strikes the screen. Everything else is a complete mystery, including whether the photon passed through the left slit or the right slit. Because the exact path of the photon is unknown, superpositionists take the peculiar view that the photon somehow passes through both slits simultaneously, which would then allow it to interfere with itself and create the striped pattern observed on the screen. But how can one photon pass through both slits?

Superpositionists argue along the following lines. If we do not know what a particle is doing, then it is allowed to do everything possible simultaneously. In the case of the photon, we do not know whether it passed through the left slit or the right slit, so we assume that it passed through both slits simultaneously. Each possibility is called a *state*, and because the photon fulfils both possibilities it is said to be in a *superposition of states*. We know that one photon left the filament and we know that one photon hit the screen on the other side of the partition, but in between it somehow split into two 'ghost photons' that passed through both slits. Superposition might sound silly, but at least it explains the striped pattern that results from Young's experiment performed with individual photons. In comparison, the old-fashioned classical view is that the photon must have passed through one of the two slits, and we simply do not know which one – this seems much more sensible than the quantum view, but unfortunately it cannot explain the observed result.

Erwin Schrödinger, who won the Nobel Prize for Physics in 1933, invented a parable known as 'Schrödinger's cat', which is often used to help explain the concept of superposition. Imagine a cat in a box. There are two possible states for the cat, namely dead or alive. Initially, we know that the cat is definitely in one particular state, because we can see that it is alive. At this point, the cat is not in a superposition of states. Next, we place a vial of cyanide in the box along with the cat and close the lid. We now enter a period of ignorance, because we cannot see or measure the state of the cat. Is it still alive, or has it trodden on the vial of cyanide and died? Traditionally we would say that the cat is either dead or alive, we just do not know which. However, quantum theory says that the cat is in a superposition of two states – it is both dead and alive, it satisfies all possibilities. Superposition occurs only when we lose sight of an object, and it is a way of describing an object during a period of ambiguity. When we eventually open the box, we can see whether the cat is alive or dead. The act of looking at the cat forces it to be in one particular state, and at that very moment the superposition disappears.

For readers who feel uncomfortable with superposition, there is the second quantum camp, who favour a different interpretation of Young's experiment. Unfortunately, this alternative view is equally bizarre. The *many-worlds interpretation* claims that upon leaving the filament the photon has two choices – either it passes through the left slit or the right slit – at which point the universe divides into two universes, and in one universe the photon goes through the left slit, and in the other universe the photon goes through the right slit. These two universes somehow interfere with each other, which accounts for the striped pattern. Followers of the many-worlds interpretation believe that whenever an object has the potential to enter one of several possible states, the universe splits into many universes, so that each potential is fulfilled in a different universe. This proliferation of universes is referred to as the *multiverse*.

Whether we adopt superposition or the many-worlds interpretation, quantum theory is a perplexing philosophy. Nevertheless, it has shown itself to be the most successful and practical scientific theory ever conceived. Besides its unique capacity to explain the result of Young's experiment, quantum theory successfully explains many other phenomena. Only quantum theory allows physicists to calculate the consequences of nuclear reactions in power stations; only quantum theory can explain the wonders of DNA; only quantum theory explains how the sun shines; only quantum theory can be used to design the laser that reads the CDs in your stereo. Thus, like it or not, we live in a quantum world.

Of all the consequences of quantum theory, the most technologically important is potentially the quantum computer. As well as destroying the security of all modern ciphers, the quantum computer would herald a new era of computing power. One of the pioneers of quantum computing is David Deutsch, a British physicist who began working on the concept in 1984, when he attended a conference on the theory of computation. While listening to a lecture at the conference, Deutsch spotted something that had previously been overlooked. The tacit assumption was that all computers essentially operated according to the laws of classical physics, but Deutsch was convinced that computers ought to obey the laws of quantum physics instead, because quantum laws are more fundamental.

Ordinary computers operate at a relatively macroscopic level, and at that level quantum laws and classical laws are almost indistinguishable. It did not therefore matter that scientists had generally thought of ordinary computers in terms of classical physics. However, at the microscopic level the two sets of laws diverge, and at this level only the laws of quantum physics hold true. At the microscopic level, quantum laws reveal their true weirdness, and a computer constructed to exploit these laws would behave in a drastically new way. After the conference, Deutsch returned home and began to recast the theory of computers in the light of quantum physics. In a paper published in 1985 he described his vision of a quantum computer operating according to the laws of quantum physics. In particular, he explained how his quantum computer differed from an ordinary computer.

Imagine that you have two versions of a question. To answer both questions using an ordinary computer, you would have to input the first

Figure 72 David Deutsch.

version and wait for the answer, then input the second version and wait for the answer. In other words, an ordinary computer can address only one question at a time, and if there are several questions it has to address them sequentially. However, with a quantum computer, the two questions could be combined as a superposition of two states and inputted simultaneously – the machine itself would then enter a superposition of two states, one for each question. Or, according to the many-worlds interpretation, the machine would enter two different universes, and answer each version of the question in a different universe. Regardless of the interpretation, the quantum computer can address two questions at the same time by exploiting the laws of quantum physics.

To get some idea of the power of a quantum computer, we can compare its performance with that of a traditional computer by seeing what happens when each is used to tackle a particular problem. For example, the two types of computer could tackle the problem of finding a number whose square and cube together use all the digits from 0 to 9 once and only once. If we test the number 19, we find that $19^2 = 361$ and $19^3 =$ 6,859. The number 19 does not fit the requirement because its square and cube include only the digits: 1, 3, 5, 6, 6, 8, 9, i.e. the digits 0, 2, 4, 7 are missing and the digit 6 is repeated.

To solve this problem with a traditional computer, the operator would have to adopt the following approach. The operator inputs the number 1 and then allows the computer to test it. Once the computer has done the necessary calculations, it declares whether or not the number fulfils the criterion. The number 1 does not fulfil the criterion, so the operator inputs the number 2 and allows the computer to carry out another test, and so on, until the appropriate number is eventually found. It turns out that the answer is 69, because $69^2 = 4{,}761$ and $69^3 = 328{,}509$, and these numbers do indeed include each of the ten digits once and only once. In fact, 69 is the only number that satisfies this requirement. It is clear that this process is time-consuming, because a traditional computer can test only one number at a time. If the computer takes one second to test each number, then it would have taken 69 seconds to find the answer. In contrast, a quantum computer would find the answer in just 1 second.

The operator begins by representing the numbers in a special way so as to exploit the power of the quantum computer. One way to represent the numbers is in terms of spinning particles – many fundamental particles possess an inherent spin, and they can either spin eastwards or westwards, rather like a basketball spinning on the end of a finger. When a particle is spinning eastwards it represents 1, and when it is spinning westwards it represents 0. Hence, a sequence of spinning particles represents a sequence of 1's and 0's, or a binary number. For example, seven particles, spinning east, east, west, east, west, west, west respectively, together represent the binary number 1101000, which is equivalent to the decimal number 104. Depending on their spins, a combination of seven particles can represent any number between 0 and 127.

With a traditional computer, the operator would then input one particular sequence of spins, such as west, west, west, west, west, west, east, which represents 0000001, which is simply the decimal number 1. The operator would then wait for the computer to test the number to see whether it fits the criterion mentioned earlier. Next the operator would input 0000010, which would be a sequence of spinning particles representing 2, and so on. As before, the numbers would have to be entered one at a time, which we know to be time-consuming. However, if we are dealing with a quantum computer, the operator has an alternative way of inputting numbers which is much faster. Because each particle is fundamental, it obeys the laws of quantum physics. Hence, when a particle is not being observed it can enter a superposition of states, which means that it is spinning in both directions at the same time, and so is representing both 0 and 1 at the same time. Alternatively, we can think of the particle entering two different universes: in one universe it spins eastwards and represents 1, while in the other it spins westwards and represents 0.

The superposition is achieved as follows. Imagine that we can observe one of the particles, and it is spinning westwards. To change its spin, we would fire a sufficiently powerful pulse of energy, enough to kick the particle into spinning eastwards. If we were to fire a weaker pulse, then sometimes we would be lucky and the particle would change its spin, and sometimes we would be unlucky and the particle would keep its westward spin. So far the particle has been in clear view all along, and we have been able to follow its progress. However, if the particle is spinning westwards and put in a box out of our view, and we fire a weak pulse of energy at it, then we have no idea whether its spin has been changed. The particle enters a superposition of eastward and westward spins, just as the cat entered a superposition of being dead and alive. By taking seven westward-spinning particles, placing them in a box, and firing seven weak pulses of energy at them, then all seven particles enter a superposition.

With all seven particles in a superposition, they effectively represent all possible combinations of eastward and westward spins. The seven particles simultaneously represent 128 different states, or 128 different numbers. The operator inputs the seven particles, while they are still in a superposition of states, into the quantum computer, which then performs its calculations as if it were testing all 128 numbers simultaneously. After 1 second the computer outputs the number, 69, which fulfils the requested criterion. The operator gets 128 computations for the price of one.

A quantum computer defies common sense. Ignoring the details for a moment, a quantum computer can be thought of in two different ways, depending on which quantum interpretation you prefer. Some physicists view the quantum computer as a single entity that performs the same calculation simultaneously on 128 numbers. Others view it as 128 entities, each in a separate universe, each performing just one calculation. Quantum computing is *Twilight Zone* technology.

When traditional computers operate on 1's and 0's, the 1's and 0's are called bits, which is short for binary digits. Because a quantum computer deals with 1's and 0's that are in a quantum superposition, they are called quantum bits, or *qubits* (pronounced 'cubits'). The advantage of qubits becomes even clearer when we consider more particles. With 250 spinning particles, or 250 qubits, it is possible to represent roughly 10^{75} combinations, which is greater than the number of atoms in the universe. If it were possible to achieve the appropriate superposition with 250 particles, then a quantum computer could perform 10^{75} simultaneous computations, completing them all in just one second.

The exploitation of quantum effects could give rise to quantum computers of unimaginable power. Unfortunately, when Deutsch created his vision of a quantum computer in the mid-1980s, nobody could quite envisage how to create a solid, practical machine. For example, scientists could not actually build anything that could calculate with spinning particles in a superposition of states. One of the greatest hurdles was maintaining a superposition of states throughout the calculation. A superposition exists only while it is not being observed, but an observation in the most general sense includes any interaction with anything external to the superposition. A single stray atom interacting with one of the spinning particles would cause the superposition to collapse into a single state and cause the quantum calculation to fail.

Another problem was that scientists could not work out how to program a quantum computer, and were therefore not sure what sort of computations it might be capable of doing. However, in 1994 Peter Shor of AT&T Bell Laboratories in New Jersey did succeed in defining a useful program for a quantum computer. The remarkable news for cryptanalysts was that Shor's program defined a series of steps that could be used by a quantum computer to factor a giant number – just what was required to crack the RSA cipher. When Martin Gardner set his RSA challenge in *Scientific American*, it took six hundred computers several months to factor a 129-digit number. In comparison, Shor's program could factor a number a million times bigger in one-millionth of the time. Unfortunately, Shor could not demonstrate his factorisation program, because there was still no such thing as a quantum computer.

Then, in 1996, Lov Grover, also at Bell Labs, discovered another powerful program. Grover's program is a way of searching a list at incredibly high speed, which might not sound particularly interesting until you realise that this is exactly what is required to crack a DES cipher. To crack a DES cipher it is necessary to search a list of all possible keys in order to find the correct one. If a conventional computer can check a million keys a second, it would take over a thousand years to crack a DES cipher, whereas a quantum computer using Grover's program could find the key in less than four minutes.

It is purely coincidental that the first two quantum computer programs to be invented have been exactly what cryptanalysts would have put at the top of their wish-lists. Although Shor's and Grover's programs generated tremendous optimism among codebreakers, there was also immense frustration, because there was still no such thing as a working quantum computer that could run these programs. Not surprisingly, the potential of the ultimate weapon in decryption technology has whetted the appetite of organisations such as America's Defense Advanced Research Projects Agency (DARPA) and the Los Alamos National Laboratory, who are desperately trying to build devices that can handle qubits, in the same way that silicon chips handle bits. Although a number of recent breakthroughs have boosted morale among researchers, it is fair to say that the technology remains remarkably primitive. In 1998, Serge Haroche at the University of Paris VI put the hype surrounding the breakthroughs into perspective when he dispelled claims that a real quantum computer is only a few years away. He said this was like painstakingly assembling the first layer of a house of cards, then boasting that the next 15,000 layers were a mere formality.

Only time will tell if and when the problems of building a quantum computer can be overcome. In the meantime, we can merely speculate as to what impact it would have on the world of cryptography. Ever since the 1970s, codemakers have had a clear lead in the race against codebreakers, thanks to ciphers such as DES and RSA. These sorts of ciphers are a precious resource, because we have come to trust them to encrypt our e-mails and guard our privacy. Similarly, as we enter the twenty-first century more and more commerce will be conducted on the Internet, and the electronic marketplace will rely on strong ciphers to protect and verify financial transactions. As information becomes the world's most valuable commodity, the economic, political and military fate of nations will depend on the strength of ciphers.

Consequently, the development of a fully operational quantum computer would imperil our personal privacy, destroy electronic commerce and demolish the concept of national security. A quantum computer would jeopardise the stability of the world. Whichever country gets there first will have the ability to monitor the communications of its citizens, read the minds of its commercial rivals and eavesdrop on the plans of its enemies. Although it is still in its infancy, quantum computing presents a potential threat to the individual, to international business and to global security.

Quantum Cryptography

While cryptanalysts anticipate the arrival of quantum computers, cryptographers are working on their own technological miracle – an encryption system that would re-establish privacy, even when confronted with the might of a quantum computer. This new form of encryption is fundamentally different from any that we have previously encountered in that it offers the hope of perfect privacy. In other words, this system would be flawless and would guarantee absolute security for eternity. Furthermore, it is based on quantum theory, the same theory that is the foundation for quantum computers. So while quantum theory is the inspiration for a computer that could crack all current ciphers, it is also at the heart of a new unbreakable cipher called *quantum cryptography*.

The story of quantum cryptography dates back to a curious idea developed in the late 1960s by Stephen Wiesner, then a graduate student at Columbia University. Sadly, it was Wiesner's misfortune to invent an idea so ahead of its time that nobody took it seriously. He still recalls the reaction of his seniors: 'I didn't get any support from my thesis advisor – he showed no interest in it at all. I showed it to several other people, and they all pulled a strange face, and went straight back to what they were already doing.' Wiesner was proposing the bizarre concept of quantum money, which had the great advantage of being impossible to counterfeit.

Wiesner's quantum money relied heavily on the physics of photons. When a photon travels through space it vibrates, as shown in Figure 73(a). All four photons are travelling in the same direction, but the angle of vibration is different in each case. The angle of vibration is known as the polarisation of the photon, and a light bulb generates photons of all polarisations, which means that some photons will vibrate up and down, some from side to side, and others at all angles in between. To simplify matters, we shall assume that photons have only four possible polarisations, which we label ↕, ↔, ⤡ and ⤢.

By placing a filter known as a Polaroid in the path of the photons, it is possible to ensure that the emerging beam of light consists of photons that vibrate in one particular direction; in other words, the photons all have the same polarisation. To some extent, we can think of the Polaroid filter as a grating, and photons as matchsticks randomly scattered onto the grating. The matchsticks will slip through the grating only if they are at the correct angle. Any photon that is already polarised in the same direction as the Polaroid filter will automatically pass through it unchanged, and photons that are polarised perpendicular to the filter will be blocked.

Unfortunately, the matchstick analogy breaks down when we think about diagonally polarised photons approaching a vertical Polaroid filter. Although matchsticks oriented diagonally would be blocked by a vertical grating, this is not necessarily the case with diagonally polarised photons approaching a vertical Polaroid filter. In fact, diagonally polarised photons are in a quantum quandary when confronted by a vertical Polaroid filter. What happens is that, half of them at random will be blocked, and half will pass through, and those that do pass through will be reoriented with a vertical polarisation. Figure 73(b) shows eight photons approaching a vertical Polaroid filter, and Figure 73(c) shows that only four of them successfully pass through it. All the vertically polarised photons have passed through, all the horizontally polarised photons have been blocked, and half of the diagonally polarised photons have passed through.

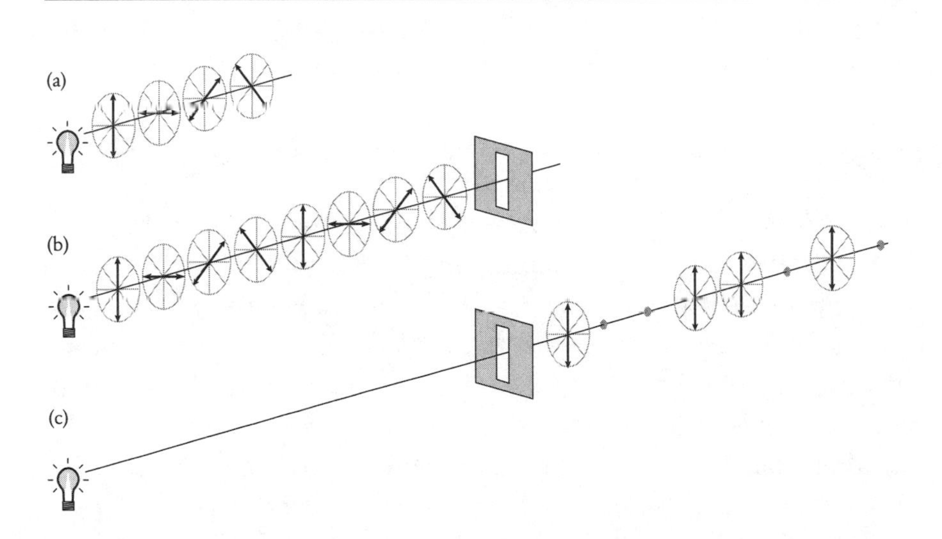

Figure 73 (a) Although photons of light vibrate in all directions, we assume for simplicity that there are just four distinct directions, as shown in this diagram. (b) The lamp has emitted eight photons, which are vibrating in various directions. Each photon is said to have a polarisation. The photons are heading towards a vertical Polaroid filter. (c) On the other side of the filter, only half the photons have survived. The vertically polarised photons have passed through, and the horizontally polarised photons have not. Half the diagonally polarised photons have passed through, and are thereafter vertically polarised.

It is this ability to block certain photons that explains how Polaroid sunglasses work. In fact, you can demonstrate the effect of Polaroid filters by experimenting with a pair of Polaroid sunglasses. First remove one lens, and close that eye so that you are looking with just the other eye through the remaining lens. Not surprisingly, the world looks quite dark because the lens blocks many of the photons that would otherwise have reached your eye. At this point, all the photons reaching your eye have the same polarisation. Next, hold the other lens in front of the lens you are looking through, and rotate it slowly. At one point in the rotation, the loose lens will have no effect on the amount of light reaching your eye because its orientation is the same as the fixed lens – all the photons that get through the loose lens also pass through the fixed lens. If you now rotate the loose lens through 90°, it will turn completely black. In this configuration, the polarisation of the loose lens is perpendicular to the polarisation of the fixed lens, so that any photons that get through the loose lens are blocked by the fixed lens. If you now rotate the loose lens by 45°, then you reach an intermediate stage in which the lenses are partially misaligned, and half of the photons that pass through the loose lens manage to get through the fixed lens.

Wiesner planned to use the polarisation of photons as a way of creating dollar bills that can never be forged. His idea was that dollar bills should each contain 20 light-traps, tiny devices that are capable of capturing and retaining a photon. He suggested that banks could use four Polaroid filters oriented in four different ways (↕, ↔ ,⤡, ⤢) to fill the 20 light-traps with a sequence of 20 polarised photons, using a different sequence for each dollar bill. For example, Figure 74 shows a bill with the polarisation sequence (⤡ ↕ ⤢ ⤢ ↔ ↕ ↕ ⤡ ↕ ⤡ ↔ ↔ ⤢ ↔ ⤡ ⤢ ↔ ⤢ ↕ ↕). Although the polarisations are explicitly shown in Figure 74, in reality they would be hidden from view. Each note also carries a traditional serial number, which is B2801695E for the dollar bill shown. The issuing bank can identify each dollar bill according to its polarisation sequence and its printed serial number, and would keep a master list of serial numbers and the corresponding polarisation sequences.

A counterfeiter is now faced with a problem – he cannot merely forge a dollar bill which carries an arbitrary serial number and a random polarisation sequence in the light-traps, because this pairing will not appear on the bank's master list, and the bank will spot that the dollar bill is a fake. To create an effective forgery, the counterfeiter must use a genuine bill as a sample, somehow measure its 20 polarisations, and then make a duplicate dollar bill, copying across the serial number and loading the light-traps in the appropriate way. However, measuring photon polarisations is a notoriously tricky task, and if the counterfeiter cannot accurately measure them in the genuine sample bill, then he cannot hope to make a duplicate.

To understand the difficulty of measuring the polarisation of photons, we need to consider how we would go about trying to perform such a measurement. The only way to learn anything about the polarisation of a photon is by using a Polaroid filter. To measure the polarisation of the photon in a particular light-trap, the counterfeiter selects a Polaroid filter and orients it in a particular way, say vertically, ↕. If the photon emerging from the light-trap happens to be vertically polarised, it will pass through the vertical Polaroid filter and the counterfeiter will correctly assume that

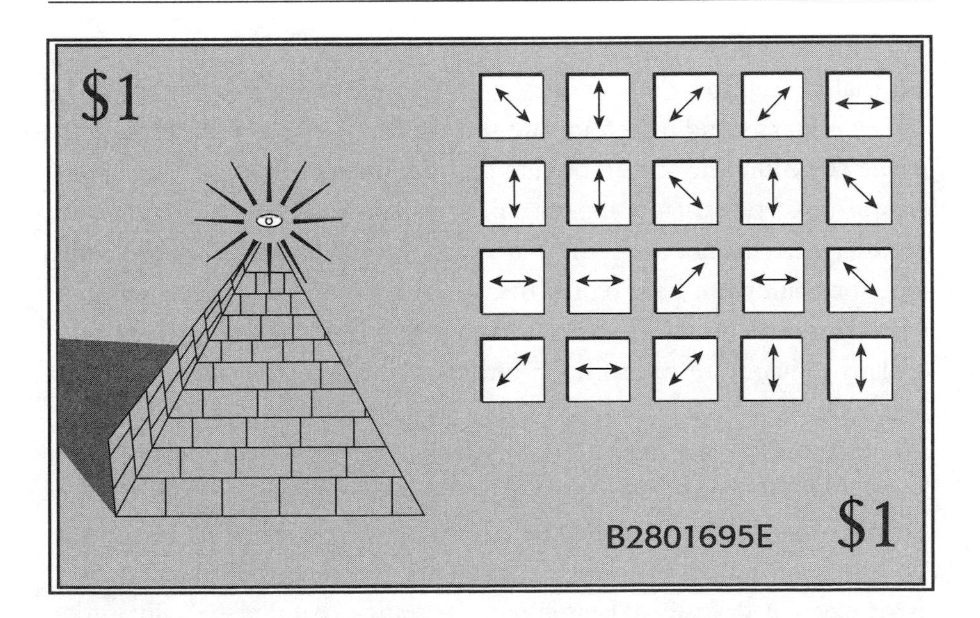

Figure 74 Stephen Wiesner's quantum money. Each note is unique because of its serial number, which can be seen easily, and the 20 light-traps, whose contents are a mystery. The light-traps contain photons of various polarisations. The bank knows the sequence of polarisations corresponding to each serial number, but a counterfeiter does not.

it is a vertically polarised photon. If the emerging photon is horizontally polarised, it will not pass through the vertical Polaroid filter, and the counterfeiter will correctly assume that it is a horizontally polarised photon. However, if the emerging photon happens to be diagonally polarised (⤡ or ⤢), it might or might not pass through the filter, and in either case the counterfeiter will fail to identify its true nature. A ⤡ photon might pass through the vertical Polaroid filter, in which case the counterfeiter will wrongly assume that it is a vertically polarised photon, or the same photon might not pass through the filter, in which case he will wrongly assume that it is a horizontally polarised photon. Alternatively, if the counterfeiter chooses to measure the photon in another light-trap by orientating the filter diagonally, say ⤡, then this would correctly identify the nature of a diagonally polarised photon, but it would fail to accurately identify a vertically or horizontally polarised photon.

The counterfeiter's problem is that he must use the correct orientation of Polaroid filter to identify a photon's polarisation, but he does not know which orientation to use because he does not know the polarisation of the photon. This catch-22 is an inherent part of the physics of photons. Imagine that the counterfeiter chooses a ⤡-filter to measure the photon emerging from the second light-trap, and the photon does not pass through the filter. The counterfeiter can be sure that the photon was not ⤡-polarised, because that type of photon would have passed through. However, the counterfeiter cannot tell whether the photon was ⤢-polarised, which would certainly not have passed through the filter, or whether it was ↕- or ↔-polarised, either of which stood a fifty-fifty chance of being blocked.

This difficulty in measuring photons is one aspect of the uncertainty principle, developed in the 1920s by the German physicist Werner Heisenberg. He translated his highly technical proposition into a simple statement: 'We *cannot* know, as a matter of principle, the present in all its details.' This does not mean that we cannot know everything because we do not have enough measuring equipment, or because our equipment is poorly designed. Instead, Heisenberg was stating that it is logically impossible to measure every aspect of a particular object with perfect accuracy. In this particular case, we cannot measure every aspect of the photons within the light-traps with perfect accuracy. The uncertainty principle is another weird consequence of quantum theory.

Wiesner's quantum money relied on the fact that counterfeiting is a two-stage process: first the counterfeiter needs to measure the original note with great accuracy, and then he has to replicate it. By incorporating photons in the design of the dollar bill, Wiesner was making the bill impossible to measure accurately, and hence creating a barrier to counterfeiting.

A naive counterfeiter might think that if he cannot measure the polarisations of the photons in the light-traps, then neither can the bank. He might try manufacturing dollar bills by filling the light-traps with an arbitrary sequence of polarisations. However, the bank is able to verify that all bills are genuine. The bank looks at the serial number, then consults its confidential master list to see which photons should be in which light-traps. Because the bank knows which polarisations to expect in each light-trap, it can correctly orient the Polaroid filter for each light-trap and perform an accurate measurement. If the bill is counterfeit, the counterfeiter's arbitrary polarisations will lead to incorrect measurements and the bill will stand out as a forgery. For example, if the bank uses a ↕-filter to measure what should be a ↕ polarised photon, but finds that the filter blocks the photon, then it knows that a counterfeiter has filled the trap with the wrong photon. If, however, the bill turns out to be genuine, then the bank refills the light-traps with the appropriate photons and puts it back into circulation.

In short, the counterfeiter cannot measure the polarisations in a genuine bill because he does not know which type of photon is in each light-trap, and cannot therefore know how to orient the Polaroid filter in order to measure it correctly. On the other hand, the bank is able to check the polarisations in a genuine bill, because it originally chose the polarisations, and so knows how to orient the Polaroid filter for each one.

Quantum money is a brilliant idea. It is also wholly impractical. To start with, engineers have not yet developed the technology for trapping photons in a particular polarised state for a sufficiently long period of time. Even if the technology did exist, it would be too expensive to implement it. It might cost in the region of $1 million to protect each dollar bill. Despite its impracticality, quantum money applied quantum theory in an intriguing and imaginative way, so despite the lack of encouragement from his thesis advisor, Wiesner submitted a paper to a scientific journal. It was rejected. He submitted it to three other journals, and it was rejected three more times. Wiesner claims that they simply did not understand the physics.

It seemed that only one person shared Wiesner's excitement for the concept of quantum money. This was an old friend by the name of Charles Bennett, who several years earlier had been an undergraduate with him at Brandeis University. Bennett's curiosity about every aspect of science is one of the most remarkable things about his personality. He says he knew at the age of three that he wanted to be a scientist, and his childhood enthusiasm for the subject was not lost on his mother. One day she returned home to find a pan containing a weird stew bubbling on the cooker. Fortunately she was not tempted to taste it, as it turned out to be the remains of a turtle that the young Bennett was boiling in alkali in order to strip the flesh from the bones, thereby obtaining a perfect specimen of a turtle skeleton. During his teenage years, Bennett's curiosity

Figure 75 Charles Bennett.

moved from biology to biochemistry, and by the time he got to Brandeis he had decided to major in chemistry. At graduate school he concentrated on physical chemistry, then went on to do research in physics, mathematics, logic and, finally, computer science.

Aware of Bennett's broad range of interests, Wiesner hoped that he would appreciate quantum money, and handed him a copy of his rejected manuscript. Bennett was immediately fascinated by the concept, and considered it one of most beautiful ideas he had ever seen. Over the next decade he would occasionally reread the manuscript, wondering if there was a way to turn something so ingenious into something that was also useful. Even when he became a research fellow at IBM's Thomas J. Watson Laboratories in the early 1980s, Bennett still could not stop thinking about Wiesner's idea. The journals might not want to publish it, but Bennett was obsessed by it.

One day, Bennett explained the concept of quantum money to Gilles Brassard, a computer scientist at the University of Montreal. Bennett and Brassard, who had collaborated on various research projects, discussed the intricacies of Wiesner's paper over and over again. Gradually they began to see that Wiesner's idea might have an application in cryptography. For Eve to decipher an encrypted message between Alice and Bob, she must first intercept it, which means that she must somehow accurately perceive the contents of the transmission. Wiesner's quantum money was secure because it was impossible to accurately perceive the polarisations of the photons trapped in the dollar bill. Bennett and Brassard wondered what would happen if an encrypted message was represented and transmitted by a series of polarised photons. In theory, it seemed that Eve would be unable to accurately read the encrypted message, and if she could not read the encrypted message, then she could not decipher it.

Bennett and Brassard began to concoct a system based on the following principle. Imagine that Alice wants to send Bob an encrypted message, which consists of a series of 1's and 0's. She represents the 1's and 0's by sending photons with certain polarisations. Alice has two possible schemes for associating photon polarisations with 1 or 0. In the first scheme, called the *rectilinear* or +-scheme, she sends ↕ to represent 1, and ↔ to represent 0. In the other scheme, called the *diagonal* or ×-scheme, she sends ⤢ to represent 1, and ⤡ to represent 0. To send a binary message, she switches between these two schemes in an unpredictable way. Hence, the binary message 1101101001 could be transmitted as follows:

Message	1	1	0	1	1	0	1	0	0	1
Scheme	+	×	+	×	×	×	+	+	×	×
Transmission	↕	⤢	↔	⤢	⤢	⤡	↕	↔	⤡	⤢

Alice transmits the first 1 using the +-scheme, and the second 1 using the ×-scheme. Hence, 1 is being transmitted in both cases, but it is represented by differently polarised photons each time.

If Eve wants to intercept this message, she needs to identify the polarisation of each photon, just as the counterfeiter needs to identify the polarisation of each photon in the dollar bill's light-traps. To measure the polarisation of each photon Eve must decide how to orient her Polaroid filter as each one approaches. She cannot know for sure which scheme Alice will be using for each photon, so her choice of Polaroid filter will be haphazard and wrong half the time. Hence, she cannot have complete knowledge of the transmission.

An easier way to think of Eve's dilemma is to pretend that she has two types of Polaroid detector at her disposal. The +-detector is capable of measuring horizontally and vertically polarised photons with perfect accuracy, but is not capable of measuring diagonally polarised photons with certainty, and merely misinterprets them as vertically or horizontally polarised photons. On the other hand, the ×-detector can measure diagonally polarised photons with perfect accuracy, but cannot measure horizontally and vertically polarised photons with certainty, misinterpreting them as diagonally polarised photons. For example, if she uses the ×-detector to measure the first photon, which is ↕, she will misinterpret it as ⤢ or ⤡. If she misinterprets it as ⤢, then she does not have a problem, because this also represents 1, but if she misinterprets it as ⤡ then she is in trouble, because this represents 0. To make matters worse for Eve, she only gets one chance to measure the photon accurately. A photon is indivisible, and so she cannot split it into two photons and measure it using both schemes.

This system seems to have some pleasant features. Eve cannot be sure of accurately intercepting the encrypted message, so she has no hope of deciphering it. However, the system suffers from a severe and apparently insurmountable problem – Bob is in the same position as Eve, inasmuch as he has no way of knowing which polarisation scheme Alice is using for each photon, so he too will misinterpret the message. The obvious solution to the problem is for Alice and Bob to agree on which polarisation scheme they will use for each photon. For the example above, Alice and Bob would share a list, or key, that reads + × + × × × + + × ×. However, we are now back to the same old problem of key distribution – somehow Alice has to get the list of polarisation schemes securely to Bob.

Of course, Alice could encrypt the list of schemes by employing a public-key cipher such as RSA, and then transmit it to Bob. However, imagine that we are now in an era when RSA has been broken, perhaps following the development of powerful quantum computers. Bennett and Brassard's system has to be self-sufficient and not rely on RSA. For months, Bennett and Brassard tried to think of a way round the key-distribution problem. Then, in 1984, the two found themselves standing on the platform at Croton-Harmon station, near IBM's Thomas J. Watson Laboratories. They were waiting for the train that would take Brassard back to Montreal, and passed the time by chatting about the trials and tribulations of Alice, Bob and Eve. Had the train arrived a few minutes early, they would have waved each other goodbye, having made no progress on the problem of key distribution. Instead, in a *eureka!* moment, they created quantum cryptography, the most secure form of cryptography ever devised.

Their recipe for quantum cryptography requires three preparatory stages. Although these stages do not involve sending an encrypted message, they do allow the secure exchange of a key which can later be used to encrypt a message.

Stage 1. Alice begins by transmitting a random sequence of 1's and 0's (bits), using a random choice of rectilinear (horizontal and vertical) and diagonal polarisation schemes. Figure 76 shows such a sequence of photons on their way to Bob.

Stage 2. Bob has to measure the polarisation of these photons. Since he has no idea what polarisation scheme Alice has used for each one, he randomly swaps between his +-detector and his ×-detector. Sometimes Bob picks the correct detector, and sometimes he picks the wrong one. If

Bob uses the wrong detector he may well misinterpret Alice's photon. Table 27 covers all the possibilities. For example, in the top line, Alice uses the rectilinear scheme to send 1, and thus transmits ↕; then Bob uses the correct detector, so he detects ↕, and correctly notes down 1 as the first bit of the sequence. In the next line, Alice does the same thing, but Bob uses the incorrect detector, so he might detect ⤢ or ⤡, which means that he might correctly note down 1 or incorrectly note down 0.

Stage 3. At this point, Alice has sent a series of 1's and 0's and Bob has detected some of them correctly and some of them incorrectly. To clarify the situation, Alice then telephones Bob on an ordinary insecure line, and tells Bob which polarisation scheme she used for each photon – but not how she polarised each photon. So she might say that the first photon was sent using the rectilinear scheme, but she will not say

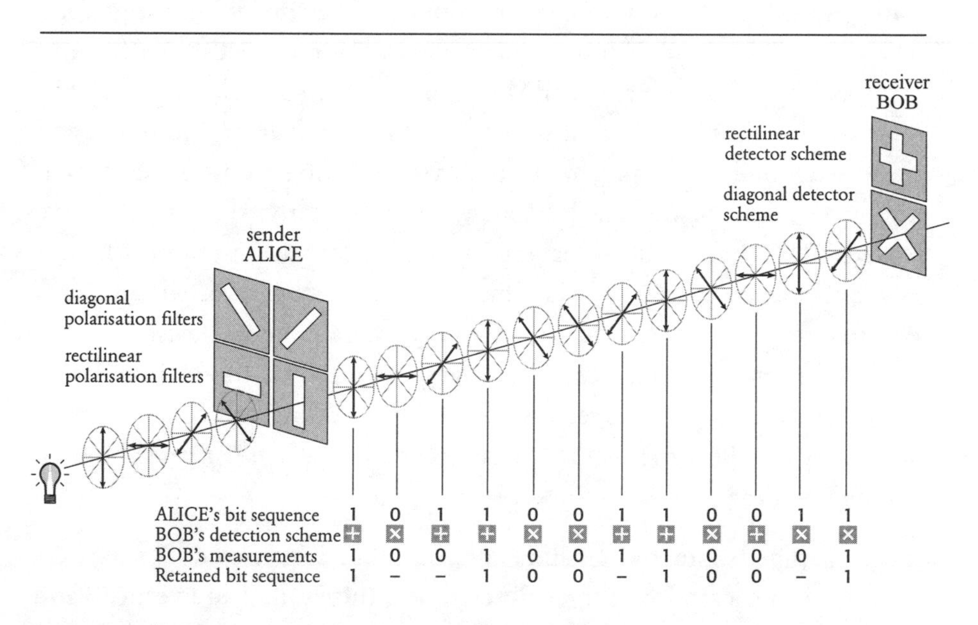

Figure 76 Alice transmits a series of 1's and 0's to Bob. Each 1 and each 0 is represented by a polarised photon, according to either the rectilinear (horizontal/vertical) or diagonal polarisation scheme. Bob measures each photon using either his rectilinear or his diagonal detector. He chooses the correct detector for the left-most photon and correctly interprets it as 1. However, he chooses the incorrect detector for the next photon. He happens to interpret it correctly as 0, but this bit is nevertheless later discarded because Bob cannot be sure that he has measured it correctly.

whether she sent ↕ or ↔. Bob then tells Alice on which occasions he guessed the correct polarisation scheme. On these occasions he definitely measured the correct polarisation and correctly noted down 1 or 0. Finally, Alice and Bob ignore all the photons for which Bob used the wrong scheme, and concentrate only on those for which he guessed the right scheme. In effect, they have generated a new shorter sequence of bits, consisting only of Bob's correct measurements. This whole stage is illustrated in the table at the bottom of Figure 76.

These three stages have allowed Alice and Bob to establish a common series of digits, such as the sequence 11001001 agreed in Figure 76. The crucial property of this sequence is that it is random, because it is derived

Table 27 The various possibilities in stage 2 of photon exchange between Alice and Bob

Alice's scheme	Alice's bit	Alice sends	Bob's detector	Correct detector?	Bob detects	Bob's bit	Is Bob's bit correct?
Rectilinear	1	↕	+	Yes	↕	1	Yes
			×	No	⤢	1	Yes
					⤡	0	No
	0	↔	+	Yes	↔	0	Yes
			×	No	⤢	1	No
					⤡	0	Yes
Diagonal	1	⤢	+	No	↕	1	Yes
					↔	0	No
			×	Yes	⤢	1	Yes
	0	⤡	+	No	↕	1	No
					↔	0	Yes
			×	Yes	⤡	0	Yes

from Alice's initial sequence, which was itself random. Furthermore, the occasions when Bob uses the correct detector are also random. The agreed sequence does not therefore constitute a message, but it could act as a random key. At last, the actual process of secure encryption can begin.

This agreed random sequence can be used as the key for a one-time pad cipher. Chapter 3 described how a random series of letters or numbers, the one-time pad, can give rise to an unbreakable cipher – not just practically unbreakable, but absolutely unbreakable. Previously, the only problem with the one-time pad cipher was the difficulty of securely distributing the random series, but Bennett and Brassard's arrangement overcomes this problem. Alice and Bob have agreed on a one-time pad, and the laws of quantum physics actually forbid Eve from successfully intercepting it. It is now time to put ourselves in Eve's position, and then we will see why she is unable to intercept the key.

As Alice transmits the polarised photons, Eve attempts to measure them, but she does not know whether to use the +-detector or the ×-detector. On half the occasions she will choose the wrong detector. This is exactly the same position that Bob is in, because he too picks the wrong detector half the time. However, after the transmission Alice tells Bob which scheme he should have used for each photon and they agree to use only the photons which were measured when Bob used the right detector. However, this does not help Eve, because for half these photons she will have measured them using the incorrect detector, and so will have misinterpreted some of the photons that make up the final key.

Another way to think about quantum cryptography is in terms of a pack of cards rather than polarised photons. Every playing card has a value and a suit, such as the jack of hearts or the six of clubs, and usually we can look at a card and see both the value and the suit at the same time. However, imagine that it is only possible to measure either the value or the suit, but not both. Alice picks a card from the pack, and must decide whether to measure the value or the suit. Suppose that she chooses to measure the suit, which is 'spades', which she notes. The card happens to be the four of spades, but Alice knows only that it is a spade. Then she transmits the card down a phone line to Bob. While this is happening, Eve tries to measure the card, but unfortunately she chooses to measure its value, which is 'four'. When the card reaches Bob he decides to measure its suit, which is still 'spades', and he notes this down. Afterwards, Alice calls Bob and asks him if he measured the suit, which he did, so Alice and Bob now know that they share some common knowledge – they both have 'spades' written on their notepads. However, Eve has 'four' written on her notepad, which is of no use at all.

Next, Alice picks another card from the pack, say the king of diamonds, but, again, she can measure only one property. This time she chooses to measure its value, which is 'king', and transmits the card down a phone line to Bob. Eve tries to measure the card, and she also chooses to measure its value, 'king'. When the card reaches Bob, he decides to measure its suit, which is 'diamonds'. Afterwards, Alice calls Bob and asks him if he measured the card's value, and he has to admit that he guessed wrong and measured its suit. Alice and Bob are not bothered because they can ignore this particular card completely, and try again with another card chosen at random from the pack. On this last occasion Eve guessed right, and measured the same as Alice, 'king', but the card was discarded because Bob did not measure it correctly. So Bob does not have to worry about his mistakes, because Alice and he can agree to ignore them, but Eve is stuck with her mistakes. By sending several cards, Alice and Bob can agree on a sequence of suits and values which can then be used as the basis for some kind of key.

Quantum cryptography allows Alice and Bob to agree on a key, and Eve cannot intercept this key without making errors. Furthermore, quantum cryptography has an additional benefit: it provides a way for Alice and Bob to find out if Eve is eavesdropping. Eve's presence on the line becomes apparent because every time that she measures a photon, she risks altering it, and these alterations become obvious to Alice and Bob.

Imagine that Alice sends ⤡, and Eve measures it with the wrong detector, the +-detector. In effect, the +-detector forces the incoming ⤡ photon to emerge as either a ↕ or a ↔ photon, because this is the only way the photon can get through Eve's detector. If Bob measures the transformed photon with his ×-detector, then he might detect ⤡, which is what Alice sent, or he might detect ⤢, which would be a mismeasurement. This is a problem for Alice and Bob, because Alice sent a diagonally polarised photon and Bob used the correct detector, yet he might have measured it incorrectly. In short, when Eve chooses the wrong detector, she will 'twist' some of the photons, and this will make Bob prone to errors, even when he is using the correct detector. These errors can be found if Alice and Bob perform a brief error-checking procedure.

The error checking is done after the three preliminary stages, by which time Alice and Bob should have identical sequences of 1's and 0's. Imagine that they have established a sequence that is 1,075 binary digits in length. One way for Alice and Bob to check that their respective sequences match would be for Alice to call Bob and read out her complete sequence to him. Unfortunately, if Eve is eavesdropping she would then be able to intercept the entire key. Checking the complete sequence is clearly unwise, and it is also unnecessary. Instead, Alice merely has to pick 75 of the digits at random and check just these. If Bob agrees with the 75 digits, it is highly unlikely that Eve was eavesdropping during the original photon transmission. In fact, the chances of Eve being on the line and not affecting Bob's measurement of these 75 digits are less than one in a billion. Because these 75 digits have been openly discussed by Alice and Bob, they must be discarded, and their one-time pad is thus reduced from 1,075 to 1,000 binary digits. On the other hand, if Alice and Bob find a discrepancy among the 75 digits, then they will know that Eve has been eavesdropping, and they would have to abandon the entire one-time pad, switch to a new line and start all over again.

To summarise, quantum cryptography is a system that ensures the security of a message by making it hard for Eve to read accurately a communication between Alice and Bob. Furthermore, if Eve tries to eavesdrop then Alice and Bob will be able to detect her presence. Quantum cryptography therefore allows Alice and Bob to exchange and agree upon a one-time pad in complete privacy, and thereafter they can use this as a key to encrypt a message. The procedure has five basic steps:

(1) Alice sends Bob a series of photons, and Bob measures them.

(2) Alice tells Bob on which occasions he measured them in the correct way. (Although Alice is telling Bob when he made the correct measurement, she is not telling him what the correct result should have been, so this conversation can be tapped without any risk to security.)

(3) Alice and Bob discard the measurements that Bob made incorrectly,

and concentrate on those measurements he made correctly in order to create an identical pair of one-time pads.

(4) Alice and Bob check the integrity of their one-time pads by testing a few of the digits.

(5) If the verification procedure is satisfactory, they can use the one-time pad to encrypt a message; if the verification reveals errors, they know that the photons were being tapped by Eve, and they need to start all over again.

Fourteen years after Wiesner's paper on quantum money had been rejected by the science journals, it had inspired an absolutely secure system of communication. Now living in Israel, Wiesner is relieved that, at last, his work is being recognised: 'Looking back, I wonder if I couldn't have made more of it. People have accused me of being a quitter, for not having tried harder to get my idea published – I guess they're right in a way – but I was a young graduate student, and I didn't have that much confidence. In any case, nobody seemed interested in quantum money.'

Cryptographers greeted Bennett and Brassard's quantum cryptography with enthusiasm. However, many experimentalists argued that the system worked well in theory, but would fail in practice. They believed that the difficulty of dealing with individual photons would make the system impossible to implement. Despite the criticism, Bennett and Brassard were convinced that quantum cryptography could be made to work. In fact, they had so much faith in their system that they did not bother building the apparatus. As Bennett once put it, 'there is no point going to the north pole if you know it's there'.

However, the mounting scepticism eventually goaded Bennett into proving that the system could really work. In 1988 he began accumulating the components he would need for a quantum cryptographic system, and took on a research student, John Smolin, to help assemble the apparatus. After a year of effort they were ready to attempt to send the first message ever to be protected by quantum cryptography. Late one evening they retreated into their light-tight laboratory, a pitch-black environment safe from stray photons that might interfere with the experiment. Having eaten a hearty dinner, they were well prepared for a long night of tinkering with the apparatus. They set about the task of trying to send polarised photons across the room, and then measuring them using a +-detector and a ×-detector. A computer called Alice ultimately controlled the transmission of photons, and a computer called Bob decided which detector should be used to measure each photon.

After hours of tweaking, at around 3 a.m., Bennett witnessed the first quantum cryptographic exchange. Alice and Bob managed to send and receive photons, they discussed the polarisation schemes that Alice had used, they discarded photons measured by Bob using the wrong detector and they agreed on a one-time pad consisting of the remaining photons. 'There was never any doubt that it would work', recalls Bennett, 'only that our fingers might be too clumsy to build it.' Bennett's experiment had demonstrated that two computers, Alice and Bob, could communicate in absolute secrecy. This was a historic experiment, despite the fact that the two computers were separated by a distance of just 30 cm.

Ever since Bennett's experiment, the challenge has been to build a quantum cryptographic system that operates over useful distances. This is not a trivial task, because photons do not travel well. If Alice transmits a photon with a particular polarisation through air, the air molecules will interact with it, causing a change in its polarisation, which cannot be tolerated. A more efficient medium for transmitting photons is via an optic fibre, and researchers have recently succeeded in using this technique to build quantum cryptographic systems that operate over significant distances. In 1995, researchers at the University of Geneva succeeded in implementing quantum cryptography in an optic fibre that stretched 23 km from Geneva to the town of Nyon.

More recently, a group of scientists at Los Alamos National Laboratory in New Mexico has once again begun to experiment with quantum cryptography in air. Their ultimate aim is to create a quantum cryptographic system that can operate via satellites. If this could be achieved, it would enable absolutely secure global communication. So far the Los Alamos group has succeeded in transmitting a quantum key through air over a distance of 1 km.

Security experts are now wondering how long it will be before quantum cryptography becomes a practical technology. At the moment there is no advantage in having quantum cryptography, because the RSA cipher already gives us access to effectively unbreakable encryption. However, if quantum computers were to become a reality, then RSA and all other modern ciphers would be useless, and quantum cryptography would become a necessity. So the race is on. The really important question is whether quantum cryptography will arrive in time to save us from the threat of quantum computers, or whether there will be a privacy gap, a period between the development of quantum computers and the advent of quantum cryptography. So far, quantum cryptography is the more advanced technology. The Swiss experiment with optic fibres demonstrates that it would be feasible to build a system that permits secure communication between financial institutions within a single city. Indeed, it is currently possible to build a quantum cryptography link between the White House and the Pentagon. Perhaps there already is one.

Quantum cryptography would mark the end of the battle between codemakers and codebreakers, and the codemakers emerge victorious. Quantum cryptography is an unbreakable system of encryption. This may seem a rather exaggerated assertion, particularly in the light of previous similar claims. At different times over the last two thousand years, cryptographers have believed that the monoalphabetic cipher, the polyalphabetic cipher and machine ciphers such as Enigma were all unbreakable. In each of these cases the cryptographers were eventually proved wrong, because their claims were based merely on the fact that the complexity of the ciphers outstripped the ingenuity and technology of cryptanalysts at one point in history. With hindsight, we can see that the cryptanalysts would inevitably figure out a way of breaking each cipher, or developing technology that would break it for them.

However, the claim that quantum cryptography is secure is qualitatively different from all previous claims. Quantum cryptography is not just effectively unbreakable, it is absolutely unbreakable. Quantum theory, the most successful theory in the history of physics, means that it is impossible for Eve to intercept accurately the one-time pad key established between Alice and Bob. Eve cannot even attempt to intercept the one-time pad key without Alice and Bob being warned of her eavesdropping. Indeed, if a message protected by quantum cryptography were ever to be deciphered, it would mean that quantum theory is flawed, which would have devastating implications for physicists; they would be forced to reconsider their understanding of how the universe operates at the most fundamental level.

If quantum cryptography systems can be engineered to operate over long distances, the evolution of ciphers will stop. The quest for privacy will have come to an end. The technology will be available to guarantee secure communications for governments, the military, businesses and the public. The only question remaining would be whether or not governments would allow us to use the technology. How would governments regulate quantum cryptography, so as to enrich the Information Age, without protecting criminals?

The Cipher Challenge: 10 Steps to $15,000

The Cipher Challenge is an opportunity to put your cipher cracking skills to the test and win a $15,000 prize. The Challenge consists of ten separate stages. The first stage is a relatively straightforward monoalphabetic cipher, and thereafter the stages will progress through the history of cryptography. In other words, the second stage contains a ciphertext that has been encrypted using one of the earliest ciphers, and the tenth stage contains one of the most modern forms of cipher. In general, each stage will be harder than the previous one.

What do you need to do in order to claim the prize?

Deciphering each of the ten ciphertexts will generate a message. In addition to the main body of each message, there will be a clearly indicated codeword. In order to claim the prize, you must collect all ten codewords. Hence, it is necessary to decipher all ten stages. Although you may tackle the stages in any order, I would recommend attacking them in the order given. In some cases, deciphering a stage will provide information that will be important for breaking the next stage.

How do you claim the prize?

To claim the prize, please send in the first two letters of each codeword, as well as your name, address and telephone number. If your letters are correct, you will be contacted within 28 days of receipt of your letter, and asked to send in the ten complete codewords. If you are the first to correctly identify all ten codewords, you will win $15,000.

All claims must be sent by recorded delivery to: The Cipher Challenge, P.O. Box 23064, London W11 3GX, UK.

The winner will be the first past the post. There is no element of chance, just skill. Please note, I will only contact competitors in the event of a correct winning entry. Furthermore, I will not be able to reply to any queries regarding the Cipher Challenge. Any updates on the challenge will be posted on the Cipher Challenge website, http://www.4thestate.co.uk/cipherchallenge

The one year prize

If the prize has not been claimed by 1 October 2000, $1500 will be awarded to whoever has made most progress soonest, which means completion of the most consecutive stages. In other words, if you have solved stages 1, 2, 3, 4 and 8, only stages 1–4 are valid in terms of competing for this prize. Before attempting to claim this prize, please check the Cipher Challenge website, which will indicate the current leader and the extent of that person's success. If you believe that you have progressed to the next stage, please send in the first two letters of the codewords from all the stages that you have deciphered, as well as your name, address and telephone number. If your letters are correct, you will be contacted within 28 days of receipt of your letter, and asked to send in the complete codewords. If your codewords are correct, then you will become the current leader and your achievement will be posted on the website. Whoever is the leader on 1 October 2000 will win $1500. This prize is quite separate from the $15,000 prize for the overall winner.

Please send claims by recorded delivery to the address already given. Once again, please note that I will only contact competitors in the event of a successful entry.

I would encourage competitors to track the state of the Challenge on the website, which will also carry other information relating to both the book and the Challenge.

OFFICIAL CONTEST RULES FOR U.S. and CANADIAN ENTRANTS

1. This contest is open to residents of the United States and Canada (except the Province of Quebec). For contest rules for the 'one year prize,' please refer to Mr. Singh's description. In addition to the one year prize, Simon Singh will select the first person to complete the Cipher Challenge successfully by January 1, 2010. If no one claiming the prize has discovered all ten code words by January 1, 2010, then the solutions will be revealed on the Cipher Challenge website, and Mr. Singh, at his discretion, will determine who has made the most progress, that is, the person who has completed the most consecutive stages. If the winner is a citizen of the United States or Canada, then the winner will be awarded $15,000.00.

2. Entries must be received by Mr. Singh no later than January 1, 2010.

3. The winner will be chosen by Mr. Singh from completed entries received. Mr. Singh's decision will be final. The winner will be notified within twenty-eight days of receipt of winning entry, or, in the event that the winner is the person who has completed the most consecutive stages by January 1, 2010, the winner will be notified on or about February 1, 2010. The winner will have thirty (30) days from the date of notice in which to accept the prize, or an alternate winner will be chosen. Mr. Singh is not responsible for late, lost, or misdirected entries.

4. The winner may be required to execute an Affidavit of Eligibility and Promotional Release. Entering the contest constitutes permission for use of the winner's name, likeness, biographical data, and contest entry, for publicity and promotional purposes, with no additional compensation.

5. Employees of Random House, Inc., its subsidiaries and affiliates, and their immediate family members are not eligible to enter this contest. This contest is open to residents of the United States and Canada who are eighteen years or older on the date of entry. Void wherever prohibited or restricted by law. All federal, state, and local regulations apply. Taxes, if any, are the winner's sole responsibility.

6. For the name of the winner, available after April 1, 2010, and until October 1, 2010, send a self-addressed, stamped envelope entirely separate from your entry to THE CIPHER CHALLENGE, P.O. Box 23064, London W11 3GX, UK.

Stage 1: Simple Monoalphabetic Substitution Cipher

BT JPX RMLX PCUV AMLX ICVJP IBTWXVR CI M LMT'R PMTN, MTN YVCJX CDXV MWMBTRJ JPX AMTNGXRJBAH UQCT JPX QGMRJXV CI JPX YMGG CI JPX HBTW'R QMGMAX; MTN JPX HBTW RMY JPX QMVJ CI JPX PMTN JPMJ YVCJX. JPXT JPX HBTW'R ACUTJXTMTAX YMR APMTWXN, MTN PBR JPCUWPJR JVCUFGXN PBL, RC JPMJ JPX SCBTJR CI PBR GCBTR YXVX GCCRXN, MTN PBR HTXXR RLCJX CTX MWMBTRJ MTCJPXV. JPX HBTW AVBXN MGCUN JC FVBTW BT JPX MRJVCGCWXVR, JPX APMGNXMTR, MTN JPX RCCJPRMEXVR. MTN JPX HBTW RQMHX, MTN RMBN JC JPX YBRX LXT CI FMFEGCT, YPCRCXDXV RPMGG VXMN JPBR YVBJBTW, MTN RPCY LX JPX BTJXVQVXJMJBCT JPXVXCI, RPMGG FX AGCJPXN YBJP RAMVGXJ, MTN PMDX M APMBT CI WCGN MFCUJ PBR TXAH, MTN RPMGG FX JPX JPBVN VUGXV BT JPX HBTWNCL. JPXT AMLX BT MGG JPX HBTW'R YBRX LXT; FUJ JPXE ACUGN TCJ VXMN JPX YVBJBTW, TCV LMHX HTCYT JC JPX HBTW JPX BTJXVQVXJMJBCT JPXVXCI. JPXT YMR HBTW FXGRPMOOMV WVXMJGE JVCUFGXN, MTN PBR ACUTJXTMTAX YMR APMTWXN BT PBL, MTN PBR GCVNR YXVX MRJCTBRPXN. TCY JPX KUXXT, FE VXMRCT CI JPX YCVNR CI JPX HBTW MTN PBR GCVNR, AMLX BTJC JPX FMTKUXJ PCURX; MTN JPX KUXXT RQMHX MTN RMBN, C HBTW, GBDX ICVXDXV; GXJ TCJ JPE JPCUWPJR JVCUFGX JPXX, TCV GXJ JPE ACUTJXTMTAX FX APMTWXN; JPXVX BR M LMT BT JPE HBTWNCL, BT YPCL BR JPX RQBVBJ CI JPX PCGE WCNR; MTN BT JPX NMER CI JPE IMJPXV GBWPJ MTN UTNXVRJMTNBTW MTN YBRNCL, GBHX JPX YBRNCL CI JPX WCNR, YMR ICUTN BT PBL; YPCL JPX HBTW TXFUAPMNTXOOMV JPE IMJPXV, JPX HBTW, B RME, JPE IMJPXV, LMNX LMRJXV CI JPX LMWBABMTR, MRJVCGCWXVR, APMGNXMTR, MTN RCCJPRMEXVR; ICVMRLUAP MR MT XZAXGGXTJ RQBVBJ, MTN HTCYGXNWX, MTN UTNXVRJMTNBTW, BTJXVQVXJBTW CI NVXMLR, MTN RPCYBTW CI PMVN RXTJXTAXR, MTN NBRRCGDBTW CI NCUFJR, YXVX ICUTN BT JPX RMLX NMTBXG, YPCL JPX HBTW TMLXN FXGJXRPMOOMV; TCY GXJ NMTBXG FX AMGGXN, MTN PX YBGG RPCY JPX BTJXVQVXJMJBCT. JPX IBVRJ ACNXYCVN BR CJPXGGC.

Stage 2: Caesar Shift Cipher

MHILY LZA ZBHL XBPZXBL MVYABUHL HWWPBZ JSHBKPBZ JHLJBZ
KPJABT HYJHUBT LZA ULBAYVU

Stage 3: Monoalphabetic Cipher with Homophones

IXDVMUFXLFEEFXSOQXYQVXSQTUIXWF*FMXYQVFJ*FXEFQUQXJFPTUFX
MX*ISSFLQTUQXMXRPQEUMXUMTUIXYFSSFI*MXKFJF*FMXLQXTIEUVFX
EQTEFXSOQXLQ*XVFWMTQTUQXTITXKIJ*FMUQXTQJMVX*QEYQVFQTHMX
LFVQUVIXM*XEI*XLQ*XWITLIXEQTHGXJQTUQXSITEFLQVGUQX*GXKIE
UVGXEQWQTHGXDGUFXTITXDIEUQXGXKFKQVXSIWQXAVPUFXWGXYQVXEQ
JPFVXKFVUPUQXQXSGTIESQTHGX*FXWFQFXSIWYGJTFXDQSFIXEFXGJP
UFXSITXRPQEUGXIVGHFITXYFSSFI*CXC*XSCWWFTIXSOQXCXYQTCXYI
ESFCX*FXCKVQFXVFUQTPUFXQXKI*UCXTIEUVCXYIYYCXTQ*XWCUUFTI
XLQFXVQWFXDCSQWWIXC*FXC*XDI**QXKI*IXEQWYVQXCSRPFEUCTLIX
LC*X*CUIXWCTSFTIXUPUUQX*QXEUQ**QXJFCXLQX*C*UVIXYI*IXKQL
QCX*CXTIUUQXQX*XTIEUVIXUCTUIXACEEIXSOQXTITXEPVJQCXDPIVX
LQ*XWCVFTXEPI*IXSFTRPQXKI*UQXVCSSQEIXQXUCTUIXSCEEIX*IX*
PWQXQVZXLFXEIUUIXLZX*ZX*PTZXYIFXSOQXTUVZUFXQVZKZWXTQX*Z
*UIXYZEEIRPZTLIXTZYYZVKQXPTZXWITUZJTZXAVPTZXYQVX*ZXLFEU
ZTHZXQXYZVKQWFXZ*UZXUZTUIXRPZTUIXKQLPUZXTITXZKQZXZ*SPTZ
XTIFXSFXZ**QJVNWWIXQXUIEUIXUIVTIXFTXYFNTUIXSOQXLQX*NXTI
KNXUQVVNXPTXUPVAIXTNSRPQXQXYQVSIEEQXLQ*X*QJTIXF*XYVFWIX
SNTUIXUVQXKI*UQXF*XDQXJFVBVXSITXUPUUQX*BSRPQXBX*BXRPBVU
BX*QKBVX*BXYIYYBXFTXEPEIXQX*BXYVIVBXFVQXFTXJFPXSIWB*UVP
FXYFBSRPQFTDFTXSOQX*XWBVXDPXEIYVBXTIFXVFSOFPEIXX*BXYBVI
*BXFTXSILFSQXQXQRPBUIV

Stage 4: Vigenère Cipher

K Q O W E F V J P U J U U N U K G L M E K J I N M W U X F Q M K J B
G W R L F N F G H U D W U U M B S V L P S N C M U E K Q C T E S W R
E E K O Y S S I W C T U A X Y O T A P X P L W P N T C G O J B G F Q
H T D W X I Z A Y G F F N S X C S E Y N C T S S P N T U J N Y T G G
W Z G R W U U N E J U U Q E A P Y M E K Q H U I D U X F P G U Y T S
M T F F S H N U O C Z G M R U W E Y T R G K M E E D C T V R E C F B
D J Q C U S W V B P N L G O Y L S K M T E F V J J T W W M F M W P N
M E M T M H R S P X F S S K F F S T N U O C Z G M D O E O Y E E K C
P J R G P M U R S K H F R S E I U E V G O Y C W X I Z A Y G O S A A
N Y D O E O Y J L W U N H A M E B F E L X Y V L W N O J N S I O F R
W U C C E S W K V I D G M U C G O C R U W G N M A A F F V N S I U D
E K O H C E U C P F C M P V S U D G A V E M N Y M A M V L F M A O Y
F N T Q C U A F V F J N X K L N E I W C W O D C C U L W R I F T W G
M U S W O V M A T N Y B U H T C O C W F Y T N M G Y T Q M K B B N L
G F B T W O J F T W G N T E J K N E E D C L D H W T V B U V G F B I
J G Y Y I D G M V R D G M P L S W G J L A G O E E K J O F E K N Y N
O L R I V R W V U H E I W U U R W G M U T J C D B N K G M B I D G M
E E Y G U O T D G G Q E U J Y O T V G G B R U J Y S

Stage 5

109 182 6 11 88 214 74 77 153 177 109 195 76 37 188
166 188 73 109 158 15 208 42 5 217 78 209 147 9 81
80 169 109 22 96 169 3 29 214 215 9 198 77 112 8 30
117 124 86 96 73 177 50 161

Stage 6

OCOYFOLBVNPIASAKOPVYGESKOVMUFGUWMLNOOEDRNCFORSOCVMTUUTY
ERPFOLBVNPIASAKOPVIVKYEOCNKOCCARICVVLTSOCOYTRFDVCVOOUEG
KPVOOYVKTHZSCVMBTWTRHPNKLRCUEGMSLNVLZSCANSCKOPORMZCKIZU
SLCCVFDLVORTHZSCLEGUXMIFOLBIMVIVKIUAYVUUFVWVCCBOVOVPFRH
CACSFGEOLCKMOCGEUMOHUEBRLXRHEMHPBMPLTVOEDRNCFORSGISTHOG
ILCVAIOAMVZIRRLNIIWUSGEWSRHCAUGIMFORSKVZMGCLBCGDRNKCVCP
YUXLOKFYFOLBVCCKDOKUUHAVOCOCLCIUSYCRGUFHBEVKROICSVPFTUQ
UMKIGPECEMGCGPGGMOQUSYEFVGFHRALAUQOLEVKROEOKMUQIRXCCBCV
MAODCLANOYNKBMVSMVCNVROEDRNCGESKYSYSLUUXNKGEGMZGRSONLCV
AGEBGLBIMORDPROCKINANKVCNFOLBCEUMNKPTVKTCGEFHOKPDULXSUE
OPCLANOYNKVKBUOYODORSNXLCKMGLVCVGRMNOPOYOFOCVKOCVKVWOFC
LANYEFVUAVNRPNCWMIPORDGLOSHIMOCNMLCCVGRMNOPOYHXAIFOOUEP
GCHK

Stage 7

M C C M M C T R U O U U U R E P U C C T C T P C C C U U P C M M P
R T C C R U P E C C M U U P C M P E P P U P U R U P P M E U P U C E
U U C U C C C M E M T U P E T P C M R C M C C U C C M P E C R T M R
U P M P M R C P M M C R U M C U U E U R P P C M O U U E U C C M U M
T U C U C U T M U U U P M U U C T C U P M M C C R P P P P M M M M E
E U M R C C C P U U E U P M U M M C C P E C U C U P C T C U E P M P
C U U E E U U U T P M M U C C T C C P P P C T P U C U C C C U R E U
T U C M E P C C E M U U U P R M M T M U C M M M C C C C C C M E P U
E C U M R E R U U U U M U R C C P M U U R U U P M U P R P P U U U U
M R C C P C P E U R M M M P U T C R U U E O U U U M C M U U R U P U
R U C M U C R U M M C U P U U M U C R E U U U P C C U R R C P R M C
T R C U U U R C T P P M U U C C U U U U M U U E P C R M E P M P U U
C C C U M M U U M C U C M C C C R T C C M E E U P T M U U M M M C C
P P T M C P T E O U U U M U U C R M C C C M C P R C R C E P M C M C
P U U C M C C O M T P R C M C P C P M C P C E R R E C C R R E C R U
P U E E P M U M T C U C E U U T P C E U M R C U U U R R U C R U U C
R P P T T C P C P C U C U M U M P E C E E R P M R M M U R U M E P M
R M M C P R U C R C P E E R P U U U U R E P C C M M E P P P R C C U
M P C C C M M E E U U P P E R U E C P U E M U C C U U C P U E P U C
M C M C U U C M M M C U P C C M M U U U C U O P U C U P M P U E C C
E U P M C E P R C T R M C C U U T E C E C C R M U C U R U C M U C R
C M P C C U O R U C T U C C M C U C M U M M T R U M C M M C P U U M
U P C C M P C U U E P C T E C T U U T C E E M T U C T E P P R U U M
U U E C M U M R U E P C U M P P O U R U C C U P U C U C U E P C M M
E C C U C E C P P C C C C O C R C R C R T U C P P T P U O C U O R U
C C C E U C P P M R R C E U U U R U R C C M T P P U R P P C T R R T
R U U P M T M U U E T R P R O E M P T P T E P R E R P T R U U U M T
R U M T P P P R U U P E O U T P T R O M U U E R M M E P U T T O T O
O M T P R M P P T M R E U R R U P M T R P P R E M U P R T R M M E O
U M M U P U U O U M E M O M E C P E U U U U C R U T T T R T U P T T
P E R E M U U R E E P E T R M P T R U U U O T R U U O O T T T O T T
E T E T O U P O M T U U O U T O E E T P T E M U U T U R C U O P T R
P O T E E M C O U U E P R M P T T T U P P R E T T R O E M U E T P O
P M T E R T E U U U P U P U U E M M O T O U M O R R C M U U U E T U

O T T E M T T C T M E T E R E U M U E E T U M E T P U T P U E T T M
P E E R T C P T O U U T R E R E T U T R E T R T R U T C M T C U U T
P O M T T P T P T O U M E O T T R P E P U T T T R T T O U M U U T P
E E C T M P P M U E C T R P U C T E U U E T P T O T P M T M C P U E
P P U P R M T P C R U R P R E M E R T U E E R O R O T O M M R C U U
E U T P T E P P E U U T P O T P P M E P E M T R E E U T U U T O T P
R E E R O P O R R M U U T M P R T T M E E E T E R U T M T O O C P E
P P M P M T P R R M E P R E U M M P R T R E E P U T T P E C T U R U
R C O P E E E O O U E M O M P T U E C E R M M M P P E P M U E M U R
T E U M R T T P U T C E R O E T M U U R O T U T T R M U E T E T T R
P R O U T U U P R E U T T R T P M T U P E E M E T E P T O E T U U T
E P T M U U E E P P T P M U P T E P R M U T T P M U M M E C R E T E
P T R T U R P M T O O U E E O T O U R U U R T U E U T P O M T P P U
R E O T C M C P R P R O O E E R U U E E R U M U U U C P P C P U E T
E R U R P O R P T P C T P E R E R M U T T R E U P R T M E C U R E P
P O U T M O T C T M P T P O E U U T O T P T O R E U E T U R M E T R
E P E E P R U C P E M M P T M U U T T E O E R M U R U U R U T P T T
E C E T O R T M T M E T T U E M U U C T O P E M U U E P U M C M U C
M T P O U C E C M T R E M C P C M C T P M M P P C M U U U U C M C C
C P T M M U C R E U U C T R R E U C U R E C P M R C E C U C E U C C
P M C T T P C R E U R M U T U P M P P M M C U T M C M C C E U U C T
U P U U U U U R C U M E P O T U U U C T E P C C P M C C T P C P U M
E R U C U M E M M R M U P C M U U C U C R U U U U C P C U P C E C M
C U U P O P C U U U U C U T T C P C M C U U C C E P U U P C M P U C
M M M P U U U E P M P P E C R C M P R E C R R U M C U E C P U P U C
E M P M U C R T U T U C R C C U P U U C U M M P U U U U E C U U C C
E C P P P R R M C M M E C C R M M R C C E C T U R M C E C C C P M M
M R P E C U U U C P P M M E C C M M R R C M U C M R C P C U C M U C
C C P C T R C U U E U C C M T E M C R C P E C C U U C U U C P E T P
C C P P T U M P C M P C M C E U C C C P C U C T C C C M T U M P T U
M E U C P P M U M P M M R E M C U M M M E R U C U C C M P U U E U C
P C E P P R U U C C U C T P U E T E R C M M M U R U U P U R P U E E
M U M U M R C U U C R M R C P T E M E C M M U C U C U U P P E T T T
M P C P M M U E M P P C U T P M C M U U P U C C P M P R C M C R P U
P M E M U U U R C O C P C U E P M R C P T M M M M C C E C U M C U U
C E C P P U C P M R M E P C U U R U C U C P R T U E R M C C R P M U

U R U U P M E U P C E C P T R U T U M C E C E P C U T C U C P E P C

C U U E T P P C P U U M C M M R O U C C P U C P P E P M E C R P C M

C U M P U C U U U E M M C U T M C U M C U E U C M U C C T P U R E U

P P C O P M P M U U M M M U U E T P U U U U U P P P P M U E C E R U

R P U R T P M P P P M E M C T U P C M E C P P C C E M U R M P T U U

R C U E P C U E C P U T C U R U C P R U M T C O C C M P U C M E P E

M P R U P P E C C P C U U C C C E U M R U U E U U E U C P C P M P U

C U M P U C U M P P R E U U U P E U P E U U C T P O T U P E T U O E

C O T T E M O T E U T E U M U P M U T P O U P E T E R P U T P R U U

U P O T T E P T R R M T C E T O R O P M T R E T R C O E T P R O E E

P T E P M M E U P E P E P U P U U R E E P E R T P E E C E P O R T U

E M E T T E P T E R M M T T E T T T P O R U M P T T E R P P U U R M

T T O M T M U M M U U T U O E P E U U O T C P E P T M R E R U R P E

T P P T T P C O R P T T T M U T R U P P T E R R E U R P R T R E T T

R C P R C U U M U P R U U U M T P R T R E T T U U U O C U M U U U U

M O T T P E M E T T E R P C T O E T U U R M E P E E O R C P E T M P

P R U T T R U U E T M O T M U U M T E R U T O T C R P M U R M U M R

M P M O O M O U O T P O R E M E M U P T O R T R R P O O U T P P P E

P M T P E O C T R R M E T O R T P E M M P E E E T R U U R U R P P U

P U R T R O U M T M R C U O T E T R C R P E E C P T E E U U E M T T

P U R U P E U O E U U M P E M U U T T E R E U M E R T T E T T T M E

U T M R T O R M E C U C U E U E P R U M T U U E R M U T R E U U P E

E M E E R C U U U T R M R T R M U U M M E P P T P R T E M T E M P E

U E T P O O O U U M O T O U T O O P E P R U U R T T T M U R T U T E

T P C O T E M T U O E T R M T E T E M M T U M O E E O O U M O P T P

R U T M R M T R T P T U U E P U U P U R R O E U E R U U O U P R T M

E T P E P P O T R M C M R U T T P U U E U R T T E E T E T U U E U E

E T U R R M E E M R E U R C T P E M U U R E P R U E O R U R U U P T

U M P E M T T P T U E M U P M O R T O O O U T P P M U U P U P E R E

R U U O U E E T U P E T E T P T T T E M R U U R T T T U T T M U P R

P R R U R U U T M T U R T C U E E O M R R T E T T M U T P P R P E P

T R E E O O T T E T R E T R U T P R U T M U U U T M U U C T U U P U

E R U E E M M U E E T T P E T M U M E T T E T T P M R E M R T P T E

T O U R T P P O E T T O M T P T E T E U T P U C U M U C U O E T U C

P E C U C M U P M U C U T T U C T U U M U C U R P U C P M C U U M U

C C E P C M M U C P T P U M U P U C M E C M P U M P P M U E M P P E

P U T E U M E P E P U P U U R M T P E M R P M M P T P O P R C R U E
P C M P P M R C C C P U C U P T U M U U P C P E M P T U U M C C C U
P C C U T U U U R C E M P E U C M R P P E P C C M M M U M P E C M T
R E R P U M P C C P T U C M C O P C U R U E C M T E C M M C C R P P
E P U U C U T M U U U C C T M C M E C P C U U U P U C U U U T C U C
C P T U U C C M M P P R E M C U U R U U M U E U U P P U C R P M R U
P C M U U E C U U C C U U R C E R R C U C P M P U U M T U U R C M P
E M U U U U C T U M T T T C U M P U M C M R T U U U C P P M E P U C
T O U P C M M C E C U M C P E C U P M T E P R U U R U R M P U P E R
C R U U C C C M P C U C M R M P M P E E P T P E M C U R C P C P U R
U T E U U E U U U P T C U C C C E M M T U T R E R E M P R R M U C C
R C U M U E P U P U E U E P M U T R U C C M U U C M M U U P M E C M
M E M U U U C M R P C M C U U C C E T P C P R R M U R R C T E C M C
M U U U U U U E C U U C U U T E P M U U R C C C U U R C U C E C P P
U C M U R C U U C R U C M C R C C C U C U M E M U U C P P P P R C R
U R U C M C P P C R M P U E P U M P O M U M M C U U U P C C C E C T
M R P U P M P O C C T P C M U U M C M C C T U C E C U U M C C M C U
E R T T R C M M U M T C P E R U U M M T R U E U M C M C C M C U U P
M U C C T P U M C U T P U M C U U U U C P P U C E T U P E R T R U U
U U M M C U M E E M C T C C P U R R U U R C P C U P C C U P M P M M
U R U U C C C E P R P U M M U T C M C M C C C U C P P C M E P C R E
M U U R C T P E M C M C C P R U C C U U U C C U U P C U U P U T R U
E E U U U E U C R P M R U U U O C P O C R P C M E C R C P C E C U U
E C P P U M P P E P C P R M P E U C P T U E M T U T T E O P R U E P
E P M T P U P T T R R E R P U E M M O P M U P R U U U M E M P P P U
T O U R O P R O P P M E T P R M T U U R P T P U U T O U U M T E P C
O E M C U U T P U U P T O T U U T T U U U R T P T R T T M O C T R U
T R O T T R O P T U M P P M U R T E U M T P E U M C M P R E P M R E
E E E U T T T E U U T M T P U R U E U U M T U P P U T T R E M T P T
R R U T U R T R U U T O T E R O T M U U U T M U P T P U U R T E R U
M M T M T T U P R P P P E M E P C M U M T R R E M U C E U P P T T T
T T P R U U U R T E E P U P U T M M T U P M R U O P E U E E T M M P
E M T P E C R E T M E O U T M E E P R E U M E M R T O T E M T O T P
T E C E P T U T R E E E M P P T P E E C P P T M U U T M U M P R M E
R E U U P T O E O P U E P T R T T E P M O U M P E U T M T T M U U U
T T P T E R M T R R U U R U U E U R T E E M U T T E P O U U E M E E

P C R U R M E T M E T O R E U U O T R T P R T T E U M M T P M M R P
E U U R E R T E O T U T R R O T O T E T T E O T U E U U E T U E T P
M U O O R T O U M C O T U E C E U U R E U U M T T E R U O T T M T E
T T E O T U T E P T R C T U U P P E R U T O U U E O R M U E M P R E
M U U P O P M O U O O T E C U O E T U C M T T P T T U U R T T M M O
P T P U C M T U U O M U M T T T O R T U P E T E T R O M T R E T T U
E U U T P P T M E U M U R U U U U R E T U T R U R R T T P P T T P O
E T E M U O T C O U E M T T M T U E U U P T U P U P T R O T U E E R
O E R O U E M C P T E R C P P T M U U M T O M C E M U T P T T T O U
T O E M T T P P C R E P O T E P P E R P O P P O T E U U U U R P U U
C P R P R M T R E U U E R M U C T O P T T U U T P M C T R M E T E M
M U O P T U U E T P P M M R M T U P R M U P R M O U P R T E U U U R
M M C O R T U M T O E T M U P M U T T P U T T E R M U U P C E T M T
U P T P P E T R U T T P O T M E C U R C P U O P M T P M C M P E P C
M M U O R R M P C M M O R C C U T C C O M C U U P R C P P P U C U U
E U P R U P M C E C T M C C U U R P P M U U E U U U U C E T U U R C
P U U R E U C E C U C C U E C U U U R C P P M C C C U P R M U C M U
C P R U P P U O M P U U U C M U U C P M U C R C P M M T C M M U O M
C M C C M U U P C C T U R U E U U U C U M T U C C M M U C T C R R U
R U M R P R U C U C E M U C C U U U E T U M C P C U R P U R C U U M
U P P C E M P P P U U M P P C C P R R C E C C R M C P P R C C R P P
M U U U R C M E P C P U C C C C U P R R U U P M C E M C U T M U C C
M E P M M P P M U U C C E M P R E U U T C P C U C M C C U C M R T P
M P C U C P P M R C M P C P E M P P P M R U U C C U U P R C E R T U
U P C U M U P U M P C R C C E P C U C C P M T R P C P C U U C R P P
R U R C C M E U U R U U M U R P E M R U C C M M U C R M C T M R P R
C U C M C U U C U M M U U U E M C T M C C M U C T C M U C M P M U T
R U R R E O C U C R C U P U C M P C E U C C E U U E P U M P T C C E
U R C U U C P U R C T P E U U M M U U U C C M M T U C R C R M R P O
U C U C U P C M P C U C T P M M U P U C U M U M C U T P P M E U U U
P U P C U U U U C M P U E M C U P C C R P P R U U M C C U C U P C P
C P C C U U U C U R C C P U R C U T U R E C R U U C M T C C C M U C
C P P P C M U C C U U U U U M M P U C R C U E C C T P C P M E E C M
U U C C C U U M C P C C C U U C U P C U P U T C M M C U M M M U M M
P U M M P T R M M P P P M R U U U C U U R E T U C P E C R P U R U R
C C C T P P M T P U P M P P M R M U R P U P U U U U U E P U C M P R

P P C C R O U U E C T U P C U P C C U U C P C P C M U E C M U T U U
P C U U T P P P C M M U P C C R U C E R T U C T E C M C U U E C R P
U M C U T C U E C C U P C U C C P U R P M M T U T P P O C U R C P C
P P M C M C C C P U P P M R U T E R M O T U M U U E M R C U U T P U
P P T T T M U O T T E R P R E T T R M T E M T E U U T T R P T T C U
T M T U P M R E U P M U E U U U U U P T E T C P U C E E C T E R M M
T M O T M P M E T R P E R O P E M E M M P R P T R U P T U O E U M P
P U R M U U E M M M P U C P U M U T M P E U U O P P U O M P T O T R
R M T P C P P P R E P E E R M R E M U T P O U E M P P E E R R M T R
T O M E P T E M U E P R T U R O O T O M U P P E R O T T P T T M P P
T P C U U U M T T U R E O P M T R E T T M E E U U O P M E R M P E T
E E R M U T T M M P E P O E T M E T E R U U O O R M E M M T R U U R
U O P R U P R P P U U U E E E T T T T P E U R E R R P U E T R U U E
O O O U E T E U U M U T U R U T R U U T O P O T U P M U R U U E R U
U U P U O O T T T P M E U E R T M O U M T P P P E O M T T U U U O E
U U E T U U E T U R P U M T M M E R R U U E T O T P T T T R P T M P
E E M T M E U U P O E T T P P P R U T E E C O U M E U U T T R T T T
R T T R T T M E P P T R T P O U T R T T O P E C R T P U T T C E M P
T O M R E T T T R E U C O T O T R P R U R P T U T E U U E P M E O T
M M U U U R R E T M O U M M P C P E T P T P R M T U P U E T E T E E
M C C T E R U R O E E P R R R R T P T U U M T P E E M C U O U U R E
C T U P P R T P P M T M U M C T T T P R R E O U T P E R U T M P U R
R U T U M O T T E E T M T R M R T O M T R R R T O P T T E R U O O M
U T P R M M P R P U E T M E U T T M P P R T P T P T T U U M R T E T
T R R O T U R U T R U U C M R C M T O C R U T P O T T P T M T E O R
R M R U E U R R T T O U R U P T U E C T E O T M T P R T P U M M R E
E E P O R P U R P R U M E M O T T R O P R U E T T U E T R O M T O U
E O P U T M T U R P T P R R T M O R E T C T M T M U E T T M R T T E
O R P C P P M M U M T T O U M T E U U R T R T R M E M U U T M T U T
R E T P M T P P M M

Stage 8

Umkehr-walze			Walze 3			Walze 2			Walze 1			Stecker-brett		Tastatur
Y	A		B	A		E	A		A	A				A
R	B		D	B		K	B		J	B				B
U	C		F	C		M	C		D	C				C
H	D		H	D		F	D		K	D				D
Q	E		J	E		L	E		S	E				E
S	F		L	F		G	F		I	F				F
L	G		C	G		D	G		R	G				G
D	H		P	H		Q	H		U	H				H
P	I		R	I		V	I		X	I				I
X	J		T	J		Z	J		B	J				J
N	K	←	X	K	←	N	K	←	L	K	←		←	K
G	L		V	L		T	L		H	L				L
O	M		Z	M		O	M		W	M		?		M
K	N	→	N	N	→	W	N	→	T	N	→		→	N
M	O		Y	O		Y	O		M	O				O
I	P		E	P		H	P		C	P				P
E	Q		I	Q		X	Q		Q	Q				Q
B	R		W	R		U	R		G	R				R
F	S		G	S		S	S		Z	S				S
Z	T		A	T		P	T		N	T				T
C	U		K	U		A	U		P	U				U
W	V		M	V		I	V		Y	V				V
V	W		U	W		B	W		F	W				W
J	X		S	X		R	X		V	X				X
A	Y		Q	Y		C	Y		O	Y				Y
T	Z		O	Z		J	Z		E	Z				Z

```
K J Q P W C A I S R X W Q M A S E U P F O C Z O Q Z V G Z G W W
K Y E Z V T E M T P Z H V N O T K Z H R C C F Q L V R P C C W L
W P U Y O N F H O G D D M O J X G G B H W W U X N J E Z A X F U
M E Y S E C S M A Z F X N N A S S Z G W R B D D M A P G M R W T
G X X Z A X L B X C P H Z B O U Y V R R V F D K H X M Q O G Y L
Y Y C U W Q B T A D R L B O Z K Y X Q P W U U A F M I Z T C E A
X B C R E D H Z J D O P S Q T N L I H I Q H N M J Z U H S M V A
H H Q J L I J R R X Q Z N F K H U I I N Z P M P A F L H Y O N M
R M D A D F O X T Y O P E W E J G E C A H P Y F V M C I X A Q D
Y I A G Z X L D T F J W J Q Z M G B S N E R M I P C K P O V L T
H Z O T U X Q L R S R Z N Q L D H X H L G H Y D N Z K V B F D M
X R Z B R O M D P R U X H M F S H J
```

Schlüssel

```
0716150413020110
```

Schriftzeichen

```
begin 644 DEBUGGER.BIN
(-&>`_EU-_/$`


end
```

Stage 9


```
begin 600 text.d
MM5P7)_8F_,H[JOF1C//L/W+)%QSK*Q37CJ-N 'W[_;CQSTW'UY0S2,\LQVG0
M@1&HY^1MHYI\>2P'F:6Y*E%X4A&$2'=L28$$..9["-ZIGA_VP(GIPK[CW3^L
M55+6OD^&=FS61(L96YG>  '59*1Q^)/C?$1/C&9PN35-HP;.>V8_/P(.:+R(
M61]'NG^UF:,#57MMQSKN[N7M>1NE;2(!RUA495Q16!;Q<*("[C*"A"@%A+=S
M8AR45+G$-#8A?29V_.6%7*6D$J_G4JX'JM^1? K@._#(B/N7-<YNU;/,JF8C
M6LD[90MVJ2'I*.G@>9U%!E(33!S^K# N7JH_Y5RYE&=J@S!>^<C3Y=PD%-RP
M9&++^"JLPOK%T)-5KI>IUA"W;7;&D(D-2/U'$3\C7 ?]B* 3*C/Y!%U >&V6
M%W85NJ:JPO(>#C1)CFEL&^H3YKR2.59XJVD??\MX+ [S?3X_F^/*1$NGH$B&
MI$L2-C'E/@OD*&5;6+P+G1S D49AO=#9\C!4D$/F;C(H#MX:\%G[K[OR+2RG
M@@SCSVG!A5%FEV!=$YD"V.2T06@>C-&)3H<:Y9BOR=V#S_>\:S8GZ.*A"$!T
M2OE-/4QWLLB<[!K8I 1Z@C9_,( #D!/G4)P2>,3!%9. Q]MV0,!F9,F1VF'@
M=!XCI_M>2?F=' ;20):%Y61[.! -W8%7M3BJUX/&!-E@A7C\(>5SZXESA$LZ
MF\_U//JGV"KKHE259927962%P-9J!*J@ DPJF]M2/>DXHA?JT"^2C7;_-9B;
MBM'CFTYUR#DOA7.J4ZW8=+3(9O>#4A+^!=4IV 6A!(PNGZ:T$O)659KNGS=>
MN"?LQ3$6F*I43Q(3_U:64V/L9$<E%">*#A9P>@(66#XDS!)-'*\JZE.,=G29
MOJLH!9.Y#+=?]!"C?2/?H5O!A]<KW^H%J "&>+EXK;II6)N6JY$%UB'BN3'F
MMS[XKP#JY(:3@V);U2,5PG 6$!46;.B/K'E7$4'MKN1]* YX^R"Q?Q+;,"./
MPL((>]UF90L7[<]9^E0*:NMBI(Q+B'>-IHF+,J0&"G0F.5L8@"_)<Y$<ZRU=
M']&L9!WD1Y<V[D:/:4J('#X(NIKKDFO@#:5O_3C%7]AC5H.? ,%;D)=7'HKE
M.(_E=(*(W5HO3RA5WP8<!ZM.K2T.:&#P\LV;!7W$ K3)/A7D&P8SVO3-?$U1
M2J10K3T>2)OVRA'Y;C<DZVV+'$VXI_ $JZ^)39,'.7MK,0*QOP9O6QRQ0F(*
M&8J9O!Z">N;S%MD%%A.SD?'^\K]"R_@XE6V# >&P.$L#$,%N"C[H:A_EPH$V
M\H)(;C0#3^C] T9Z0=,9UQ(3N^3D],9PVM<AJ.T:('(.=1FB;NBV_YS!\7ON
M?-T%5B;2J^TORBWA^Z$B'$K8LC;'A+>@87(6!8Q%FRS=^;Y*0$FC">;I!NI*
@#%0SNY_0_-EK1>;84QMT0/(KQQ2LL+R##K:I=NK7.OT

end
```

Stage 10

Shorter message:

```
10052 30973 22295 13534 12990 66921 15454 81904 58209 26472 18119
11542 99190 01294 87266 20201 55809 80932 92390 96710 64341 91354
27685 27572 48495 78859 80627 33369 29356 36094 85523
```

Longer message:

```
begin 600 text.d
M.4#)>S I:R!!4)NA+\%T%V/(AW!7HHDPS$;T[\E!RWA?,J8:X#D[!:XF,A>K
MXT9$Q)37\IOMG6KL-$6?A!#FZ2Y)N+4%*.^2K!SP?Z2'8O7LZ]QP \T=QG-*
MAMJA;Q@3H[8^U/L<ILL%TA0J9M*F@8F?H:76%<33JOESAP=@3:(\:8NBGFM0
M,MP3B^CP%/D8DICZ$VO(7IS(DTJRZ&#Y- 7I\-#VI0">J@+O!CT.+6B9K$J%
4:EAB9%1#;(P+I>1!#<+2+;(7.W<

end
```

Appendix A

The Opening Paragraph of *A Void* by Georges Perec, translated by Gilbert Adair

Today, by radio, and also on giant hoardings, a rabbi, an admiral notorious for his links to masonry, a trio of cardinals, a trio, too, of insignificant politicians (bought and paid for by a rich and corrupt Anglo-Canadian banking corporation), inform us all of how our country now risks dying of starvation. A rumour, that's my initial thought as I switch off my radio, a rumour or possibly a hoax. Propaganda, I murmur anxiously – as though, just by saying so, I might allay my doubts – typical politicians' propaganda. But public opinion gradually absorbs it as a fact. Individuals start strutting around with stout clubs. 'Food, glorious food!' is a common cry (occasionally sung to Bart's music), with ordinary hard-working folk harassing officials, both local and national, and cursing capitalists and captains of industry. Cops shrink from going out on night shift. In Mâcon a mob storms a municipal building. In Rocadamour ruffians rob a hangar full of foodstuffs, pillaging tons of tuna fish, milk and cocoa, as also a vast quantity of corn – all of it, alas, totally unfit for human consumption. Without fuss or ado, and naturally without any sort of trial, an indignant crowd hangs 26 solicitors on a hastily built scaffold in front of Nancy's law courts (this Nancy is a town, not a woman) and ransacks a local journal, a disgusting right-wing rag that is siding against it. Up and down this land of ours looting has brought docks, shops and farms to a virtual standstill.

Appendix B

Some Elementary Tips for Frequency Analysis

(1) Begin by counting up the frequencies of all the letters in the ciphertext. About five of the letters should have a frequency of less than 1 per cent, and these probably represent j, k, q, x and z. One of the letters should have a frequency greater than 10 per cent, and it probably represents e. If the ciphertext does not obey this distribution of frequencies, then consider the possibility that the original message was not written in English. You can identify the language by analysing the distribution of frequencies in the ciphertext. For example, typically in Italian there are three letters with a frequency greater than 10 per cent, and nine letters have frequencies less than 1 per cent. In German, the letter e has the extraordinarily high frequency of 19 per cent, so any ciphertext containing one letter with such a high frequency is quite possibly German. Once you have identified the language you should use the appropriate table of frequencies for that language for your frequency analysis. It is often possible to unscramble ciphertexts in an unfamiliar language, as long as you have the appropriate frequency table.

(2) If the correlation is sympathetic with English, but the plaintext does not reveal itself immediately, which is often the case, then focus on pairs of repeated letters. In English the most common repeated letters are ss, ee, tt, ff, ll, mm and oo. If the ciphertext contains any repeated characters, you can assume that they represent one of these.

(3) If the ciphertext contains spaces between words, then try to identify words containing just one, two or three letters. The only one-letter words in English are a and I. The commonest two-letter words are of, to, in, it, is, be, as, at, so, we, he, by, or, on, do, if, me, my, up, an, go, no, us, am. The most common three-letter words are the and and.

(4) If possible, tailor the table of frequencies to the message you are trying to decipher. For example, military messages tend to omit pronouns and articles, and the loss of words such as I, he, a and the will reduce the frequency of some of the commonest letters. If you know you are tackling a military message, you should use a frequency table generated from other military messages.

(5) One of the most useful skills for a cryptanalyst is the ability to identify words, or even entire phrases, based on experience or sheer guesswork. Al-Khalīl, an early Arabian cryptanalyst, demonstrated this talent when he cracked a Greek ciphertext. He guessed that the ciphertext began with the greeting 'In the name of God'. Having established that these letters corresponded to a specific section of ciphertext, he could use them as a crowbar to prise open the rest of the ciphertext. This is known as a crib.

(6) On some occasions the commonest letter in the ciphertext might be E, the next commonest could be T, and so on. In other words, the frequency of letters in the ciphertext already matches those in the frequency table. The E in the ciphertext appears to be a genuine e, and the same seems to be true for all the other letters, yet the ciphertext looks like gibberish. In this case you are faced not with a substitution cipher, but with a transposition cipher. All the letters do represent themselves, but they are in the wrong positions.

Cryptanalysis by Helen Fouché Gaines (Dover) is a good introductory text. As well as giving tips, it also contains tables of letter frequencies in different languages, and provides lists of the most common words in English.

Appendix C

The So-called Bible Code

In 1997 *The Bible Code* by Michael Drosnin caused headlines around the world. Drosnin claimed that the Bible contains hidden messages which could be discovered by searching for equidistant letter sequences (EDLSs). An EDLS is found by taking any text, picking a particular starting letter, then jumping forward a set number of letters at a time. So, for example, with this paragraph we could start with the 'M' in Michael and jump, say, five spaces at a time. If we noted every fifth letter, we would generate the EDLS **mesahirt** . . . .

Although this particular EDLS does not contain any sensible words, Drosnin described the discovery of an astonishing number of Biblical EDLSs that not only form sensible words, but result in complete sentences. According to Drosnin, these sentences are Biblical predictions. For example, he claims to have found references to the assassinations of John F. Kennedy, Robert Kennedy and Anwar Sadat. In one EDLS the name of Newton is mentioned next to gravity, and in another Edison is linked with the light bulb. Although Drosnin's book is based on a paper published by Doron Witzum, Eliyahu Rips and Yoav Rosenberg, it is far more ambitious in its claims, and has attracted a great deal of criticism. The main cause of concern is that the text being studied is enormous: in a large enough text, it is hardly surprising that by varying both the starting place and the size of the jump, sensible phrases can be made to appear.

Brendan McKay at the Australian National University tried to demonstrate the weakness of Drosnin's approach by searching for EDLSs in *Moby Dick*, and discovered thirteen statements pertaining to assassinations of famous people, including Trotsky, Gandhi and Robert Kennedy. Furthermore, Hebrew texts are bound to be particularly rich in EDLSs, because they are largely devoid of vowels. This means that interpreters can insert vowels as they see fit, which makes it easier to extract predictions.

Appendix D

The Pigpen Cipher

The monoalphabetic substitution cipher persisted through the centuries in various forms. For example, the pigpen cipher was used by Freemasons in the 1700s to keep their records private, and is still used today by schoolchildren. The cipher does not substitute one letter for another, rather it substitutes each letter for a symbol according to the following pattern.

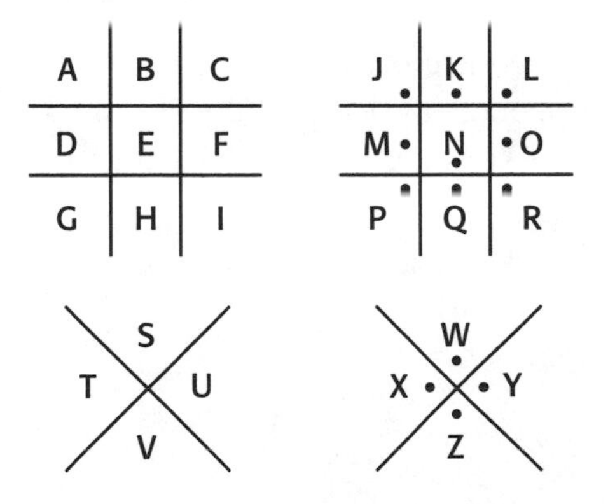

To encrypt a particular letter, find its position in one of the four grids, then sketch that portion of the grid to represent that letter. Hence:

a =

b =

:

:

z =

If you know the key, then the pigpen cipher is easy to decipher. If not, then it is easily broken by:

Appendix E

The Playfair Cipher

The Playfair cipher was popularised by Lyon Playfair, first Baron Playfair of St Andrews, but it was invented by Sir Charles Wheatstone, one of the pioneers of the electric telegraph. The two men lived close to each other, either side of Hammersmith Bridge, and they often met to discuss their ideas on cryptography.

The cipher replaces each pair of letters in the plaintext with another pair of letters. In order to encrypt and transmit a message, the sender and receiver must first agree on a keyword. For example, we can use Wheatstone's own name, CHARLES, as a keyword. Next, before encryption, the letters of the alphabet are written in a 5×5 square, beginning with the keyword, and combining the letters I and J into a single element:

C H A R L
E S B D F
G I/J K M N
O P Q T U
V W X Y Z

Next, the message is broken up into pairs of letters, or digraphs. The two letters in any digraph should be different, achieved in the following example by inserting an extra x between the double m in hammersmith, and an extra x is added at the end to make a digraph from the single final letter:

Plaintext	meet me at hammersmith bridge tonight
Plaintext in digraphs	me-et-me-at-ha-mx-me-rs-mi-th-br-id-ge-to-ni-gh-tx

Encryption can now begin. All the digraphs fall into one of three categories – both letters are in the same row, or the same column, or neither. If both letters are in the same row, then they are replaced by the letter to the immediate right of each one; thus mi becomes NK. If one of the letters is at the end of the row, it is replaced by the letter at the beginning; thus ni becomes GK. If both letters are in the same column, they are replaced by the letter immediately beneath each one;

thus ge becomes OG. If one of the letters is at the bottom of the column, then it is replaced by the letter at the top; thus ve becomes CG.

If the letters of the digraph are neither in the same row nor the same column, the encipherer follows a different rule. To encipher the first letter, look along its row until you reach the column containing the second letter; the letter at this intersection then replaces the first letter. To encipher the second letter, look along its row until you reach the column containing the first letter; the letter at this intersection replaces the second letter. Hence, me becomes GD, and et becomes DO. The complete encryption is:

Plaintext in digraphs	me	et	me	at	ha	mx	me	rs	mi	th	br	id	ge	to	ni	gh	tx
Ciphertext	GD	DO	GD	RQ	AR	KY	GD	HD	NK	PR	DA	MS	OG	UP	GK	IC	QY

The recipient, who also knows the keyword, can easily decipher the ciphertext by simply reversing the process: for example, enciphered letters in the same row are deciphered by replacing them by the letters to their left.

As well as being a scientist, Playfair was also a notable public figure (Deputy Speaker of the House of Commons, postmaster general, and a commissioner on public health who helped to develop the modern basis of sanitation) and he was determined to promote Wheatstone's idea among the most senior politicians. He first mentioned it at a dinner in 1854 in front of Prince Albert and the future Prime Minister, Lord Palmerston, and later he introduced Wheatstone to the Under Secretary of the Foreign Office. Unfortunately, the Under Secretary complained that the system was too complicated for use in battle conditions, whereupon Wheatstone stated that he could teach the method to boys from the nearest elementary school in 15 minutes. 'That is very possible', replied the Under Secretary, 'but you could never teach it to attachés.'

Playfair persisted, and eventually the British War Office secretly adopted the technique, probably using it first in the Boer War. Although it proved effective for a while, the Playfair cipher was far from impregnable. It can be attacked by looking for the most frequently occurring digraphs in the ciphertext, and assuming that they represent the commonest digraphs in English: th, he, an, in, er, re, es.

Appendix F

The ADFGVX Cipher

The ADFGVX cipher features both substitution and transposition. Encryption begins by drawing up a 6 × 6 grid, and filling the 36 squares with a random arrangement of the 26 letters and the 10 digits. Each row and column of the grid is identified by one of the six letters A, D, F, G, V or X. The arrangement of the elements in the grid acts as part of the key, so the receiver needs to know the details of the grid in order to decipher messages.

	A	**D**	**F**	**G**	**V**	**X**
A	8	p	3	d	1	n
D	l	t	4	o	a	h
F	7	k	b	c	5	z
G	j	u	6	w	g	m
V	x	s	v	i	r	2
X	9	e	y	0	f	q

The first stage of encryption is to take each letter of the message, locate its position in the grid and substitute it with the letters that label its row and column. For example, 8 would be substituted by AA, and p would be replaced by AD. Here is a short message encrypted according to this system:

Message	attack at 10 pm
Plaintext	a t t a c k a t 1 0 p m
Stage 1 Ciphertext	DV DD DD DV FG FD DV DD AV XG AD GX

So far this is a simple monoalphabetic substitution cipher, and frequency analysis would be enough to crack it. However, the second stage of the ADFGVX is a transposition, which makes cryptanalysis much harder. The transposition depends on a keyword, which in this case happens to be the word MARK, and which must be shared with the receiver. Transposition is carried out according to the following recipe. First, the letters of the keyword are written in the top row of a fresh grid. Next, the stage 1 ciphertext is written underneath it in a series of rows, as shown below. The columns of the grid are then re-arranged so that the letters of the keyword are in alphabetical order. The final ciphertext is achieved by going down each column and then writing out the letters in this new order.

M	A	R	K
D	V	D	D
D	D	D	V
F	G	F	D
D	V	D	D
A	V	X	G
A	D	G	X

Re-arrange columns so that the letters of the keyword are in alphabetical order →

A	K	M	R
V	D	D	D
D	V	D	D
G	D	F	F
V	D	D	D
V	G	A	X
D	X	A	G

Final Ciphertext V D G V V D D V D D G X D D F D A A D D F D X G

The final ciphertext would then be transmitted in Morse code, and the receiver would reverse the encryption process in order to retrieve the original text. The entire ciphertext is made up of just six letters (i.e. A, D, F, G, V, X), because these are the labels of the rows and columns of the initial 6 × 6 grid. People often wonder why these letters were chosen as labels, as opposed to, say, A, B, C, D, E and F. The answer is that A, D, F, G, V and X are highly dissimilar from one another when translated into Morse dots and dashes, so this choice of letters minimises the risk of errors during transmission.

Appendix G

The Weaknesses of Recycling a One-time Pad

For the reasons explained in Chapter 3, ciphertexts encrypted according to a one-time pad cipher are unbreakable. However, this relies on each one-time pad being used once and only once. If we were to intercept two distinct ciphertexts which have been encrypted with the same one-time pad, we could decipher them in the following way.

We would probably be correct in assuming that the first ciphertext contains the word the somewhere, and so cryptanalysis begins by assuming that the entire message consists of a series of the's. Next, we work out the one-time pad that would be required to turn a whole series of the's into the first ciphertext. This becomes our first guess at the one-time pad. How do we know which parts of this one-time pad are correct?

We can apply our first guess at the one-time pad to the second ciphertext, and see if the resulting plaintext makes any sense. If we are lucky, we will be able to discern a few fragments of words in the second plaintext, indicating that the corresponding parts of the one-time pad are correct. This in turn shows us which parts of the first message should be the.

By expanding the fragments we have found in the second plaintext, we can work out more of the one-time pad, and then deduce new fragments in the first plaintext. By expanding these fragments in the first plaintext, we can work out more about the one-time pad, and then deduce new fragments in the second plaintext. We can continue this process until we have deciphered both plaintexts.

This process is very similar to the decipherment of a message enciphered with a Vigenère cipher using a key that consists of a series of words, such as the example in Chapter 3, in which the key was CANADABRAZILEGYPTCUBA.

Appendix H

The *Daily Telegraph* Crossword Solution

ACROSS

1. Troupe
4. Short Cut
9. Privet
10. Aromatic
12. Trend
13. Great deal
15. Owe
16. Feign
17. Newark
22. Impale
24. Guise
27. Ash
28. Centre bit
31. Token
32. Lame dogs
33. Racing
34. Silencer
35. Alight

DOWN

1. Tipstaff
2. Olive oil
3. Pseudonym
5. Horde
6. Remit
7. Cutter
8. Tackle
11. Agenda
14. Ada
18. Wreath
19. Right nail
20. Tinkling
21. Sennight
23. Pie
25. Scales
26. Enamel
29. Rodin
30. Bogie

Appendix I

Exercises for the Interested Reader

Some of the greatest decipherments in history have been achieved by amateurs. For example, Georg Grotefend, who made the first breakthrough in interpreting cuneiform, was a schoolteacher. For those readers who feel the urge to follow in his footsteps, there are several scripts that remain a mystery. Linear A, a Minoan script, has defied all attempts at decipherment, partly due to a paucity of material. Etruscan does not suffer from this problem, with over 10,000 inscriptions available for study, but it has also baffled the world's greatest scholars. Iberian, another pre-Roman script, and the futhark runes from Scandinavia are equally unfathomable.

The most intriguing ancient European script appears on the unique Phaistos Disc, discovered in southern Crete in 1908. It is a circular tablet dating from around 1700 BC bearing writing in the form of two spirals, one on each side. The signs are not handmade impressions, but were made using a variety of stamps, making this the world's oldest example of typewriting. Remarkably, no other similar document has ever been found, so decipherment relies on very limited information – there are 242 characters divided into 61 groups. However, a typewritten document implies mass production, so the hope is that archaeologists will eventually discover a hoard of similar discs, and shed light on this intractable script.

One of the great challenges outside Europe is the decipherment of the Bronze Age script of the Indus civilisation, which can be found on thousands of seals dating from the third millennium BC. Each seal depicts an animal accompanied by a short inscription, but the meaning of these inscriptions has so far evaded all the experts. In one exceptional example the script has been found on a large wooden board with giant letters 37 cm in height. This could be the world's oldest billboard. It implies that literacy was not restricted to the elite, and raises the question as to what was being advertised. The most likely answer is that it was part of a promotional campaign for the king, and if the identity of the king can be established, then the billboard could provide a way into the rest of the script.

Appendix J

The Mathematics of RSA

What follows is a straightforward mathematical description of the mechanics of RSA encryption and decryption.

(1) Alice picks two giant prime numbers, p and q. The primes should be enormous, but for simplicity we assume that Alice chooses $p = 17$, $q = 11$. She must keep these numbers secret.

(2) Alice multiplies them together to get another number, N. In this case $N = 187$. She now picks another number e, and in this case she chooses $e = 7$.

(e and $(p - 1) \times (q - 1)$ should be relatively prime, but this is a technicality.)

(3) Alice can now publish e and N in something akin to a telephone directory. Since these two numbers are necessary for encryption, they must be available to anybody who might want to encrypt a message to Alice. Together these numbers are called the public-key. (As well as being part of Alice's public-key, e could also be part of everybody else's public-key. However, everybody must have a different value of N, which depends on their choice of p and q.)

(4) To encrypt a message, the message must first be converted into a number, M. For example, a word is changed into ASCII binary digits, and the binary digits can be considered as a decimal number. M is then encrypted to give the ciphertext, C, according to the formula

$$C = M^e \pmod N$$

(5) Imagine that Bob wants to send Alice a simple kiss: just the letter X. In ASCII this is represented by 1011000, which is equivalent to 88 in decimal. So, $M = 88$.

(6) To encrypt this message, Bob begins by looking up Alice's public-key, and discovers that $N = 187$ and $e = 7$. This provides him with the encryption formula required to encrypt messages to Alice. With $M = 88$, the formula gives

$$C = 88^7 \pmod{187}$$

(7) Working this out directly on a calculator is not straightforward, because the display cannot cope with such large numbers. However, there is a neat trick for calculating exponentials in modular arithmetic. We know that, since $7 = 4 + 2 + 1$,

$$88^7 \pmod{187} = [88^4 \pmod{187} \times 88^2 \pmod{187} \times 88^1 \pmod{187}] \pmod{187}$$

$$88^1 = 88 = 88 \pmod{187}$$

$$88^2 = 7{,}744 = 77 \pmod{187}$$

$$88^4 = 59{,}969{,}536 = 132 \pmod{187}$$

$$88^7 = 88^1 \times 88^2 \times 88^4 = 88 \times 77 \times 132 = 894{,}432 = 11 \pmod{187}$$

Bob now sends the ciphertext, $C = 11$, to Alice.

(8) We know that exponentials in modular arithmetic are one-way functions, so it is very difficult to work backwards from $C = 11$ and recover the original message, M. Hence, Eve cannot decipher the message.

(9) However, Alice can decipher the message because she has some special information: she knows the values of p and q. She calculates a special number, d, the decryption key, otherwise known as her private-key. The number d is calculated according to the following formula

$$e \times d = 1 \pmod{(p\text{-}1) \times (q\text{-}1)}$$

$$7 \times d = 1 \pmod{16 \times 10}$$

$$7 \times d = 1 \pmod{160}$$

$$d = 23$$

(Deducing the value of d is not straightforward, but a technique known as Euclid's algorithm allows Alice to find d quickly and easily.)

(10) To decrypt the message, Alice simply uses the following formula,

$M = C^d \pmod{187}$

$M = 11^{23} \pmod{187}$

$M = [11^1 \pmod{187} \times 11^2 \pmod{187} \times 11^4 \pmod{187} \times 11^{16} \pmod{187}] \pmod{187}$

$M = 11 \times 121 \times 55 \times 154 \pmod{187}$

$M = 88 = X$ in ASCII.

Rivest, Shamir and Adleman had created a special one-way function, one that could be reversed only by somebody with access to privileged information, namely the values of p and q. Each function can be personalised by choosing p and q, which multiply together to give N. The function allows everybody to encrypt messages to a particular person by using that person's choice of N, but only the intended recipient can decrypt the message because the recipient is the only person who knows p and q, and hence the only person who knows the decryption key, d.

Glossary

ASCII American Standard Code for Information Interchange, a standard for turning alphabetic and other characters into numbers.

asymmetric-key cryptography A form of cryptography in which the key required for encrypting is not the same as the key required for decrypting. Describes public-key cryptography systems, such as RSA.

Caesar-shift substitution cipher Originally a cipher in which each letter in the message is replaced with the letter three places further on in the alphabet. More generally, it is a cipher in which each letter in the message is replaced with the letter *x* places further on in the alphabet, where *x* is a number between 1 and 25.

cipher Any general system for hiding the meaning of a message by replacing each letter in the original message with another letter. The system should have some built-in flexibility, known as the key.

cipher alphabet The rearrangement of the ordinary (or plain) alphabet, which then determines how each letter in the original message is enciphered. The cipher alphabet can also consist of numbers or any other characters, but in all cases it dictates the replacements for letters in the original message.

ciphertext The message (or plaintext) after encipherment.

code A system for hiding the meaning of a message by replacing each word or phrase in the original message with another character or set of characters. The list of replacements is contained in a codebook. (An alternative definition of a code is any form of encryption which has no built-in flexibility, i.e. there is only one key, namely the codebook.)

codebook A list of replacements for words or phrases in the original message.

cryptanalysis The science of deducing the plaintext from a ciphertext, without knowledge of the key.

cryptography The science of encrypting a message, or the science of concealing the meaning of a message. Sometimes the term is used more generally to mean the science of anything connected with ciphers, and is an alternative to the term cryptology.

cryptology The science of secret writing in all its forms, covering both cryptography and cryptanalysis.

decipher To turn an enciphered message back into the original message. Formally, the term refers only to the intended receiver who knows the key required to obtain the plaintext, but informally it also refers to the process of cryptanalysis, in which the decipherment is performed by an enemy interceptor.

decode To turn an encoded message back into the original message.

decrypt To decipher or to decode.

DES Data Encryption Standard, developed by IBM and adopted in 1976.

Diffie–Hellman–Merkle key exchange A process by which a sender and receiver can establish a secret key via public discussion. Once the key has been agreed, the sender can use a cipher such as DES to encrypt a message.

digital signature A method for proving the authorship of an electronic document. Often this is generated by the author encrypting the document with his or her private-key.

encipher To turn the original message into the enciphered message.

encode To turn the original message into the encoded message.

encrypt To encipher or encode.

encryption algorithm Any general encryption process which can be specified exactly by choosing a key.

homophonic substitution cipher A cipher in which there are several potential substitutions for each plaintext letter. Crucially, if there are, say, six potential substitutions for the plaintext letter a, then these six characters can only represent the letter a. This is a type of monoalphabetic substitution cipher.

key The element that turns the general encryption algorithm into a specific method for encryption. In general, the enemy may be aware of the encryption algorithm being used by the sender and receiver, but the enemy must not be allowed to know the key.

key distribution The process of ensuring that both sender and receiver have access to the key required to encrypt and decrypt a message, while making sure that the key does not fall into enemy hands. Key distribution was a major problem in terms of logistics and security before the invention of public-key cryptography.

key escrow A scheme in which users lodge copies of their secret keys with a trusted third party, the escrow agent, who will pass on keys to law enforcers only under certain circumstances, for example if a court order is issued.

key length Computer encryption involves keys which are numbers. The key length refers to the number of digits or bits in the key, and thus indicates the biggest number that can be used as a key, thereby defining the number of possible keys. The longer the key length (or the greater the number of possible keys), the longer it will take a cryptanalyst to test all the keys.

monoalphabetic substitution cipher A substitution cipher in which the cipher alphabet is fixed throughout encryption.

National Security Agency (NSA) A branch of the U.S. Department of Defense, responsible for ensuring the security of American communications and for breaking into the communications of other countries.

one-time pad The only known form of encryption that is unbreakable. It relies on a random key that is the same length as the message. Each key can be used once and only once.

plaintext The original message before encryption.

polyalphabetic substitution cipher A substitution cipher in which the cipher alphabet changes during the encryption, for example the Vigenère cipher. The change is defined by a key.

Pretty Good Privacy (PGP) A computer encryption algorithm developed by Phil Zimmermann, based on RSA.

private-key The key used by the receiver to decrypt messages in a system of public-key cryptography. The private-key must be kept secret.

public-key The key used by the sender to encrypt messages in a system of public-key cryptography. The public-key is available to the public.

public-key cryptography A system of cryptography which overcomes the problems of key distribution. Public-key cryptography requires an asymmetric cipher, so that each user can create a public encryption key and a private decryption key.

quantum computer An immensely powerful computer that exploits quantum theory, in particular the theory that an object can be in many states at once (superposition), or the theory that an object can be in many universes at once. If scientists could build a quantum computer on any reasonable scale, it would jeopardise the security of all current ciphers except the one-time pad cipher.

quantum cryptography An unbreakable form of cryptography that exploits quantum theory, in particular the uncertainty principle – which states that it is impossible to measure all aspects of an object with absolute certainty. Quantum cryptography guarantees the secure exchange of a random series of bits, which is then used as the basis for a one-time pad cipher.

RSA The first system that fitted the requirements of public-key cryptography, invented by Ron Rivest, Adi Shamir and Leonard Adleman in 1977.

steganography The science of hiding the existence of a message, as opposed to cryptography, which is the science of hiding the meaning of a message.

substitution cipher A system of encryption in which each letter of a message is replaced with another character, but retains its position within the message.

symmetric-key cryptography A form of cryptography in which the key required for encrypting is the same as the key required for decrypting. The term describes all traditional forms of encryption, i.e. those in use before the 1970s.

transposition cipher A system of encryption in which each letter of a message changes its position within the message, but retains its identity.

Vigenère cipher A polyalphabetic cipher which was developed around 1500. The Vigenère square contains 26 separate cipher alphabets, each one a Caesar-shifted alphabet, and a keyword defines which cipher alphabet should be used to encrypt each letter of a message.

Acknowledgements

While writing this book I have had the privilege of meeting some of the world's greatest living codemakers and codebreakers, ranging from those who worked at Bletchley Park to those who are developing the ciphers that will enrich the Information Age. I would like to thank Whitfield Diffie and Martin Hellman, who took the time to describe their work to me while I was in sunny California. Similarly, Clifford Cocks, Malcolm Williamson and Richard Walton were enormously helpful during my visit to cloudy Cheltenham. In particular, I am grateful to the Information Security Group at Royal Holloway College, London, who allowed me to attend the M.Sc. course on information security. Professor Fred Piper, Simon Blackburn, Jonathan Tuliani, and Fauzan Mirza all taught me valuable lessons about codes and ciphers.

While I was in Virginia, I was fortunate to be given a guided tour of the Beale treasure trail by Peter Viemeister, an expert on the mystery. Furthermore, the Bedford County Museum and Stephen Cowart of the Beale Cypher and Treasure Association helped me to research the subject. I am also grateful to David Deutsch and Michele Mosca of the Oxford Centre for Quantum Computation, Charles Bennett and his research group at IBM's Thomas J. Watson Laboratories, Stephen Wiesner, Leonard Adleman, Ronald Rivest, Paul Rothemund, Jim Gillogly, Paul Leyland and Neil Barrett.

Derek Taunt, Alan Stripp and Donald Davies kindly explained to me how Bletchley Park broke Enigma, and I was also helped by the Bletchley Park Trust, whose members regularly give enlightening lectures on a variety of topics. Dr Mohammed Mrayati and Dr Ibrahim Kadi have been involved in revealing some of the early breakthroughs in Arab cryptanalysis, and were kind enough to send me relevant documents. The periodical *Cryptologia* also carried articles about Arabian cryptanalysis, as well as many other cryptographic subjects, and I would like to thank Brian Winkel for sending me back-issues of the magazines.

I would encourage readers to visit the National Cryptologic Museum near Washington, D.C. and the Cabinet War Rooms in London, and I hope that you will be as fascinated as I was during my visits. Thank you to the curators and librarians of these museums for helping me with my research. When I was pressed for time, James Howard, Bindu Mathur, Pretty Sagoo, Anna Singh and Nick Shearing all helped me to uncover important and interesting articles, books and documents, and I am grateful to them for their efforts. Thanks also go to Antony Buonomo at www.vertigo.co.uk who helped me to establish my website.

As well as interviewing experts, I have also depended on numerous books and articles. The list of further reading contains some of my sources, but it is neither a complete bibliography nor a definitive reference list. Instead, it merely includes material that may be of interest to the general reader. Of all the books I have come across during my research, I would like to single out one in particular: *The Codebreakers* by David Kahn. This book documents almost every cryptographic episode in history, and as such it is an invaluable resource.

Various libraries, institutions and individuals have provided me with photographs. All the sources are listed in the picture credits, but particular thanks go to Sally McClain, for sending me photographs of the Navajo code talkers; Professor Eva Brann, for discovering the only known photo of Alice Kober; Joan Chadwick, for sending me a photo of John Chadwick; and Brenda Ellis, for allowing me to borrow photos of James Ellis. Thanks also go to Hugh Whitemore, who gave me permission to use a quote from his play *Breaking the Code*, based on Andrew Hodges' book *Alan Turing – The Enigma*.

On a personal note, I would like to thank friends and family who put up with me over the two years while I was writing this book. Neil Boynton, Dawn Dzedzy, Sonya Holbraad, Tim Johnson, Richard Singh and Andrew Thompson all helped me to keep sane while I was struggling with convoluted cryptographic concepts. In particular, Bernadette Alves supplied me with a rich mixture of moral support and perceptive criticism. Travelling back in time, thanks also go to all the people and institutions that have shaped my career, including Wellington School, Imperial College and the High Energy Physics Group at Cambridge University; Dana Purvis, at the BBC, who gave me my first break in television; and Roger Highfield, at the *Daily Telegraph*, who encouraged me to write my first article.

Finally, I have had the enormous good fortune to work with some of the best people in publishing. Patrick Walsh is an agent with a love of science, a concern for his authors and a boundless enthusiasm. He has put me in touch with the kindest and most capable publishers, most notably Fourth Estate, whose staff endure my constant stream of queries with great spirit. Last, but certainly not least, my editors, Christopher Potter, Leo Hollis and Peternelle van Arsdale, have helped me to steer a clear path through a subject that twists and turns its way across three thousand years. For that I am tremendously grateful.

Further Reading

The following is a list of books aimed at the general reader. I have avoided giving more detailed technical references, but several of the texts listed contain a detailed bibliography. For example, if you would like to know more about the decipherment of Linear B (Chapter 5), then I would recommend *The Decipherment of Linear B* by John Chadwick. However, if this book is not detailed enough, then please refer to the references it contains.

There is a great deal of interesting material on the Internet relating to codes and ciphers. In addition to the books, I have therefore listed a few of the websites that are worth visiting.

General

Kahn, David, *The Codebreakers* (New York: Scribner, 1996).

A 1,200-page history of ciphers. The definitive story of cryptography up until the 1950s.

Newton, David E., *Encyclopedia of Cryptology* (Santa Barbara, CA: ABC-Clio, 1997).

A useful reference, with clear, concise explanations of most aspects of ancient and modern cryptology.

Smith, Lawrence Dwight, *Cryptography* (New York: Dover, 1943).

An excellent elementary introduction to cryptography, with more than 150 problems. Dover publishes many books on the subject of codes and ciphers.

Beutelspacher, Albrecht, *Cryptology* (Washington, D.C.: Mathematical Association of America, 1994).

An excellent overview of the subject, from the Caesar cipher to public-key cryptography, concentrating on the mathematics rather than the history. It is also the cryptography book with the best sub-title: *An Introduction to the Art and Science of Enciphering, Encrypting, Concealing, Hiding, and Safeguarding, Described Without any Arcane Skullduggery but not Without Cunning Waggery for the Delectation and Instruction of the General Public.*

Chapter 1

Gaines, Helen Fouché, *Cryptanalysis* (New York: Dover, 1956).
A study of ciphers and their solution. An excellent introduction to cryptanalysis, with many useful frequency tables in the appendix.

Al-Kadi, Ibraham A., 'The origins of cryptology: The Arab contributions', *Cryptologia*, vol. 16, no. 2 (April 1992), pp. 97–126.
A discussion of recently discovered Arab manuscripts, and the work of al-Kindī.

Fraser, Lady Antonia, *Mary Queen of Scots* (London: Random House, 1989).
A highly readable account of the life of Mary Queen of Scots.

Smith, Alan Gordon, *The Babington Plot* (London: Macmillan, 1936).
Written in two parts, this book examines the plot from the points of view of both Babington and Walsingham.

Steuart, A. Francis (ed.), *Trial of Mary Queen of Scots* (London: William Hodge, 1951).
Part of the Notable British Trials series.

Chapter 2

Standage, Tom, *The Victorian Internet* (London: Weidenfeld & Nicolson, 1998).
The remarkable story of the development of the electric telegraph.

Franksen, Ole Immanuel, *Mr Babbage's Secret* (London: Prentice-Hall, 1985).
Contains a discussion of Babbage's work on breaking the Vigenère cipher.

Franksen, Ole Immanuel, 'Babbage and cryptography. Or, the mystery of Admiral Beaufort's cipher', *Mathematics and Computer Simulation*, vol. 35, 1993, pp. 327–67.
A detailed paper on Babbage's cryptological work, and his relationship with Rear-Admiral Sir Francis Beaufort.

Rosenheim, Shawn, *The Cryptographic Imagination* (Baltimore, MD: Johns Hopkins University Press, 1997).
An academic assessment of the cryptographic writings of Edgar Allan Poe and their influence on literature and cryptography.

Poe, Edgar Allan, *The Complete Tales and Poems of Edgar Allan Poe* (London: Penguin, 1982).
Includes 'The Gold Bug'.

Viemeister, Peter, *The Beale Treasure: History of a Mystery* (Bedford, VA: Hamilton's, 1997).
An in-depth account of the Beale ciphers written by a respected local historian. It includes the entire text of the Beale pamphlet, and is most easily obtained directly from the publishers; Hamilton's, P.O. Box 932, Bedford, VA, 24523, USA.

Chapter 3

Tuchman, Barbara W., *The Zimmermann Telegram* (New York: Ballantine, 1994).
A highly readable account of the most influential decipherment in the First World War.

Yardley, Herbert O., *The American Black Chamber* (Laguna Hills, CA: Aegean Park Press, 1931).
A racy history of cryptography, which was a controversial best-seller when it was first published.

Chapter 4

Hinsley, F.H., *British Intelligence in the Second World War: Its Influence on Strategy and Operations* (London: HMSO, 1975).
The authoritative record of intelligence in the Second World War, including the role of Ultra intelligence.

Hodges, Andrew, *Alan Turing: The Enigma* (London: Vintage, 1992).
The life and work of Alan Turing. One of the best scientific biographies ever written.

Kahn, David, *Seizing the Enigma* (London: Arrow, 1996).
Kahn's history of the Battle of the Atlantic and the importance of cryptography. In particular, he dramatically describes the 'pinches' from U-boats which helped the codebreakers at Bletchley Park.

Hinsley, F.H., and Stripp, Alan (eds), *The Codebreakers: The Inside Story of Bletchley Park* (Oxford: Oxford University Press, 1992).
A collection of illuminating essays by the men and women who were part of one of the greatest cryptanalytic achievements in history.

Smith, Michael, *Station X* (London: Channel 4 Books, 1999).
The book based on the British Channel 4 TV series of the same name, containing anecdotes from those who worked at Bletchley Park, otherwise known as Station X.

Harris, Robert, *Enigma* (London: Arrow, 1996).
A novel revolving around the codebreakers at Bletchley Park.

Chapter 5

Paul, Doris A., *The Navajo Code Talkers* (Pittsburgh, PA: Dorrance, 1973).
A book devoted to ensuring that the contribution of the Navajo code talkers is not forgotten.

McClain, S., *The Navajo Weapon* (Boulder, CO: Books Beyond Borders, 1994).
A gripping account that covers the entire story, written by a woman who has spent much time talking to the men who developed and used the Navajo code.

Pope, Maurice, *The Story of Decipherment* (London: Thames & Hudson, 1975).
A description of various decipherments, from Hittite hieroglyphs to the Ugaritic alphabet, aimed at the layperson.

Davies, W.V., *Reading the Past: Egyptian Hieroglyphs* (London: British Museum Press, 1997).
Part of an excellent series of introductory texts published by the British Museum. Other authors in the series have written books on cuneiform, Etruscan, Greek inscriptions, Linear B, Maya glyphs, and runes.

Chadwick, John, *The Decipherment of Linear B* (Cambridge: Cambridge University Press, 1987).
A brilliant description of the decipherment.

Chapter 6

Data Encryption Standard, FIPS Pub. 46–1 (Washington, D.C.: National Bureau of Standards, 1987).
The official DES document.

Diffie, Whitfield, and Hellman, Martin, 'New directions in cryptography', *IEEE Transactions on Information Theory*, vol. IT-22 (November 1976), pp. 644–54.
The classic paper that revealed Diffie and Hellman's discovery of key exchange, opening the door to public-key cryptography.

Gardner, Martin, 'A new kind of cipher that would take millions of years to break', *Scientific American*, vol. 237 (August 1997), pp. 120–24.
The article which introduced RSA to the world.

Hellman, M.E., 'The mathematics of public-key cryptography', *Scientific American*, vol. 241 (August 1979), pp. 130–39.
An excellent overview of the various forms of public-key cryptography.

Diffie, Whitfield, 'The first ten years of public-key cryptography', *Proceedings of the IEEE*, vol. 76, no. 5 (May 1988), pp. 560–77.
Another excellent overview of public-key cryptography.

Chapter 7

Zimmermann, Philip R., *The Official PGP User's Guide* (Cambridge, MA: MIT Press, 1996).
A friendly overview of PGP, written by the man who developed it.

Garfinkel, Simson, *PGP: Pretty Good Privacy* (Sebastopol, CA: O'Reilly & Associates, 1995).
An excellent introduction to PGP and the issues surrounding modern cryptography.

Bamford, James, *The Puzzle Palace* (London: Penguin, 1983).
Inside the National Security Agency, America's most secret intelligence organisation.

Koops, Bert-Jaap, *The Crypto Controversy* (Boston, MA: Kluwer, 1998).
An excellent survey of the impact of cryptography on privacy, civil liberty, law enforcement and commerce.

Diffie, Whitfield, and Landau, Susan, *Privacy on the Line* (Cambridge, MA: MIT Press, 1998).
The politics of wire-tapping and encryption.

Chapter 8

Deutsch, David, *The Fabric of Reality* (London: Allen Lane, 1997).
Deutsch devotes one chapter to quantum computers, in his attempt to combine quantum physics with the theories of knowledge, computation and evolution.

Bennett, C. H., Brassard, C., and Ekert, A., 'Quantum Cryptography', *Scientific American*, vol. 269 (October 1992), pp. 26–33.
A clear explanation of the evolution of quantum cryptography.

Deutsch, D., and Ekert, A., 'Quantum computation', *Physics World*, vol. 11, no. 3 (March 1998), pp. 33–56.
One of four articles in a special issue of *Physics World*. The other three articles discuss quantum information and quantum cryptography, and are written by leading figures in the subject. The articles are aimed at physics graduates and give an excellent overview of the current state of research.

Internet Sites

The Mystery of the Beale Treasure
http://www.roanokeva.com/ttd/stories/beale.html
A collection of sites relating to the Beale ciphers. The Beale Cypher and Treasure Association is currently in transition, but it hopes to be active again by the year 2000.

Bletchley Park
http://www.cranfield.ac.uk/ccc/bpark/
The official website, which includes opening times and directions.

The Alan Turing Homepage
  http://www.turing.org.uk/turing/

Enigma emulators
  http://www.attlabs.att.co.uk/andyc/enigma/enigma_j.html
  http://www.izzy.net/~ian/enigma/applet/index.html
  Two excellent emulators that show how the Enigma machine works. The former allows you to alter the machine settings, but it is not possible to track the electrical path through the scramblers. The latter has only one setting, but has a second window that shows the scramblers moving and the subsequent effect on the electrical path.

Phil Zimmermann and PGP
  http://www.nai.com/products/security/phil/phil.asp

Electronic Frontier Foundation
  http://www.eff.org/
  An organisation devoted to protecting rights and promoting freedom on the Internet.

Centre for Quantum Computation
  http://www.qubit.org/

Information Security Group, Royal Holloway College
  http://isg.rhbnc.ac.uk/

National Cryptologic Museum
  http://www.nsa.gov:8080/museum/

American Cryptogram Association (ACA)
  http://www.und.nodak.edu/org/crypto/crypto/
  An association which specialises in setting and solving cipher puzzles.

Cryptologia
  http://www.dean.usma.edu/math/resource/pubs/cryptolo/index.htm
  A quarterly journal devoted to all aspects of cryptology.

Cryptography Frequently Asked Questions
  http://www.cis.ohio-state.edu/hypertext/faq/usenet/cryptography-faq/top.html

RSA Laboratories' Frequently Asked Questions About Today's Cryptography
  http://www.rsa.com/rsalabs/faq/html/questions.html

Yahoo! Security and Encryption Page
  http://www.yahoo.co.uk/Computers_and_Internet/Security_and_Encryption/

Crypto Links
  http://www.ftech.net/~monark/crypto/web.htm

Picture credits

Line illustrations by Miles Smith-Morris.
Hieroglyphs reproduced by kind permission of British Museum Press.
Linear B characters reproduced by kind permission of Cambridge University Press.

Figure 1 Scottish National Portrait Gallery, Edinburgh; Figure 6 Ibrahim A. Al-Kadi and Mohammed Mrayati, King Saud University, Riyadh; Figure 9 Public Record Office, London; Figure 10 Scottish National Portrait Gallery, Edinburgh; Figure 11 Cliché Bibliothèque Nationale de France, Paris, France; Figure 12 Science and Society Picture Library, London; Figures 20 and 25 *The Beale Treasure – History of a Mystery* by Peter Viemeister; Figure 26 David Kahn Collection, New York; Figure 27 Bundesarchiv, Koblenz; Figure 28 National Archive, Washington DC; Figure 29 General Research Division, The New York Public Library, Astor, Lenox and Tilden Foundations; Figures 31 and 32 Luis Kruh Collection, New York; Figure 38 David Kahn Collection; Figures 39 and 40 Science and Society Picture Library, London; Figures 41 and 42 David Kahn Collection, New York; Figure 43 Imperial War Museum, London; Figures 44 and 45 Private collection of Barbara Eachus; Figure 47 Godfrey Argent Agency, London; Figure 50 Imperial War Museum, London; Figure 51 Telegraph Group Limited, London; Figures 52 and 53 National Archive, Washington DC; Figures 54 and 55 British Museum Press, London; Figure 56 Louvre, Paris © Photo RMN; Figure 58 Department of Classics, University of Cincinnati; Figure 59 Private collection of Eva Brann; Figure 60 Source unknown; Figure 61 Private collection of Joan Chadwick; Figure 62 Sun Microsystems; Figure 63 Stanford, University of California; Figure 65 RSA Data Security, Inc.; Figure 66 Private collection of Brenda Ellis; Figure 67 Private collection of Clifford Cocks; Figures 68 and 69 Private collection of Malcolm Williamson; Figure 70 Network Associates, Inc.; Figure 72 Penguin Books, London; Figure 75 Thomas J. Watson Laboratories, IBM.

Index

Page folios that refer to pictures and figures appear in italics